INTRODUCTION TO SECURITY

INTRODUCTION TO SECURITY

THIRD EDITION

GION GREEN

BUTTERWORTH PUBLISHERS
Boston • London
Sydney • Wellington • Durban • Toronto

All references in this book to personnel of male gender are used for convenience only and shall be regarded as including both males and females.

Library of Congress Cataloging in Publication Data

Green, Gion.
Introduction to security.

Bibliography: p.
Includes index.
1, Industry—Security measures. 2. Retail trade—Security measures. 3. Electronic data processing departments—Security measures. I. Title.
HV8290.G74 1981 658.4'7 80-29131
ISBN 0-409-95036-X

Butterworth Publishers
80 Montvale Avenue
Stoneham, MA 02180

10 9 8 7 6 5

Printed in the United States of America

CONTENTS

PREFACE AND ACKNOWLEDGEMENTS TO THE THIRD EDITION

The field of security is still in the early stages of growth and evolution. Many changes have occurred over the past several years, and I'm sure many more are in store for the profession in the years ahead. Even the terms used to label the field—let alone the definitions—have undergone changes. Is the security professional involved with security, safety, loss prevention or risk management? The correct answer is "all of the above." As the field becomes more professional, the duties of the security practitioner encompass a far wider range of responsibilities. No longer is security saddled with the night watchman image.

The problem of writing an introductory text is in determining where the emphasis should be placed. Should it be on security methods and hardware, serving management, or on loss prevention theory? For an introductory text to be effective, it should cover the total picture, giving the reader a glimpse of the various, diverse components which make up the security function. This has been the intent behind past editions of INTRODUCTION TO SECURITY, as it is behind this third edition.

The history and purpose of security, as well as the methods and hardware utilized by the profession, are covered in the first several chapters. Loss prevention theory is developed in such chapters as "Internal Theft Controls," "Retail Security," and "Cargo Security." In "Risk Analysis and the Security Survey" and "Insurance," the reader is introduced to the principles of threat assessment and risk management. Basic principles of security management appear in "The Proprietary Security Organization." Safety is covered in the chapters "Fire Prevention and Emergency Planning" and "OSHA and Safety Planning."

The third edition of INTRODUCTION TO SECURITY contains three new chapters. An essential chapter, "Security and the Law," treats the basics of law and discusses its effect on the security profession. Another addition, "Government Security and Organizations," briefly outlines the role of security within the federal government. The final new chapter, "Institutional Security: Hospitals, Banks, Museums," not only familiarizes the reader with various security career opportunities, but also compares the loss prevention principles and methods of different institutions.

The third edition has been thoroughly updated and existing chapters have been slightly expanded. Some examples of topics new to this edition are space protection systems, emergency planning in the event of labor disputes, establishing a safety and loss control program, fidelity bonding, background history and reference checking, and the problems related to the arrest, apprehension, and prosecution of shoplifters.

An illustration program has been added to this edition, in hopes that these illustrations will not only incorporate new material but will also increase the reader's understanding of the text.

I wish to express my gratitude to Dr. Kenneth Fauth at Northern Michigan University; Raymond Lavertue, Sr., at Bristol Community College; George Moore at Northern Virginia Community College in Woodbridge; and David Schachtsiek at David Allan and Associates for the excellent suggestions they offered when reviewing this edition. I also want to thank all of those instructors who took the time to complete the questionnaire which the publisher prepared. The comments from the following were particularly helpful:

Richard A. Barlow, Los Angeles Southwest College
Roland Benedetti, St. John's University
Elmer Criswell, Harrisburg Area Community College
Robert J. Fischer, Western Illinois University
Wolcott S. Gaines, Southern Maine Vocational Technical Institute
Thomas Kissane, Iona College
Francis W. Norwood, Middle Tennessee State University
Tom Pruette, Parkersburg Community College
Vernon Rich, University of Texas at Tyler
Rudolph Strasswimmer, Mercy College
John J. Sullivan, Mercy College

Without the direction of so many people, this new edition of INTRODUCTION TO SECURITY would not have been possible.

PREFACE AND ACKNOWLEDGEMENTS TO THE REVISED (SECOND) EDITION

The enormous growth of security in the past two decades has raised many questions that have only recently begun to come into focus. The heavy demands upon the resources of public police agencies created by the sharp increase in criminal activity have to some degree diverted their attention and finite resources from traditional efforts in the protection of private property, leaving that responsibility more and more to the proprietor. In turn, this has placed private security in the position of playing an increasingly active role in many areas of crime and loss prevention.

It is the purpose of this book to provide a broad overview of private security in its practical application, and to suggest certain theoretical approaches to some of its problems. After considering the history and organization of security, the text details the fundamental principles of risk assessment, physical protection, systems of defense, internal security, fire prevention, emergency planning, safety and insurance protection.

Industrial security is a reference point throughout the general discussion of the security function and basics of defense. Additionally, the authors have chosen to examine a selected sample of various industries to illustrate the application of security systems, procedures and techniques to particular problems. Included are retail security, one of the prime targets of criminal attack; hospital security, representative of the protection and loss prevention function in a public service facility or institution; cargo security, which affects virtually all businesses across the board, and consumers as well, and which provides occasion for examining the special problems of protecting goods when they are not under control in a fixed location; and computer security, which is of ever-growing importance to the entire business and industrial community, as well as to government.

In preparing this new, Revised Edition, the authors have reorganized the text to provide a more comprehensive discussion of the security function and its organizational aspects prior to the detailed study of specific problems. New chapters have been included on defining security's role; on the proprietary security organization; and on risk analysis and the security survey. There is now a separate chapter on fire protection and emergency planning, and a new chapter on career opportunities in loss prevention. The chapter on accident prevention and safety has also been expanded, reflecting the growing concern through the 1970s for safety and health in the workplace. Review questions have also been included for all chapters.

The book is designed, first of all, as an introductory text for those students who are interested in a career in security; for those presently active in the field who are pursuing additional academic knowledge; and for those students concentrating on business administration who wish to examine the basic principles of an organizational function which is rapidly becoming an organic part of virtually every business. In the opinion of the authors, such a study is a requisite to an understanding of the needs of the modern business structure.

Security managers and others with a professional interest in the subject should also find the text of value. Its general review of fundamental principles should provide an effective yardstick against which to measure particular experience. And its detailed re-examination of security practices may prove useful to security professionals in evaluating existing programs and systems.

In the view of the authors, to meet the current needs for security services, as well as the increased demands which can be foreseen in the future, private security must be prepared to evaluate its position and to improve the quality of its performance. Nothing less will meet the challenges faced by a free society in a changing world. Every effort must be made to professionalize an industry, now in a state of dynamic growth, which plays such a significant role in our social and economic structure. Universal standards of training and performance must be sought. Above all, security must search for and develop—not least of all through formal security education—innovative personnel with managerial talents to assume the future leadership of this vital industry.

The authors are indebted to the legion of protection professionals who have added their knowledge and experience to the growing body of security literature, and we are specifically grateful to those leading security specialists who contributed information and insight to the chapter on career opportunities. Finally, we would particularly like to acknowledge the contribution of Alvin L. Van Der Most, Occupational Safety and Health, Aircraft Division of the Northrop Corporation, for his review and suggestions for additional material in the chapter on accident prevention and safety; and of Dr. John M. Carroll, Professor of Computer Science, University of Western Ontario, for reviewing the chapter on computer security.

PART I
INTRODUCTION

Chapter 1

ORIGINS AND DEVELOPMENT OF SECURITY

Security, in its semantic and philosophical sense, implies a stable, relatively predictable environment in which an individual or group may pursue its ends without disruption or harm, and without fear of such disturbance or injury. The concept of security in an organizational sense, as a means by which this safety and stability can be achieved, has evolved gradually throughout the history of Western civilization, shaped by a wide variety of institutional and cultural patterns.

In examining the origins and development of security, it is both obvious and instructive to observe that security holds a mirror up, not to nature, but to society and its institutions. Thus, in medieval England, there were programs to clear brush and other concealment on either side of the king's roads as a precaution against highway robbers; for night thieves, there were night watchmen. In modern America, these rudimentary security measures find their counterparts in the cleared areas adjoining perimeter fences and buildings, in security patrols and intrusion alarms.

Throughout history it is possible to trace the emerging concepts of security as a response to and a reflection of a changing society, mirroring not only its social structure but also its economic conditions, its conception of law and perception of crime, and its morality. Nowhere is this more evident, or more relevant to the development of modern security, than in the history of England.

SECURITY IN ENGLAND

Feudalism and Security

In early England, feudalism provided security for both the individual and the group to a very high degree. The Anglo-Saxons brought with them to England a predisposition to accepted mutual responsibility for civil and military protection of individuals. They also brought with them a strong affinity for the feudal contract whereby an overlord guaranteed the safety of persons and property, and provided arms and treasures to vassals who administered the work of serfs bound to the land.

In a world of constant warfare between men of power, security could be found in no other way. Stability lay in the system and in the power and cleverness of the lord. Group security lay in group solidarity. Formal systems of security that developed over the years of the Middle Ages were largely confirmations of those systems toward which the people of this society had gravitated naturally.

Post-Norman Reforms

Post-Norman England, beginning with King John, saw the introduction of concepts declaring the supremacy of law over arbitrary edict, thus developing a base of confidence in the continuity of the system and its institutions. Above all, there was a formal declaration of the individuals rights and responsibilities as between the state and its subjects and between subjects themselves.

Judicial reforms during this era saw the emergence of local juries, circuit judges, coroners to restrain the power of local sheriffs, and justices of the peace appointed to hear and determine criminal cases. The movement also began which would eventually see the complete separation of courts and the exercise of the rule of law from the whims and power of the king.

Concurrently, there were measures specifically aimed at the enforcement of public order. The Statute of Winchester (also known as the Statute of Westminster) in 1285 revived and reorganized the old institutions of national police and national defense. It described the "duty of watch and ward," which enjoined every man to pursue and bring to justice felons whenever "hue and cry" was raised.

Every district was made responsible for crimes committed within its bounds; the gates of all towns were required to be closed at nightfall, and all strangers were required to give an account of themselves to the magistrates. Interestingly, one "security" practice already mentioned required brushwood and other concealments to be cleared for a space of 200 feet on either side of the king's highways to protect travelers against attack by robbers.

Attempts were made to control vice and crime at the local level, and boroughs enacted their own ordinances to that end. Since organized agencies for the enforcement of such laws were virtually nonexistent, however, these efforts had a limited success. Privately established night watches and patrols were often the citizens' only protection against direct assault.

Exploration and Change

The development of systems of protection and enforcement appeared to come with greater rapidity and sophistication from the 14th through the 18th century. Seeds for this development were planted during the social revolution that heralded the end of the remaining elements of the feudal structure in the latter half of the 13th century.

Security was one thing in a largely rural society controlled by kings and feudal barons; it was another thing entirely in a world swept by enormous changes. The voyages of exploration which opened up new markets and trade routes created a new and increasingly important merchant class, whose activities came to dominate the port cities and trading centers. Concurrently, acts of enclosure and consolidation drove displaced small tenants off the land, and they migrated to the cities in great numbers.

By the beginning of the 16th century, the social patterns of the Middle Ages were being broken down. Increased urbanization of the population had created conditions of considerable hardship. Poverty and crime increased rapidly. No public law enforcement agencies existed that could restrain the mounting wave of crime and violence, and no agencies existed that could alleviate the root causes of the problem.

Different kinds of police agencies were privately formed. Individual merchants hired men to guard their property. Merchant associations also created the *merchant police* to guard shops and warehouses. *Night watchmen* were employed to make their rounds. *Agents* were engaged to recover stolen property, and *parochial police* were hired by the people of the various parishes into which the major cities were divided.

Attention then turned to the reaffirmation of laws to protect the common good. Although the Court of Star Chamber had been abolished in 1641, its practices were not officially proscribed until 1689, when Parliament agreed to crown William and Mary if they would reaffirm the ancient rights and privileges of the people. They agreed, and Parliament ratified the Bill of Rights—which for all time limited the power of the king as well as affirming and protecting the inalienable rights of the individual.

The 18th Century

By the 18th century, it is possible to discern both the shape of efforts toward communal security and the kinds of problems that would continue to plague an increasingly urban society for the next 200 years.

In 1737, for instance, a new aspect of individual rights came to be acknowledged; for the first time tax revenues were used for the payment of a night watch. This was a significant development in security practice, since it was a precedential step that established for the first time the use of tax revenues for common security purposes.

Eight years later, Parliament authorized a special committee to study security problems. The study resulted in a program employing various existing private security forces to extend the scope of their protection. The resulting heterogeneous group, however, was too much at odds. It proved ineffective in providing any satisfactory level of protection.

In 1748, Henry Fielding, magistrate and author (most notably, of the unforgettable *Tom Jones*), proposed a permanent, professional and adequately paid security force. His invaluable contributions included a foot patrol to make the streets safe, a mounted patrol for the highways, the famous "Bow Street Runners" or special investigators, and police courts.

It is interesting to note that Fielding also wrote an ironic novel called *Jonathan Wild—The Story of a Great Man.* Its hero, Jonathan Wild, was a real person in 18th century London, perhaps the most notorious fence, thief and master criminal of modern times. He was so real, in fact, that an account of his activities occupies eight pages of a staff report prepared in 1972 for the Select Committee on Small Business of the U.S. Senate, more than 200 years after Wild was hanged at Tyburn.

How is it that the spectre of Jonathan Wild still haunts those bodies charged with finding means to minimize crime? In many ways, Wild's career typified the problems of security—or, more specifically, theft control—in the 18th century.

For many centuries, the English common law almost totally ignored the receiver of stolen goods. As the Senate Committee report observes, "Because Jonathan Wild was such an extraordinary criminal, it is easy to lose sight of the fact that, first, he was at base a receiver, and second, that his whole organization was geared to facilitate that primary enterprise."[1] But mere receiving of stolen goods, even with knowledge, did not make the receiver an accessory in the eyes of the law.

Perhaps this attitude of the common law can be explained, in part, by the relative unimportance of dealings in stolen property in the early stages of the development of the law of crimes. Until the 17th century, the amount of movable property available for theft was probably limited, and opportunities to dispose of this property, other than by personal consumption, rather restricted. Lacking a professional police force, the attention of the community, and the law, was primarily directed toward apprehending offenders rather than tracing and recovering stolen property. The victims of property crimes were left to rely upon their own ingenuity, bolstered by several shaky legal remedies, to secure the return of their plundered goods and chattles.

It was not until the late 17th century that the legislature moved for the first time to combat the problem of the receiver of stolen goods. In 1691, under a statute enacted during the reign of William and Mary, the receiver was made subject to prosecution, but only as an accessory after the fact. The tradition remained throughout the 18th and early 19th century, that the receiver was an accessory rather than a principal to the crime. The weakness of the law in its attitude toward property crimes, as much as the lack of effective law enforcement, combined to make possible Jonathan Wild's legendary career.

The Impact of Industrial Expansion

The Industrial Revolution began to gather momentum in the latter half of the 18th century. By 1801, the poet William Blake, of apocalyptic vision, was writing disapprovingly of "these dark, Satanic mills". Like the migrations off the land 200 years earlier, people again flocked to the cities—not pushed this time as they had been by enclosure and dispossession, but lured by promises of work and wages.

The already crowded cities were choked with this new influx of wealth seekers. What they found were long hours, crippling work and miserly wages. Men and women—even very young children—worked in unsafe factories. Disease periodically swept the crowded quarters. Family life, heretofore the root of all stability, was virtually destroyed in this environment. Thievery, crimes of violence, and juvenile deliquency were the order of the day. All the ills of such a structure, as we can see in analogous situations today, overtook the emerging industrial centers.

Little was done to alleviate the growing problems. Indeed, the prevailing philosophy of the time argued *against* doing anything. In 1776 Adam Smith gained a large and appreciative audience for his *Wealth of Nations*. In it, he contended that labor was the source of wealth; and it was by freedom of labor, by allowing the worker to pursue his own interest in his own way, that the public wealth would best be promoted. Any attempt to force labor into artificial channels, to shape by laws the course of commerce, to promote special branches of industry in particular countries, would be not only wrong to the worker and merchant, but harmful to the wealth of the state.

In this new age in which such statements of *laissez faire* were generally accepted, industrial centers became the spawning grounds for crimes of all kinds. At one time, counterfeiting was so common that it was estimated that more counterfeit money than government issue was in circulation. Over 50 false mints were found in London alone.

The backlash to such a high crime rate was inevitable and predictable. Penalties were increased to deter potential criminals. At one time, over 150 capital offenses existed, ranging from picking pockets to serious crimes of violence. Yet, no visible decline in crime resulted. It was, for all purposes, a "society which lacked any effective means of enforcing the criminal laws in general. A Draconian code of penalties which prescribed the death penalty for a host of crimes failed to balance the absence of efficient enforcement machinery."[2]

Private citizens resorted to carrying arms for protection, and they continued to band together to hire special police to protect their homes and businesses.

Sir Robert Peel and the "Bobbies"

In 1822, Sir Robert Peel became Home Secretary. He had an abiding interest in creating a strong, unified professional police force. This interest had emerged earlier when, as Secretary for Irish Affairs, he had reformed the Irish constabulary—members of which were thereafter referred to as "Peelers." As Home Secretary, Peel initiated the criminal law reform bill and he reorganized the metropolitan police force, also referred to as "Peelers" or, more commonly, "Bobbies." He also attempted to decentralize police efforts and to develop the responsibility of each community for its own security.

Unfortunately, not all of Sir Robert's efforts met with success. Neither the Police Act of 1835, establishing city and borough police forces, nor the County Act of 1839, setting up country police, nor various other acts passed in mid-century, created adequate police operations. Private guard forces continued in use to recover stolen property, as well as to provide protection for private persons and businesses.

Nevertheless, based on Peel's thoroughgoing reforms and revisions, the metropolitan police force became a model for law enforcement agencies in years to come, not only in England but also in America. Modern policing, it is often said, was born with the "Bobby."

SECURITY IN AMERICA

Security practices in the early days of colonial America followed the patterns that colonists had been familiar with in England. The need for mutual protection in a new and alien land drew them together in groups much like those of earlier centuries.

As the settlers moved west in Massachusetts, along the Mohawk Valley in New York, and into central Pennsylvania and Virginia, the need for protection against hostile Indians was the principal security interest. Settlements generally consisted of a central fort or stockade surrounded by the farms of the inhabitants. If hostilities threatened, an alarm was sounded, and the members of the community left their homes for the protection of the fort, where all able-bodied persons were involved in its defense. In such circumstances, a semi-military flavor often characterized security measures, including guard posts and occasional patrols.

Protection of people and property in established towns again followed English traditions. Sheriffs were elected as chief security officers in colonial Virginia and Georgia; constables were appointed in New England. Watchmen were hired to patrol the streets at night. As the Report of the Task Force on Private Security notes, "These watchmen remained familiar figures and constituted the primary security measures until the establishment of full-time police forces in the mid-1800's."[3]

Such watchmen, it should be pointed out, were without training, had no legal authority, were either volunteers or were paid a pittance, and were generally held in low regard — circumstances which bear a remarkable similarity to observations in the Rand Report on private security in 1971.[4]

Police Powers in the New Nation

After the Revolution, the United States made sporadic attempts to establish adequate security systems, but no unifying philosophy or principle was developed that could act as a guide in their formation. The absence of underlying principles for police and security was further aggravated by the multiplicity of overlapping jurisdictions in the new nation. Yet, these were an inevitable and generally desirable result of the principles upon which the country was founded. The principle of states' rights had a profound and continuing impact upon law enforcement.

The Constitution was carefully drawn and ratified as an instrument threading a narrow line between the rights of the several states and the powers vested in the federal government. The fear of excessive and, perhaps, arbitrary power in a central government, like that experienced under the English crown, persuaded the Constitutional Congress to provide the federal body only those powers required to administer the broad needs of the land and to delegate the rest to the individual states.

The issue was, of course, the cause of battle lines being drawn in the Civil War. It has always, before and since, been a subject of much debate and discussion. Just where, and in what way, the police powers of the federal government should supersede those of other jurisdictions within its whole has been defined slowly over the years by numerous cases argued before the Supreme Court. The line of separation defining these powers may always be in flux; certainly it is subject to reexamination to this day.

Police power is bestowed upon the state to permit interference with normal personal and property rights in the interest of conducting government in the general public interest. Implicit in such necessity is the intervention of government in the settlement of disputes; collection of revenues; enforcement of criminal law; the regulation of industry, commerce, and agriculture; the administration of agencies providing for the public safety; and so on.

In this regard, Chief Justice Roger B. Taney said in 1847, "What are the police powers of a state? They are nothing more or less than the powers of government inherent in every sovereignty to the extent of its dominions. That is to say—the power to govern men and things within the limits of its domain."

The exercise of various police powers through the enforcement of criminal laws or administrative action may, like the taxing power, interfere with property, but it does so not to collect revenue but to promote the public welfare. Justification for such action is derived from common law in such maxims as "*salus populi suprema lex*" (the public safety is the supreme law).

The balance between the police powers of either the state or federal government and the guarantees of due process embodied in both the Fifth and Fourteenth Amendments have also been the subject of much attention by the courts. The end result of such issues was that, although the exercise of a broad range of police powers was affirmed in various jurisdictions, the fragmentation created by these entities made for difficulty in the formation of efficient police systems.

Today, there are roughly 40,000 police jurisdictions in the United States. It is inevitable that police efficiency and effectiveness must vary considerably throughout the country.

Development of Public Law Enforcement

The development of police and security forces seems to follow no predictable pattern other than that such development was traditionally in response to public pressure for action.

Outside of the establishment of night watch patrols in the 17th century, little effort to establish formal security agencies was made until the beginnings of a police department were established in New York City in 1783. Detroit followed in 1801, and Cincinnati in 1803. Chicago established a police department in 1837, San Francisco in 1846, Los Angeles in 1850, Philadelphia in 1855, and Dallas in 1856.

New York, influenced by the recent success of the police reforms of Sir Robert Peel, adopted his general principles in 1833. By and large, however, police methods in departments across the country were rudimentary. Departments as a whole were inefficient, ill-trained and corrupt.

A significant factor in such conditions was the "spoils system." Andrew Jackson had, if not created, at least articulated the system by which the party in power showered preference of all kinds, including police jobs, on its loyal followers. Such disregard for professionalism in favor of political expediency served only to decrease the already low public esteem for such agencies.

Not until the passage of the Civil Service Act of 1883 was any effective attempt made to upgrade the level of police protection, in spite of rising urban crime rates and frequent revelations of police corruption and inefficiency.

Early Federal Law Enforcement Agencies

In 1864, the U.S. Treasury Department organized its law enforcement agency — the second such federal service to be formed. The Post Office Department had set up its investigative service in 1828.

In 1870, the United States Department of Justice began operations, and the Border Patrol was formed in 1882. Although the Department of Justice organized an internal agency, the Bureau of Investigation, in 1908 to investigate federal crimes, it did not effectively blossom into the national police agency known as the F.B.I. until 1924.

Growth of Private Security

The slow development of public law enforcement agencies, both state and federal, combined with the steady escalation of crime problems in an increasingly urban and industrialized society, created security needs which were met by what might be called the first professional private security responses in the second half of the 19th century.

In the 1850's *Allan Pinkerton*, a cooper from Scotland and the Chicago Police Department's first detective, established one of the oldest, and currently the largest, private security operation in the United States. His North West Police Agency, formed in 1855, provided security and conducted investigations of crimes for various railroads. Two years later, the Pinkerton Protection Patrol began to offer a private watchman service for railroad yards and industrial concerns.

The services this agency provided were important to its clients largely because public law enforcement agencies were either inadequate for the job or because their jurisdictions were such that they could not take necessary action. At the outset of the Civil War, General George McClellan, formerly of the Illinois Central Railroad and a client of Pinkerton's, hired the agency as an intelligence gathering arm of the Union Army which had no espionage apparatus at that time.

Brinks, Inc. was founded by Washington Perry Brink in Chicago in 1859 as a freight and package delivery service. More than 30 years later, in 1891, he transported his first payroll — the beginning of armored car and courier service. By 1900 Brinks had a fleet of 85 wagons in the field.[5]

William J. Burns, a former Secret Service investigator and head of the Bureau of Investigation (forerunner of the F.B.I.), started the William J. Burns Detective Agency in 1909. It became the sole investigating agency for the American Banking Association and grew to become the second largest (to Pinkerton's) contract guard and investigative service in the United States. For all intents and purposes, Pinkerton's and Burns' were the only national investigative bodies concerned with non-specialized crimes in the country until the advent of the F.B.I.

Another 19th century pioneer in this field was *Edwin Holmes,* who offered the first burglar alarm service in the country in 1858. Baker Industries initiated a fire-control and detection equipment business in 1909.

From the 1870's only private agencies had provided contract security services to industrial facilities across the country. In many cases, particularly around the end of the 19th century and during the Depression of the 1930's, the services were, to say the least, controversial. Both the Battle of Homestead in 1892, in which workers striking that plant were shot and beaten by security forces, and the strikes in the automobile industry in the middle 1930's are examples of excesses from over-zealous security operatives in relatively recent history.

Railway Police

During the late 19th century, various Railway Police Acts enacted by most states authorized the railroads to establish their own security forces with full powers to police equipment and property. This was essential for the security of the railroads, since their property extended through so many jurisdictions that security would have been impossible without a proprietary force, in the absence of any federal agency empowered to cut across the various jurisdictions involved.

The growth of private railroad police in the late 19th and early 20th centuries was phenomenal. By 1914, over 12,000 such police were in operation, and by 1921 the Association of American Railways provided a coordinated security service to all of its members. While the railroad police were granted some police powers, they always have been, in fact, private security forces.

World War II and After

Fears over industrial sabotage and espionage had given brief impetus to private security during the first World War, but this demand slackened during the years between the two great wars, except for the strike-breaking activities of the Depression years.

The dramatic development in the use of private protective forces during World War II, in the interest of national security, was quite another story. Defense contractors were required by the federal government to meet comprehensive security standards for the protection of classified information and defense-related production. Before the end of the war more than 200,000 men and women had been sworn in as security personnel. Their training was often undertaken by local police. In the urgency of the war years, whatever level of competence they achieved, these new personnel were thrust into every conceivable kind of security position. Ultimately they were assigned to every facility across the country which was even remotely connected with production for the war effort.

Thus, when the war ended, not only had private security personnel established their effectiveness in vast new areas, but they constituted a large work force, looking for jobs, trained and ready for a level of protective service which business and industry had never known—or needed—before.

The power politics of the Cold War years continued to bring the need for improved, more sophisticated security agencies and operations into sharp focus. The initial motivation in this increasing interest had been in protecting the nation against acts of sabotage and the theft of classified information, but the steady increase in crimes of violence as well as crimes against property continued to direct attention to the need for more and better countermeasures.

The 1940's can be viewed as a watershed period in the history of private security. There was no going back. The era of modern security had begun.

SECURITY TODAY

The preceding overview of security concepts and practices as they have emerged throughout English and American history is acknowledged sketchy; little else is possible in such brief space. However, one of the significant facts to be drawn from such a review is the realization that the concepts of modern private security and modern police systems are surprisingly new. Public law enforcement in anything like its present form has been in existence only slightly more than 100 years! And private security forces, whose history is in fact longer, have only recently come into wide general use. Although some private security firms such as Pinkerton's, Burns, Globe Security Systems (originally Globe Detective Service), Baker Industries, Brinks and others have been in business for more than half a century prior to World War II, such services were nowhere near as large or commonly employed as they are today. With few exceptions, proprietary or in-house security forces hardly existed before the defense-related "plant protection" boom of the early 1940's. The impetus for modern private security effectively began in that decade—and it has come of age only in the third quarter of the 20th century.

With this dynamic growth have come profits, problems and increasing professionalism. Each is a significant part of the picture of security today.

Crime Trends

Over the past 25 years, the United States has become the victim of what Arthur J. Bilek, chairman of the Task Force on Private Security of the National Advisory Committee on Criminal Justice Standards and Goals, has called "a crime epidemic." The F.B.I.'s annual Uniform Crimed Reports document the steady increase of crimes of all kinds, with the greatest escalation in crimes related to property.

U.S. News and World Report in 1974 estimated crime-related losses in the business community annually to be $21.3 billion.[6] The United States Chamber of Commerce, in a 1976 report, estimated "ordinary crimes" against business at $23.6 billion during 1975.[7] In another report, the Department of Commerce estimated retail crime costs for the same year at $6.5 billion and a 1979 report estimated these retail crime costs to have escalated to $12 billion.[8] *Industry Week* magazine reported in 1975 that internal crime costs business between $15 billion and $50 billion annually.[9] And the United States Chamber of Commerce, in a handbook on "White Collar Crime," has estimated the cost to business of crimes in this category at "not less than $40 billion."[10]

These figures on the extent of crime against business dramatize the absence of consistent hard data indicating the exact size of the problem today. Variations of billions of dollars in estimates are the result of estimates based on "educated guesses," interpolation, and adjustments for inflation, among others. However, some progress has been made in the last decade in coming to grips with the problem of defining the size of the problem.

Obviously these figures vary, principally because satisfactory measures of many crimes against business and industry have not yet been found, but also because much internal crime, in particular, is never reported to the police. Reasons for not reporting to the police are several: to avoid bad publicity, because internal disciplinary action has already been taken, or to avoid embarrassing management by exposing to the public its lack of security controls. Nevertheless, such questions as may exist concern only the *degree*,, not the fact of the dramatic escalation of crime against business in our society.

As the brief history of security in the past has indicated, there is always an intimate link between cultural and social change and crime, as there is between crime and the security measures adopted to combat the threat. A bewildering variety of causes, both social and economic, are cited for rising crime in this era. Among them are an erosion of family and religious restraints, the trend toward "permissiveness," the increasing anonymity of business at every level of commerce, the decline of feelings of worker loyalty toward the company, high unemployment, the Vietnam War and its attendant turmoil, and a general decline in morality, accompanied by the pervasive attitude that there is no such thing as right and wrong, there is only what "feels good."

It is far beyond the scope of this book to attempt to analyze or even to catalog all of the factors involved in the trend toward increasing crime, even were we to restrict such a study to crimes against business and industry. What is important here is to make clear note of the fact of such increases—and their impact upon society's attempt to protect itself.

Most significant is the realization that "the sheer magnitude of crime in our society prevents the criminal justice system by itself from adequately controlling the preventing crime."[11] In spite of their steady growth, both in costs and in numbers of personnel, public law enforcement agencies have increasingly been compelled to be reactive and to concentrate more of their activity on the maintenance of public order and the apprehension of criminals. As the Private Security Task Force Report observes, "U.S. Bureau of the Census statistics reflect some 12.4 million commercial and business establishments in the United States. The approximate 500,000 local law enforcement personnel in this country can not possibly provide protection for all of these establishments."[12]

Security by the Numbers

Faced with alarming increases in crime on the one hand, and overstrained public law enforcements agencies on the other, society has turned increasingly to private security for protection. *U.S. News and World Report* estimated that total expenditures for private security in 1974 came to $6 billion[13] A variety of sources consistently put the cost of security products and services in 1975 at approximately $3 billion, a figure raised to $4.8 billion in 1978. During the 15-year period from 1958 to 1973, the sales of contractual security more than quadrupled, leaping from $428 million to nearly $2 billion.[14] *Security Letter* reports contractual security revenues of $3.5 billion in 1978. In general, expenditures for private security by the business community, for both equipment and personnel, are rising at an annual rate of between 10 and 12 percent.

The Rand Report, released in 1971, suggests that estimates of the size of private security forces in the national media are exaggerated. While agreeing that private security grew in the 1960's at an annual rate of 11 percent, the Report contends that public law enforcement grew at an even faster rate and would rapidly begin to outgrow private security during the 1970's, as shown in Figure 1-1.

It is interesting to look at the Rand Report's projection for 1975. It is accurate in its view of the trend, but it has one flaw. Both the totals for the 1960's and the projection for 1975 were based upon Bureau of Labor and Census Bureau figures. Significantly, as the Report concedes,[15] these data report ***primary occupation only***, and thus, do not reflect the large numbers of private security personnel who are part-time workers or "moonlighters." As just one example of the difference this makes, a survey conducted by the Private Security Task Force in 1975 in the city of St. Louis, Missouri, found the total number of licensed private security personnel to be 2,977 while commissioned public police officers totaled 2,177. But *in addition to* the licensed private security professional, nearly half of the public police officers (about 1,000) also had approved secondary employment in private security, thus bringing the total number of people actually engaged in private security to approximately 4,000—or twice the number of officers engaged in public law enforcement.[16] Not only do parttime workers form a large part of the overall proprietary security forces in many parts of the country, but "part-time security workers range from 20 to 50 percent of the total employees in the contract security industry."[17] Clearly, the real figures for private security employment, both in terms of "bodies" and man-hours, could be higher than those suggested in the Rand Report.

Other research reached similar conclusions. The Report of the Task Force on Private Security cites research by the Law Enforcement Assistance Administration (LEAA) which indicated that there were more than one million persons employed in the private security industry in the United States, compared with approximately

Figure 1-1. Growth trends in protection services. Source: James S. Kakalik and Sorrel Wildhorn, *The Private Police Industry: Its Nature and Extent,* Vol. 2 (Rand Corporation), p. 33.

	1950	**1960**	**1967**	**1975 (est.)**	***1978**
Public police and other public law enforcement	199,000	260,000-266,000-	363,000	489,000	**802,000
Total guards and private security personnel	282,000	302,000-357,000	398,000	444,000	548,000

*Security Letter Vol. x, No. 14. Part II

**This figure can be used only as an approximation since it also includes firemen.

650,000 persons engaged in all levels of local, state and national law enforcement.[18] In light of current research, these figures now appear to be distorted.

Growing Pains and Government Involvement

Inevitably, the explosive growth of the security industry in the second half of this century has not been without its problems, leading to rising concerns for the quality of selection, training and performance of security personnel. "Horror stories" involving armed and untrained guards have drawn mass media attention. Both the Rand Report and the Private Security Task Force Report call attention to a serious lack of adequate training at all levels of private security. And within the industry itself, there has developed increasing pressure for improved standards, higher pay and greater professionalism.

Of the many forces acting to affect the business of security, the federal government seems at the moment to be most significant. Although a breakdown of expenditures for crime prevention by all levels of government shows that about 75 percent is expended by local governments, with state and federal shares equal to about 12 percent each, the federal government's role is increasing.

Basic to this deepening involvement at the federal level in crime prevention is the Law Enforcement Assistance Administration (LEAA), which is funded to encourage cooperation between local law enforcement agencies and to promote research, development programs and studies, generally, to improve the criminal justice system.

Established by the Omnibus Act of 1968, the LEAA grew out of the President's Commission on Law Enforcement and Administration of Justice, established in 1965. Its principal role is to allocate funds to state and local governments for the improvement of law enforcement procedures. Additionally, it budgets funds for the collection and dissemination of crime statistics and, through its National Institute of Law Enforcement and Criminal Justice, develops new approaches, techniques, systems, equipment and devices designed to improve law enforcement.

The Rand Report, frequently referenced in this chapter, was prepared under an LEAA grant. The five volumes of this report, generated by a 16-month study, sought to describe the nature and extent of the private security industry; to examine its problems, costs and benefits to society, and to suggest guidelines for its future development and regulation. It was this report which first drew national attention to some of the problems of the private security industry, particularly in the areas of inadequate training and abuses of authority.

The National Advisory Committee on Criminal Justice Stnadards and Goals was formed by the LEAA in 1975, with the purpose of developing and publishing standards and goals for every area of the criminal justice system and crime prevention. Task forces were formed to investigate such areas of concern as the police, courts, corrections, the juvenile justice system, organized crime, terrorism, and private security. Most relevant to our purposes is the Task Force on Private Security, chaired by Arthur J. Bilek.

The Report of the Task Force on Private Security, also frequently cited in this discussion, was released in 1976. It represents a "giant step" in the articulation of uniform standards for the selection and training of personnel, for the development of a code of ethics, for research and standards in areas of physical security systems and equipment, and for improved interaction between private and public law enforcement in their common goal of crime prevention. It is through the promulgation of such uniform standards and goals, along with increased public debate, that the necessary professionalism will come to the giant new industry of private security.

A New Professionalism

It is a new professionalism that private security is moving toward today. That urging comes from leaders in the field as well as from outsiders. Responsible security executives have begun to express their own concerns over the lack of regulation of the industry, a situation that leads to cutting corners and lowering standards to meet the competitive threat. Russell L. Colling, past president of the International Association for Hospital Security, has observed that "in order to upgrade and regulate the security function, good licensing laws are required."[19] Saul Astor, president of Management Safeguards, Inc., and for more than 20 years a prominent spokesman for security and loss prevention, writes, "A code of ethics is most important for the security services industry; without it, our work is an hypocrisy and we have no *raison d'etre* that would in any way contribute to the betterment of our communities. We would merely be automatons reacting to repetitive stimuli, unworthy indeed of the status of a professional."[20] In defining the desired professionalism, Astor cites the need for a code of ethics and for "credentials," including education and training, experience, and membership in a professional society.

This thrust toward professionalism is observed in the proliferation of active private security trade organizations. It is promoted by such organizations as the American Society for Industrial Security (ASIS), which has a membership of more than 14,000 security executives; the International Association for Hospital Security (IAHS); the National Association of School Security Directors (NASSD); the Security Equipment Industry Association (SEIA), and many others. It finds its voice in an emerging library of professional security literature, both magazines and books. And it looks to its future in the rapid development in recent years of college-level courses and degree programs in security.

Despite the many efforts to professionalize the field of private security, there are still many who feel that major obstacles need to be overcome. The most persistent one has to do with the training and education of the security guard. Current standards, codes of ethics, and educational courses which ASIS has introduced in its Certified Protection Program (CPP) are directed at security management personnel, not security guards. Many guards are underpaid, undertrained, undersupervised, and not regulated. Minimal standards do exist in some places, but there is still a reluctance to train, educate, and adequately compensate the guard force. Business practices of making a product for profit can make it difficult for companies to see the need for paying for costly security programs. Thus, they often opt for the lowest priced "solution", whether or not it affords real protection. Fortunately, this kind of thinking is undergoing a change as industry realizes that the adage "you get what you pay for" very definitely applies to the quality of security. This realization will, in turn, add pressure for the industry to upgrade the position of security guard.

SUMMARY

Security, as we have seen, finds its sturdiest historical roots in both the concepts of justice and the social structure of early England. It is possible to trace the evolution of ideas of law and of mutual security through the history of England and across the sea to colonial America.

In America, the development of public law enforcement was strongly affected by the fragmentation made necessary in a nation made up of separate state jurisdictions. This fact contributed to the rise of specialized private police and investigative agencies in the 1800's. Modern police departments, influenced by the example of Sir Robert Peel and the Metropolitan Police in London, England, began to develop into something like their present form in the second half of the 19th century.

Individual security services, such as armored car and alarm services, continued to develop in the first half of the 20th century, but the need for security, especially in plant protection, was greatly intensified during World War II and in the immediate postwar years. And, as crime flourished in the next two decades, both contractual and proprietary security grew at an unprecedented rate, aided significantly by dramatic advances in electronic technology.

Today, in the words of the Private Security Task Force Report, "as a result of ever-rising crime rates, coupled with the enormous demands placed upon public law enforcement agencies and their lack of adequate resources to deal with these demands, private security has become a multi-billion-dollar-a-year industry, and the number of private security personnel surpasses that of public law enforcement in many localities. Moreover, present crime and financial statistics indicate that the industry will continue to experience significant growth in future years."[21]

REVIEW QUESTIONS

1. What events in medieval England brought about the creation and use of private night watches and patrols?

2. What factors or conditions of the times made it possible for Jonathan Wild to become so extraordinarily successful?
3. How did "state's rights" contribute to the use of private security agencies in 19th Century America?
4. How did World War II affect the growth of modern private security?
5. What is LEAA and what has been its impact on the private security industry?

PART II
THE SECURITY FUNCTION

Chapter 2

DEFINING SECURITY'S ROLE

As we have seen, during the 19th and early 20th centuries public police operated only on a local basis. They had neither the resources nor the authority to extend their investigations or pursuit of criminals beyond the sharply circumscribed boundaries in which they performed their duties. When the need arose to reach beyond these boundaries, or to cut through several of these jurisdictions, law enforcement was undertaken by such private security forces as the Pinkerton Agency, railway police, or the Burns Detective Agency.

As the police sciences developed, public agencies began to assume a more significant role in the investigation of crime and, through increased cooperation among government agencies, the pursuit of suspected criminals. Concurrent with this evolution of public law enforcement, private agencies shifted their emphasis away from investigation and toward crime prevention. This led to an increasing use of guard services to protect property and to maintain order. Today, in terms of numbers, guard forces are by far the predominant element in private security.

But what other protective measures are available? Who proves them? Who is responsible for planning and executing these procedures? Where do the roles of private and public police overlap and where do they diverge? What are the particular hazards for which private security is now held responsible, and how is it determined that these threats are sufficient to justify the adoption of protective procedures?

To answer these questions it is necessary to define private security and its role more exactly.

What Is Private Security?

Although the term "private security" has been used in previous pages without question, there is no universal agreement on a definition, or even upon the suitability of the term itself. Cogent arguments have been made, for example, for substituting the term "loss prevention" for security.

The Rand Report defines private security to include all protective and loss-preventive activities not performed by law enforcement agencies. Specifically,

> The terms private police and private security forces and security personnel are used generically in this report to include all types of private organizations and individuals providing all types of security-related services, including investigation, guard, patrol, lie detection, alarm, and armored transportation.[22]

The Private Security Task Force takes exception to this definition on several grounds. The Task Force argues that "quasi-public police" should be excluded from consideration, on the grounds that they are paid out of public funds, even though they may be performing what are essentially private security functions. The Task Force also makes the distinction that private security personnel must be employees of a "for-profit" organization or firm, as opposed to a non-profit or governmental agency. The complete Task Force definition states:

> Private security includes those self-employed individuals and privately funded business entities and organizations providing security-related services to specific clientele for a fee, for the individual or entity that retains or employs them, or for themselves, in order to protect their persons, private property, or interests from varied hazards.[23]

The Task Force argues that the profit motive, and the source of those profits, are basic elements of private security. While this definition might be suitable for the specific purposes of the Report, it hardly seems acceptable as a general definition. Many airports, hospitals, and schools, to name only three types of institutions, employ private security forces without the "for-profit" orientation, yet it would be difficult to contend, for example, that the members of the International Association for Hospital Security are not private security personnel.

Neither the profit nature of the organization being protected, nor even the source of funds by which personnel are paid, holds up as a useful distinction. A night watchman at a public school is engaged to protect a non-profit installation and is paid out of public funds. His function, however, is clearly different from that of the public law enforcement officer. He is—and is universally accepted as—a private security guard.

How, then, should private security be defined for the purposes of this text?

The opening lines of this text suggest that security "implies a stable, relatively predictable environment in which an individual or group may pursue its ends without disruption or harm, and without fear of such disturbance or injury."

Such security can be effected by military forces, public law enforcement agencies, by the individual or organization concerned, or by organized private enterprises. Where the protective services are provided by personnel who are not only paid out of public funds but also charged with the *general* responsibility for the public welfare, their function is that of public police. Where the services are provided for the protection of *specific* individuals or organizations, they normally fall into the area of private security.

It should be noted that public and private police may, in certain circumstances, perform the same functions for the same individuals or organizations. A law enforcement officer might, in some circumstances, be assigned to protect a threatened individual; a private bodyguard frequently is hired to perform the same protective function. Public police commonly perform patrol functions which include checking the external premises of stores or manufacturing facilities. But patrol is also one of the major activities of private security. The activity itself, then, is not always differentiating. Private security functions are essentially *client-oriented*; public law enforcement functions are *society-* or *citizen-oriented.*

Another key distinction is the possession and exercise of police powers, that is the power of arrest. The vast majority of private security personnel have no police powers; they act as private citizens. In some jurisdictions, ''special officer'' status is granted, in most cases by statute or ordinance, which includes limited power of arrest in specified areas or premises. The limitations on the exercise of special police powers, and the fact that their activities are client-oriented and client-controlled (as opposed to being directed primarily by public law enforcement agencies) make it reasonable to include such personnel as part of the private security industry. (This discussion omits the situation of the law enforcement officer who is ''moonlighting'' as a part-time private security guard, since his police powers derive from his public rather than his private role.)

Finally, the role of private security can be differentiated from public law enforcement by the predominent purpose of each. Private security is concerned principally with crime or loss *prevention.* By contrast, public police direct most of their efforts toward the enforcement of laws and the apprehension of criminals *after a crime has been committed.* The success of private security efforts is directly related to the absence of crime or loss; the success of public law enforcement tends to be measured by successful apprehension and prosecution.

For our purposes, then, private security can be defined as *those individuals, organizations and services, other than public law enforcement agencies, which are engaged primarily in the prevention of crime, loss, or harm to specific individuals, organizations, or facilities.*

Types of Hazards

The hazards against which private security seeks to provide protection are commonly divided into *man-made* and *natural* hazards.

Natural hazards may include fire, windstorm, flood, earthquake and other disasters, building collapse, equipment failure, accidents, and safety hazards. It should be noted that fire and accidents are also quite often man-made, intentionally or unintentionally.

Man-made hazards may include crimes against the person, theft and pilferage, fraud and embezzelment, espionage and sabotage, civil disturbances, bomb threats, fire and accidents (as noted above) and, in some situations, overt attack.

The degree of exposure to specific hazards will vary for different facilities. The threat of fire or explosion is much greater in a chemical plant; the potential of loss from shoplifting or internal theft is greater in a retail store. Each organization or facility must, ideally, be protected against the full range of hazards, but in practice, a particular protection system will emphasize some hazards more than others.

In some organizations, the whole area of accident prevention and safety has taken on such importance, primarily because of state and federal occupational safety and health legislation, that this responsibility has become a full-time objective in itself, in charge of a Director of Safety. Security can then devote its energies to other areas or loss. Similarly, some large industrial facilities have full-time fire departments. In most situations, however, both fire and accident prevention are part of the responsibility of the security department.

Security Functions

Security practices and procedures cover a broad spectrum of activities designed to eliminate or reduce the full range of potential hazards (loss, damage or injury). These protection measures may include, but are by no means limited to, all of the following:

1. Building and perimeter protection, by means of
   barriers, fences, walls, gates
   protected openings
   lighting
   surveillance (guards).
2. Intrusion and access control, by means of
   door and window security
   locks and keys
   security containers (files, safes, vaults)
   visitor and employee identification programs
   package controls
   parking and traffic controls
   inspections
   guard posts and patrols.
3. Alarm and surveillance systems.
4. Fire prevention and control, including
   evacuation and fire response programs
   extinguishing systems
   alarm systems.
5. Emergency and disaster planning.

6. Prevention of theft and pilferage, by means of
   personnel screening
   background investigations
   procedural controls
   polygraph and PSE (Psychological Stress Evaluator) investigations.
7. Accident prevention and safety.
8. Enforcement of crime- or loss-related rules, regulations and policies.

In addition to these basic loss-preventive functions, security services in some situations might also provide armored car and armed courier service, bodyguard protection, management consulting, security consulting, and other specific types of protection.

These services may be *proprietary,* or in-house, in which case the security force is hired and controlled directly by the protected organization; or it may be a *contract* security service, in which case the company contracts with a specialized firm which engages to provide designated security services for a fee. Contract security employees are actually employees of the contract security firm. Most security functions may be provided by either proprietary or contract forces or services; and in practice it is often common to find a combination of such services used.

Since the focus on this text is primarily on the proprietary security organization, it is of value to briefly consider the uses of contract security services here.

CONTRACT SECURITY SERVICES

A 1975 study by the Morton Research Corporation placed the number of contract security firms in the United States in 1973 at 4,182. In the six-year period from 1967 to 1973, there had been an increase in the number of protective service establishments of 1,624.[24] Security services in 1978 totaled approximately 5,500 companies doing an estimated $3.5 billion in business, primarily providing guard, investigative, central station alarm, and armored car and courier services.

It has been said that six large, publicly owned firms dominate the industry, accounting for approximately half of all revenues generated by contract security services. The rapid rise in the number of smaller firms suggests that they are claiming a larger share of the total market. However, based on 1978 data presented by *Security Letter*, the eight leading national firms provide 33 percent of the employment for the security industry.[25]

Many firms, particularly the smaller ones, specialize in the type of services offered to a client. The larger the firm the more likely it is to provide a full range of security services as necessary.

The major categories of these services are guard forces, patrols, investigative services, alarm response, and armored car delivery and courier services.

Guard Service

Guard supply represents the major service provided in the industry today. Only part of a guard's job is crime-related. Whenever possible or necessary, he is required to prevent major crimes and to report those that have been committed. But his major role may be to direct traffic, to screen persons desiring access to a facility, and generally, to enforce company rules. In many modern applications, his role is less regulatory than helpful. He may direct or escort persons to their destination within a facility, he may act as a receptionist or as a source of information, or he may be primarily concerned with safety.

Since many guards are concerned for only a small percentage of time with crime-related activity, there is some effort in various quarters to adjust the guard's appearance to fit his role by outfitting him in blazer and slacks rather than the uniform with its police or coercive connotation.

A guard is, however, a guard. Even if, in his particular assignment, he is never confronted with criminal activity, he is still charged with certain responsibilities in that area, and he is additionally responsible to protect the interests of his employer on the employer's property.

Private guards differ from public police both in their legal status and in that they perform in areas where the public police cannot legally or practically operate. The public police have no authority to enforce private regulations, nor have they the obligation to investigate the unsubstantiated possibility of crime (such as employee theft) on private property.

The job of the private guard is to provide specific services under the direction and control of a private employer, one who feels that he needs to exercise controls or supervision over his own property or goods, or to provide additional services that the public police as a practical matter simply cannot provide.

Patrol Service

Private patrol services offer a periodic inspection of various premises by one or more patrolmen operating either on foot or in a patrol car.

The tour of such patrols may cover several locations of a single client, or it may include several establishments owned by different clients within a limited neighborhood. Inspections of patrolled premises may be visual perusal from the outside, or they may require entering the premises for a more thorough inspection. Typically, the arrangement made with a client specifies that a certain number of inspections will be made within a given period of time or with a specified frequency.

The patrolman differs from the guard in that he operates through a tour covering various locations, whereas the guard stands a fixed post or walks a limited area. The patrol service is more economical, since the guard mans a post for the full period during which a danger exists. But the patrol has the possible disadvantage of being circumvented by an intruder who knows that there will be some period of time between inspections of a given premise.

Investigative Services

The private investigator is a gatherer of facts—in essence, a research man who spends the greatest portion of his time collecting background information for pre-employment checks or personnel, background checks of applicants for insurance or credit, and investigation of insurance claims. Much of his work is non-crime-related, although he may become involved either part- or full-time in undercover investigations of employee theft or in detecting shoplifting.

Investigations in divorce-related matters are declining as divorce laws are becoming more liberalized. Tracing missing persons or investigating criminal matters on behalf of the accused are a very small part of a typical investigator's work. In fact, the Pinkerton Agency will not handle any matter involving the defense of persons under prosecution by the public law enforcement agencies, nor matters pertaining to domestic or marital investigation.

Although there are situations, such as the long-term relationship between the American Banking Association and the Burns Agency, where private investigations are called upon to *supplement* the work of public police, the great majority of private investigative work is *complementary* to the public law enforcement effort.

Alarm Response Service

Central station alarm systems consist essentially of alarm sensors located in the protected premises, and a communication line from the sensors to a privately owned central station alarm board which is monitored and responded to by private security personnel.

In some cases, central station systems do not dispatch personnel to respond to an alarm, but merely relay the alarm received to public police headquarters. But in most such systems, the alarm is relayed and someone is also sent to the scene.

Alarms connected to a central station are usually designed to detect intrusion, but they can also be used to monitor industrial processes or conditions.

Certainly, central station coverage of a facility is cheaper than full-time security employees performing essentially the same function. A drawback is that the current false-alarm rate for virtually all intrusion systems is still very high, resulting in a growing resistence to the use of direct connection systems to police headquarters. Central station operators have some flexibility in checking the validity of an alarm before notifying the police, so they may, to some degree, reduce the incidence of false alarms demanding police response.

In cases where central station personnel actively investigate the intrusion, even taking steps to apprehend a suspect before the arrival of the police, they *supplement* the public police effort.

Armored Delivery Service

Armored delivery services provide for the safe transfer of money, valuables, or any goods the employer may wish to move from one location to another. By far, the widest use of this service is to transport cash and negotiables from a receiving point to a bank or other depository. Payrolls, cash receipts, or cash supplies for the daily business are the principal traffic of the armored delivery service.

Personnel employed by such services are not concerned with the general security of the premises they serve; their responsibility is confined to the safe transport of sensitive items as directed by the customer.

Courier services perform a similar function in the safe transport of valuables. They are distinguished from armored car services principally by using means other than special armored vehicles.

Other Services

In addition to the services described, private security firms also provide such services as crowd control, canine patrol, bodyguard and escort service, executive protection, polygraph examination and psychological stress evaluation, security consulting, and other related loss-preventative assistance to business and industry. Bodyguard service, in particular, has increased rapidly in the mid-1970's, in response to the rising incidence of executive kidnappings and hostage situations.

CONTRACT VS. PROPRIETARY SERVICES

Although some researchers, including the authors of the Rand Report previously mentioned, have perceived more rapid growth in contract security services than in proprietary security, the more recent studies by the Private Security Task Force "concluded that the growth of proprietary security has paralleled that of contractual security."[26] Although not always as visible as other forms of security, proprietary forces have, in fact, experienced equal if not greater growth.

Since the various contract functions described, with the exception of armored car delivery service, can be undertaken as proprietary *or* inhouse activities, how then is the choice to be made between the two types of services?

The Rand Report discusses the pros and cons of the issue in a presentation which is a synthesis of views from both in-house and contract services, various major users of such services, and material from several periodicals dealing with the subject. Some of their conclusions are reflected in the following discussion of the relative merits of the two approaches to security services.

Various factors will determine the decision to hire proprietary or contract guards in any given situation. These factors will include the location to be guarded, the size of the force required, its mission, the length of time the guards will be needed, and the quality of personnel required.

Advantages of Contract Services

Cost. Few experts disagree that contract guards are less expensive than a proprietary unit. Contract guards will earn from $3.10 to $3.50 per hour (with the supplier firm charging from $5.00 to $5.25 per hour); in-house guards typically earn more because of the general wage rate of the facility employing them. In many cases that wage level has been established by collective bargaining.

Contract guards receive fewer fringe benefits, and their services can be provided more economically by large contract firms by virtue of savings in costs of hiring, training, and insurance because of volume. Short-term guard service on the proprietary basis can create such large start-up costs that the effort is impractical.

Liability insurance, payroll taxes, uniforms and equipment, and the time involved in training, sick leave and vacations are all extra cost factors that must be considered in establishing a proprietary force.

Administration. Establishing in-house guard service requires the development and administration of a recruitment program, personnel screening procedures, and training programs. It will also involve the direct supervision of all guard personnel. Hiring contract guards solves the administrative problems of scheduling and substituting manpower when someone is sick or terminates his employment.

There is little question that the administrative chores are substantially decreased when a contract service is employed. At the same time, the contracting customer is obliged to check the supplier's performance of contracted services on an ongoing basis, and he must additionally insist on a satisfactory level of quality at all times. To this extent, management of the client firm is not totally relieved of administrative responsibilities.

Manpower. During any periods when the need for guards changes in any way, it is necessary to lay off existing guards or take on additional manpower. Such changes may come about fairly suddenly or unexpectedly.

In-house forces rarely have this flexibility in manpower. If they have extra men available for emergency use, such men are an unnecessary expense when they are idle. Similarly, if there is a temporary decrease in the need for guards, it would hardly be efficient to dismiss extra men only to rehire additional guards a short time later when the situation changed again.

With their larger pools of manpower, contract firms can use their personnel with a high degree of efficiency. By proper scheduling, they can provide extra men on short notice. Firms with relatively small needs in guard personnel will find this availability problem significant unless they have access to a ready pool of trained guards. Almost invariably, such a pool involves a contract service.

Unions. Guard users in favor of non-union guards support their position by arguing that such guards are not likely to go out on strike, they are less apt to sympathize or support striking employees, and they can be paid less because they receive few, if any, fringe benefits.

Since 90 percent of all unionized guards are proprietary personnel, anyone subscribing to the arguments listed here would clearly favor hiring contract guards. Only 10 percent to 25 percent of the guards employed by the three largest contract guard agencies (Pinkerton, Burns, Wackenhut) are unionized.

Impartiality. It is often suggested that contract guards can more readily and more effectively enforce regulations than in-house personnel. The rationale is that contract guards are paid by a different employer and, because of their relatively low seniority, have few opportunities to form close associations with other employees of the client. This produces a more consistently impartial performance of duty.

Expertise. When a client hires a guard service, he also hires the management of that service to guide him in his overall security program. This can prove valuable even to a firm that is already sophisticated in security administration. A different view from a competitive supplier trying to create good will with his client can always be illuminating.

Advantages of Proprietary Guards

Quality of Personnel. Proponents of proprietary guard systems argue that the higher pay and fringe benefits offered by employers, as well as the higher status of in-house guards, attracts higher quality personnel. Such employees have been more carefully screened, and they show a lower rate of turnover.

Control. Many managers feel that they have a much greater degree of control over personnel when they are directly on the firm's payroll. The presence of contract supervisors between guards and client management can interfere with the rapid, accurate flow of information either up or down.

An in-house force can be trained to suit the needs of the facility, and the progress and effectiveness of training can be better observed in this context. The individual performance of each member of the force can also be evaluated more readily.

Loyalty. In-house guards are reported to develop a keener sense of loyalty to the firm they are protecting than do contract guards. The latter, who may be shifted from one client to another, and who have a high turnover rate, simply do not have the opportunity to generate any sense of loyalty to the specific—often temporary—client-employer.

Prestige. Many managers simply prefer to have their own men on the job. They feel that the firm gains prestige by building its own security force to its own specifications, rather than renting one on the outside.

Conclusions

Obviously, in weighing the various factors on either side of this debate, the prudent manager will carefully study the quality and performance of the guard firms available to service his facility. He will assure himself of their standards of personnel, training and supervision. He will make a careful analysis of the comparative costs for proprietary or contract services, and he will make an estimate as to their relative effectiveness in his particular application.

In situations where the demand for guards fluctuates considerably, a contract service is probably indicated. If a fairly large, stable guard force is required, an in-house organization might be favored.

As an indication of the experience of a sample of industrial firms, a survey undertaken by the American Society for Industrial Security revealed that approximately 40 percent of the respondents used contract services for 50 percent or more of their guard needs. Of all users of some contract security, over 50 percent responded that they did so for reasons of economy or to avoid the administrative headaches of labor and personnel problems. Generally speaking, their experience with such service was rated as from fair to good.[27]

REVIEW QUESTIONS

1. What does the term 'private security' mean to you?
2. What are the differences between proprietary and contractual guard services?
3. What are the basic services typically performed by contractual security personnel?
4. What are the advantages and disadvantages of using contractual security services?
5. What are the advantages and disadvantages of using proprietary security services?

Chapter 3

THE PROPRIETARY SECURITY ORGANIZATION

Security problems become apparent in virtually every area of a given company's activities. The need to deal forcefully and systematically with these problems has become increasingly evident to the industrial and commercial community, and steps have been taken by greater and greater numbers of these organizations to create a security effort as an organic element of the corporate structure, rather than turning to outside (contract) security services, or to minimal efforts at physical security.

Where and how the in-house or proprietary security department operates within the organizational framework, and how this relates to the total security system of individual concerns, depends upon the needs of that organization. General principles will apply throughout much of the business community, but specific applications must be tailored to the problems faced by each enterprise. Our concern here, then, is with those considerations which have broad application in the organization of the security function.

Determining the Need

In evaluating the need to install or expand the company security function, the immediate urgency for increased security must be considered, along with the status, growth and prior performance of the security effort. The peculiarities of the company itself in the context of intra-company relationships, whether by design or natural evolution, must be a factor. The potential for growth of the company and the attendant growth of staff activities should also be considered.

Ultimately, management will have to determine the costs and the projected effectiveness of the security function. Then it will have to face the big question as to whether security can be truly and totally integrated into the organization at all. If, upon analysis, it is found that the existing structure would, in some way, suffer from the addition of new organizational functions, alternatives to the integrated proprietary security department must be sought.

These alternatives usually consist only of the application and supervision of physical security measures. This inevitably results in the fragmentation of protective systems in the various areas requiring security. However, these alternatives are sometimes effective, especially in those firms whose overall risk and vulnerability are low. But as the crime rate continues to climb, and as criminal methods of attack and the underground network of distribution continue to become more sophisticated, anything less than total integration will become increasingly more inadequate.

Once management has recognized that existing problems—real or potential—make the introduction or enlargement of security a necessity for continued effective operation, it is obliged to exert every effort to create an atmosphere in which security can exert its full efforts to accomplish stated company objectives. Any equivocation by management at this point can only serve to weaken or to ultimately undermine the security effectiveness that might be obtained by a clearer statement of total support and directives resulting in intra-company cooperation with security efforts.

SECURITY'S PLACE IN THE ORGANIZATION

The degree and nature of the authority vested in the security manager become matters of the greatest importance when such a function is fully integrated into the organization. Any evaluation of the scope of authority required by security to perform effectively must consider a variety of factors, both formal and informal, that exist in the structure.

Definition of Authority

It is management's responsibility to establish the level of authority at which security may operate in order to accomplish its mission. It must have authority to deal with the establishment of security systems. It must be able to conduct inspections of performance in many areas of the company. It must be in a position to evaluate performance and risk throughout the company.

All such authority relationships, of course, should be clearly established in order to facilitate the transmission of directives and the necessary response to them. It should be noted, however, that these relationships take many forms in any company, not infrequently including an assumption of a role by a member of the organization who becomes accepted as a designated executive simply by past compliance and by custom. In such cases, where management does not move to curtail or redefine his authority, he continues in such a posture indefinitely, whatever his formal status might be. It is management's responsibility to continually reassess the chains of authority in the interests of efficient operation.

Organizational structure generally distinguishes between *line* and *staff* relationships. Line executives are those who are delegated authority in the direct chain of command to accomplish specific organizational objectives. Staff personnel generally provide an advisory or service function to a line executive.

In general, the security manager can be considered to serve a staff function. Traditionally, this means that, as the head of a specialized operation, he is responsible to a senior executive or (in the fully integrated organization) to the president of the company. His role is that of an advisor. Theoretically, it is the president who, through his authority, implements the activities suggested by his advisors.

This is not always the practice. By the very nature of his expertise, the security manager has authority delegated by the senior executive to whom he reports. In effect, he is granted a part of the authority of his line superior. This is known as *functional authority*.

Such authority rarely appears on a table of organization, since it is delegated and can be modified or withdrawn by the superior. In the case of security, this functional authority may consist of advising operating personnel on security matters, or it may and should develop into more complete functional responsibility to formulate policies and issue directives prescribing procedures to be followed in any area affecting the security of the company.

Most department managers cooperate with security directives readily, since they lack the specialized knowledge required of upper echelon security personnel and they are generally unfamiliar with the requirements of effective supervision of security systems and procedures. It is nonetheless important that the security manager operate with the utmost tact and diplomacy in matters which may have an effect, however small, on the conduct of personnel or procedures in other departments. Every effort should be made to consult with the executive in charge of any such affected department before issuing instructions that implement security procedures.

Levels of Authority

Obviously, there are many mixtures of authority levels at which the security manager operates.

His functional authority may encompass a relatively limited area, prescribed by broad outlines of basic company policy.

In matters of investigation, he may be limited to a staff function in which he may advise and recommend or even assist in conducting the investigation, but he would not have direct control or command over the routines of employees.

It is customary for the security manager to exercise line authority over preventive activities of the company. In this situation he commands the guards who in turn command the employees in all matters over which he has jurisdiction.

The security manager will, of course, have full line authority over the conduct of his own department, within which he, too, will have staff personnel as well as those to whom he has delegated functional authority.

Reduced Losses and Net Profit

Although security is a staff function, it could be viewed as a line operation—and one day may be. An effective security program, intelligently managed, can, by reducing losses, maximize profits just as surely as can a merchandising or a production function.

With crime rates soaring, currently costing all business approximately $40 billion annually, every business is targeted for losses. And all these losses come off the net profit.

Many managers, particularly those in retail establishments, push as hard as possible for an increase in gross sales. They frequently brush aside any words of caution about inventory shrinkage due to internal and external theft. New records in gross sales are their goals. Unfortunately, companies doing a gross annual business of from $50 million to $75 million have filed for bankruptcy.

It is the net that keeps business and industry alive. The gross may be a splashy figure. It may provide some excitement for the proud manager. But the net is the bottom line—and anything that eats away at that slender lifeline seriously endangers the organization.

Retailers, who are the hardest hit by criminal attack, may operate at a net profit of two percent. It has been estimated that the national average loss due to crime is anywhere from .7 percent to as high as 10 percent of sales. A store netting two percent and doing a gross weekly business of $50,000 will net a little over $50,000 a year. But if that store is hit at one percent, the owner will end up with $25,000 instead of the $50,000 really earned.

An effective security operation could cut those losses by as much as 90 percent. The savings between the investment in security and the additional earnings realized from reduced losses are net profit.

In this light, security can be seen as a vital function in the profit picture of any company. Any operation that can minimize losses and maximize net profits should clearly take its place as an independent organizational function, reporting to the highest level of executive authority.

Non-Integrated Structures

In spite of the obvious advantages to integrating the security function into the organization as an organic function, many firms continue to locate this vital operation as a reporting activity of some totally unrelated department. Since, in many cases, the operation grew out of some security need that arose in a particular area, there is a tendency for that area to assume administrative control over it—and to maintain that control long after security has begun to extend its operational interests beyond departmental lines into various activities of the company.

In this way, security was traditionally attached to the financial function of the organization, since financial control was usually the most urgent need in a company otherwise unprepared to provide internal security. The disadvantages of such an arrangement are severe enough to endanger the effectiveness of security's efforts.

Functional authority cannot be delegated beyond the authority of the delegant. What this means is that, when the security manager gets his authority from the comptroller, it cannot extend past the comptroller's area of responsibility. If the comptroller can extend the role of security by dispensation of the chief executive, the line of authority becomes clouded and cumbersome.

Most business experts agree that functional authority should not be used to direct the activities of anyone more than one level down from the delegate in order to preserve the integrity of line functions. Clearly, the assignment of security under the financial officer is a clumsy arrangement representing bad management practice.

Relation to Other Departments

Every effort should be made to incorporate security into the organizational functions. It must be recognized, however, that by so doing management creates a new function that, like personnel and finance, among others, cuts across departmental lines and enters into every activity of the company.

Security considerations should, ideally, be as much a presence in every decision at every level as are cost decisions. This will not mean that security factors will always take precedence over matters of production or merchandising, for example, any more than specific price factors will always determine decisions in these same areas. But security should always be considered. If its recommendations are overridden from time to time—sometimes a wise decision, where the cost of disruption involved in overcoming certain risks is greater than the risk itself—this will be done with full knowledge of the risks involved.

Obviously, the management of the security function and its goals must be compatible with the aims of the organization. This means that security must provide for continued protection of the organization without significant interference with its essential activities. Security must preserve the atmosphere in which the company's activities are carried on by developing systems that will protect those activities in much their existing condition, rather than attempting to alter them to conform to certain abstract standards of security. When the overall objectives of any organization are bent and shaped to accommodate the efforts of any of the particular functions designed to help achieve those aims, the total corporate effort inevitably becomes distorted and suffers accordingly.

Organizationally, the relationship between security and other departments *should* present no difficulty. The interface serves to solve potentially disruptive or damaging problems shared by both functions. The company's goals are achieved by the elimination of all such problems. In practice, however, this harmony is not always found. Resentment and a sense of loss of authority can interfere with the cooperative intra-departmental relationships that are so vital to a company's progress.

Such conflicts will be minimized where security's authority is clearly defined and understood.

The Security Manager's Role

Directing our attention to the generalization of the security operation and the manager's role in it, we can find many common elements that are significant. In its organizational functions, security encompasses four basic activities with varying degrees of emphasis. These are:

Managerial—Includes those classic management functions common to managers of all departments within any organization. Among these are planning, organizing, employing, leading, supervising, and innovating.

Administrative—Involves budget and fiscal supervision, office administration, establishment of policies governing security matters and development of systems and procedures, developing of training programs for security personnel and security education of all other employees, providing communication and liaison between departments in security related matters.

Preventive—Includes supervision of guards, patrols, fire and safety personnel, inspections of restricted areas, regular audits of performance, appearance, understanding and competence of security personnel, control of traffic, condition of all security equipment such as alarms, lights, fences, doors, windows, locks, barriers, safes and communication equipment.

Investigative—Involves security clearances, investigation of all losses or violations of company regulations, inspections, audits, liaison with public police and fire agencies, classified documents.

It is important to remember that these latter three functions must be carried out to further the organizational needs of security. It follows that, in order to perform effectively, the security manager must be thoroughly conversant with all of the techniques and technologies inherent in such functions. But, in order to achieve the stated goals or the projected ends of the organization, he must be sufficiently skilled in his managerial duties to effectively plan, guide and control the performance of his department.

He cannot remain, as has been true so often in the past, merely a "security expert"—a technician with a high enough degree of empirical or pragmatic information to qualify him to undertake certain basic preventive or investigative tasks. The more he involves himself personally in such jobs, the more he will neglect his managerial functions. Security's role in the operation will suffer accordingly.

It is important to remember that companies which recognize the need and the efficiency of incorporating security as an organic part of their enterprise have begun the process of creating a new organizational function that will, along with such traditional functions as marketing, production, finance and personnel, play a significant role in the daily, as well as the projected, destiny of the company.

In this light, it is clear that the security manager will function as an indispensable member of the staff. His role will extend far beyond the time-honored one of principal-in-charge of burglar alarms and package inspections, to which he has so often been relegated. This is not to suggest that there is any trend toward development of a power base for security management, but rather that many enlightened, modern company managers have assigned a higher priority to integrated security systems in an effort to encourage the growth of this function as an essential element of the firm's survivial.

ORGANIZING THE SECURITY FUNCTION

While the organization and administration of a security department is a subject in itself, beyond the scope of this general introduction to security, it is nevertheless important and necessary to take an overview of the security organization, looking briefly at both the function and the staff required to implement it.

From a management point of view, organizing the security effort involves:

1. Planning
2. Establishing controls
3. Organizing the security department
4. Hiring personnel
5. Training
6. Supervision
7. "Selling" security
8. Departmental review and evaluation

Planning

It is an extraordinarily common mistake to put the cart before the horse in security planning—that is, to create a department, hire personnel, and then look around for something to do, on the premise that crime is rising, losses almost certainly exist, and something, therefore, must be done about them.

In reality, need comes first. A hazard must exist before it becomes practical to establish an organized effort to prevent or minimize it. The first step in security planning is a detailed analysis of potential areas of loss, their probability and their gravity in terms of corporate goals. Only then can the specific objectives of the security function be defined.

This relationship of corporate goals to security planning is suggested in Figure 3-1. To express this relationship in a simplified way, if a company's goal is higher profits and the widespread prevalence of employee theft is eating away those profits, a primary objective of the security function should be to reduce employee theft and thus to contribute to the corporate goal of increased profits.

Analyzing risks is discussed in detail in a later chapter, along with the security survey. In addition to this threat assessment, then, planning involves establishing objectives, allocating resources within prescribed or authorized budgetary limitations, and determining what should be done, how it should be done, and how soon it should be set into operation.

Establishing Controls

Security planning, including threat assessment, will result in a determination of the degree of security required in all areas of a company. Decisions must also be made as to the *means* by which such security can be most efficiently, effectively and economically achieved. New policies and procedures may be formulated, physical aids to security may be ordered, and the size and deployment of security personnel will be determined. All of these factors must be balanced in the consideration of the protec-

Figure 3-1. Establishing objectives in security planning.

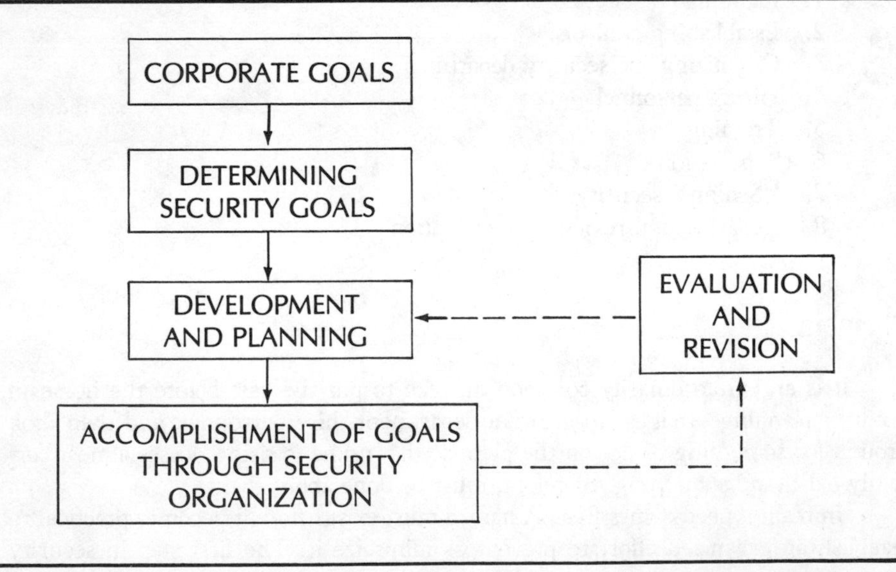

tion of the facility to arrive at a formula providing the most protection at the least expense.

Controls must be established over procedures such as shipping, receiving and warehousing, inventory, cash handling, auditing, accounting, etc. Since all of these functions are performed in other departments, the most effective and efficient method of implementing such controls is by the presentation of a control or accountability system to the department manager and allowing him to express his views and to make counter-suggestions. There is no reason to suppose that a totally satisfactory control procedure cannot be reached in this spirit of mutuality. Only when such controls break down or prove to be inadequate should the security manager or his deputy step in to handle the matter directly.

As will be discussed in later chapters, loss-preventive controls would also cover all physical protection devices, including interior and exterior barriers of all kinds, alarm and surveillance systems, and communication systems.

Identification and traffic patterns are other necessary controls. Identification implies the recognition of authorized versus unauthorized personnel, and traffic in this context includes all movement of personnel, visitors, vehicles, goods and materials.

Organizing the Security Department

An organization, as such, is *people,* so in considering the organizational structure of a security department we are referring to the assignment of duties and responsibilities to people in a command relationship in order to achieve defined goals.

It is necessary first to identify tasks, and then to develop the organization required to discharge those tasks. To put this another way, the goals or objectives of the department are divided into practical work units, and within those units specific jobs are defined.

A simplified table of organization for a small industrial security department of 20 persons might take the form charted in Figure 3-2. Even such a small organization requires careful description of specific duties and responsibilities, from the manager down to the guard on patrol, with clearly defined reporting levels and a clearly understood hierarchy of command. In this sample, for instance, the security manager himself would have more extensive "line" duties than would be the case in a larger department. He would be more directly involved in day-to-day operations (such as investigations), whereas in a larger department he might be occupied entirely with planning, advising, communications, public relations, and other administrative duties, leaving operations to his subordinates.

The particular security organization, like any other organizational structure, must be designed to meet particular needs. For this reason, it is impossible to suggest a "model organization" for the individual security department, even within the same type of enterprise (such as manufacturing plants or retail stores). The specific risks, the size of the company, the physical environment, the budget—all of these

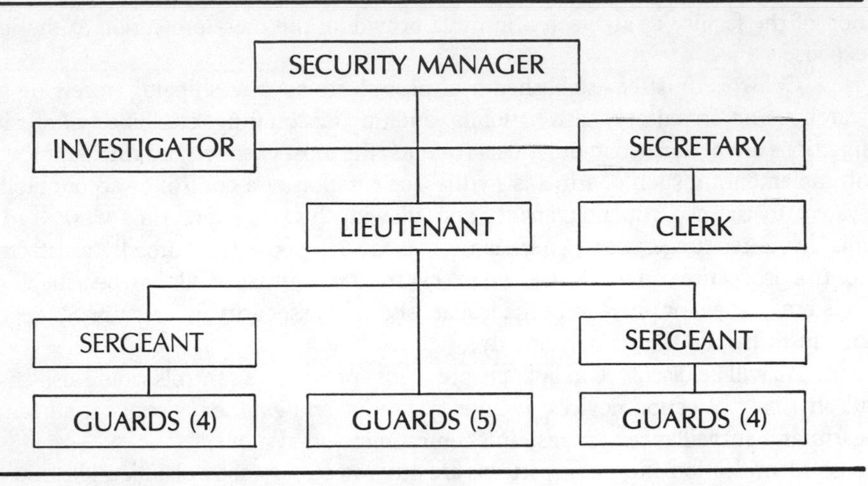

Figure 3-2. Small security department organization.

affect and to a great extent dictate the nature of the security response and thus of the organization needed to carry it out. One company's ideal organizational structure will not fit another's, except by chance.

This does not mean that the individual manager cannot benefit from the practices of others. However, he should not try to adopt any other "security package." He must adapt standard practices to suit his particular situation.

Some common matters of concern in any organizational structure involve delegation of authority, span of control, and the question of how many personnel are required.

Delegation of authority becomes necessary in any organization involving more than a handful of people. Delegation separates the ultimate and the operating responsibility. In our sample, the security manager delegates responsibility for supervising guard force operations to his lieutenant, who in turn delegates the operating responsibility to the sergeants on the first and third shifts, since the lieutenant obviously cannot work three full shifts himself.

For delegation of authority to work, the responsibility must be truly delegated—it cannot be "given" and then routinely overridden. Once the manager has determined that the lieutenant is capable of supervising the guard function in our example, he should allow the lieutenant to exercise that responsibility. And at each stage of the organizational ladder, the subordinate to whom authority is delegated must *accept* that responsibility; otherwise the entire command structure breaks down.

The degree to which a manager or supervisor is able to delegate responsibility, rather than trying to do everything himself, is a good measure of his managerial ability. Conversely, it has been said that the single most common management failing—in all organizations, not just security—is the inability or unwillingness to delegate responsibility and the authority necessary to carry it out. The result, inevitably, is a bottleneck at the manager's level, where one person must do or approve everything. The corollary result is a weakening of the entire chain of command below the manager's level.

Span of control refers to the number of personnel over which any individual can exercise direct supervision effectively. In the small security department illustrated, there is a sergeant over four guards on both the first and third shifts. The lieutenant himself supervises the five guards on the second shift. Many would regard this as a relatively ideal situation. However, giving a supervisor too little to do can sometimes be as damaging as giving him too much. The effective span of control for a given situation will depend on the complexity of duties, the number of problems, the geographical area, and many other factors. In some situations, especially where duties are routine and of a similar nature, it would be satisfactory to have one supervisor over 10 or 12 guards. Beyond that number, however, the span of control becomes so wide as to be seriously questioned.

How many security personnel are required is generally proportional to the size of the facility, expressed both in terms of square footage or acreage and the number of employees involved. Small businesses of 20 or 30 employees rarely require, and even more rarely can afford, the luxury of a security service of any size. At some point, however, as we consider larger and larger facilities, there is a need for such personnel. This can only be determined by the individual needs of a particular firm as demonstrated by a survey, and as further permitted by the funds available.

Where the security needs of a firm indicate the use of security personnel, they can be the single most important element in the security program. Since they are also the largest single item of expense, they must be used with the greatest efficiency possible. Only careful individual analysis of the needs of each facility will determine the optimum number and use of security personnel. For example, premises with inadequate perimeter barriers would need a larger security force than one with an effective barrier. In determining security needs, therefore, it is important that all protective elements be considered as a supportive whole.

One rule of thumb that deserves mention concerns the number of personnel required to cover a single post around the clock, providing coverage for three eight-hour shifts. The number is not three, but 4½ or five persons, to allow for vacation time, sick leave, terminations, or training. In the larger organization there is greater flexibility in the deployment of manpower, and 4½ persons might provide sufficient coverage. In the small organization five guards would be needed as a minimum to cover that single post 24 hours a day on a sustaining basis.

Hiring Security Personnel

The selection of security personnel must be preceded by a careful analysis of personnel needs to implement plans previously drawn. Job descriptions must be developed and labor markets must be explored.

Whatever specifications are arrived at, it is important that security personnel must be emotionally mature and stable people who can, in addition to their other skills and training, relate to other people under many conditions, including those of stress. It is also important to look for those persons whose potential is such that they may be expected to advance into the managerial ranks.

In considering the selection of personnel, it is useful to examine briefly the kinds of responsibilities they may be expected to assume.

Duties of Security Personnel. The duties of the security officer are many and varied, but among them are common elements that can serve as a guide to every security manager.

1. He protects the buildings and grounds to which he is assigned, including the contents, occupants and visitors.
2. He enforces rules and regulations governing the facility.
3. He directs traffic, both foot and vehicular.
4. He maintains order on his post and helps those persons requiring assistance or information.
5. He familiarizes himself with all special and general orders, and carries them out to the letter.
6. He supervises and enforces applicable systems of identifying personnel and vehicles, conducts package and vehicle inspection, and apprehends those persons entering or leaving the facility without required authorization.
7. He conducts periodic prescribed inspections of all areas at designated times to ascertain their condition of security and safety.
8. He acts for management in maintaining order and reports any incidents which disrupt such order.
9. He reports incidents of employees engaged in horseplay, loitering, or in violation of clearly stated policy. He reports all sickness or accidents involving employees.
10. He instantly sounds the alarm and responds to fires.
11. He logs and turns in any lost or unclaimed property. In the event any property is reported stolen, he checks the recovered property log first before proceeding in the matter.
12. He makes full reports to his supervisor on all unusual circumstances.

Posts and Patrols. Security personnel may be assigned to a variety of posts, but these fall into just a few categories. They may be assigned to a fixed post, to a patrol detail, or to reserve.

Fixed Posts may be gate houses, building lobbies, or guarding a particularly sensitive or dangerous location. *Patrol duty* involves walking or riding a given route to observe the condition of the facility. The perimeter is an important patrol, as are warehouse areas, or open yard storage areas. *Reserves* are those persons standing by in the event assistance is needed by security personnel on fixed posts or patrol duty. The scope of their special orders varies from company to company, but a list of those things that might be required will give the flavor of the tour of duty in an industrial facility.

Security personnel on patrol will make their tours on routes or in areas assigned by the supervisor in charge. They must be fully aware of all policies and procedures governing their tour as well as those that govern the area patrolled.

1. Make sure that the area is secure from intrusion and all gates and other entrances, as prescribed, are closed and locked. In interior spaces, they must check to see that all doors, windows, skylights and vents, as prescribed, are locked and secure against intrusion as well as possible damage from the weather.
2. Turn off lights, fans, heaters and other electrical equipment when its operation is not indicated.
3. Check for unusual conditions, including accumulations of trash or refuse, blocking of fire exits, access to fire-fighting equipment, etc. Any such conditions, if not immediately correctable, must be reported immediately.
4. Check for unusual sounds and investigate their source. Such sounds might indicate an attempted entry, the movement of unauthorized personnel, the malfunctioning of machinery, or any other potentially disruptive problem.
5. Check any unusual odors and report them immediately, if the source is not readily discovered. Such odors frequently indicate leakage or fire.
6. Check for damage to doors, tracks or weight guards. In cases where doors have been held open by wedges, tiebacks or other devices, these should be removed and their presence reported at the end of the tour of duty.
7. Check for running water in all areas, including wash rooms.
8. Check to see that all fire-fighting equipment is in its proper place and that access to it is in no way obstructed.
9. Check whether all processes in the area of the patrol are operating as prescribed.
10. Check the storage of all highly flammable substances, such as gasoline, kerosene, and volatile cleaning fluids, to assure that they are properly covered and properly secured against ignition.
11. Check for cigar or cigarette butts. Report the presence of such butts in "no smoking" areas.
12. Report the discovery of damage or any hazardous conditions, whether or not they can be corrected.
13. Exercise responsible control over watchman and fire alarm keys and keys to those spaces as may be issued.

14. Report all conditions which are the result of violations of security or safety policy. Repeated violations of such policies will require investigation and correction.

To carry out such assignments, it is essential that security personnel meet high standards of character and loyalty. They must be in good enough physical condition to undergo arduous exertion in the performance of their duties. They must have adequate eyesight and hearing and have full and effective use of their limbs. In some circumstances exceptions may be made, but these would be for assignments to posts requiring little or no physical exertion or dexterity. They must be of stable character and should be capable of good judgment and resourcefulness.

All applicants for such positions must be carefully investigated. Since they will frequently handle confidential material as well as items of value, and will, in general, occupy positions of great trust, they must be of the highest character. Each applicant should be fingerprinted and checked through local and federal agencies where legally permissible. A background investigation of the applicant's habits and associates should also be conducted. Signs of instability or patterns of irresponsibility should disqualify him.

All of these recommendations set high standards for the security officer—but nothing less will satisfy the emerging professionalism of the security function.

Training

Development and training of security personnel must be a continuing concern of management. Indeed, the lack of adequate training in the past has been the major criticism leveled against private security both within the industry and from outside. The Rand Report's description of the typical private guard as "an aging white male, poorly educated, usually untrained and very poorly paid"[28] has been widely quoted. A survey of private security personnel undertaken in the preparation of that Report in 1971 found that 65 percent of all respondents had received no training prior to actually beginning work (this figure dropped to 33 percent for security guards); 19 percent were put to work on their own on the first day; only seven percent received more than eight hours of initial prework training; and the remainder received a small amount of on-the-job training by a superior or fellow employee. Perhaps even more disturbing, of those who carried firearms (49 percent of the sample), only 19 percent received initial firearms training, and only 10 percent received periodic firearms retraining.[29]

The data cited were developed from an extremely small sample of inhouse and contract security personnel (only 275) and have too often been swallowed whole. Nevertheless, as the Report of the Task Force on Private Security observed five years later, "every major research project reviewed and every study conducted for this report point to a serious lack of personnel training at all levels of private security."[30]

It is clear that adequate training can and must be an important aspect of security planning in the proprietary organization. (The need is as great in contract security services, of course, where the problem is compounded by competitive pressures of the marketplace. The onus for low training standards must be borne in great part by employers whose overriding consideration in selecting security services is the lowest bid.) Proficiency in security is largely a product of the combination of experience and a thorough training program designed to improve the officer's skills and knowledge and to keep him current with the field. The Private Security Task Force's recommendations included:

1. A minimum of eight hours of formal pre-assignment training, and
2. Basic training of a minimum of 32 hours within three months of assignment, of which a maximum of 16 hours can be supervised on-the-job training.[31]

The merits of training will be reflected in the security officer's attitude and performance, improved morale and increased incentive. Training also provides greater opportunities for promotion and a better understanding on the part of the officer of his relationship to management and the objectives of his job.

It should not be presumed that a former law enforcement officer requires no training. He does. In order for him to be successful in security he must develop new skills and, not incidentally, forget some of his previous training.

A training program would cover a wide variety of subjects and procedures, some of them varying according to the nature of the organization being served. Among them might be the following:

1. Company orientation and indoctrination.
2. Company and security department policies, systems and procedures.
3. Operation of each department.
4. Background in applicable law (citizen's arrest, search and seizure, individual rights, rules of evidence).
5. Report writing.
6. General and special orders.
7. Discipline.
8. Self defense.
9. First aid.
10. Pass and identification systems.
11. Package and vehicle search.
12. Communications procedures.
13. Techniques of observation.
14. Operation of equipment.
15. Professional standards, including attitudes toward employees.

Properly trained security personnel will be cheerful, cooperative and tolerant in their dealings within the company. They will be patient. They will understand that they are not members of a law enforcement agency but employees who have the job of providing security. If they are goodhumored, tactful, patient and professional, their fellow employees will learn to respect them and to look to them for assistance in many ways—frequently beyond the scope of their duties. They should be encouraged to provide whatever assistance they can, since this can only increase employee respect for them as individuals as well as professional security people, thereby encouraging cooperation. And employee cooperation in the security task is vital to its success.

Supervision

In additon to planning, establishing controls, organizing a department, hiring personnel, and training, the security manager's responsibility ties include supervision of security. And it is in the handling of this function that the entire security program will provide to be effective or inadequate.

The security manager must maintain close supervision over communications within his own departmental structure. It is essential that he communicate downward in expressing departmental directives and polices. It is equally important that he receive regular communication up the organizational ladder from his subordinates. He must regularly study and analyze the channels of communication to be certain that the input he receives is accurate, relevant, timely, concise and informative.

Additionally, the security manager must set up a system of supervision of all departmental personnel to establish means for reviewing performance and instituting corrective action when it is necessary.

He must, above all, lead. His qualities of leadership will, in and of themselves, prove ultimately to be the most effective supervisory approach.

Security personnel are, in many respects, the most effective security device available. They rarely turn in false alarms. They can react to irregular occurrences. They can follow a thief and arrest him. They can detect and respond. They can prevent accidents and put out fires. In short, they are human—and they can perform as no machine can.

But, as humans, they are subject to human frailities. Security personnel must be adequately supervised in the performance of their duties. It is important to be sure that policy is followed, that each member of the security force is thoroughly familiar with policy, and that the training and indoctrination program is adequate to communicate all the necessary information to each member of security. Each guard must be disciplined for a violation of policy, and at the same time, management must see to it that he now knows the policy that was violated.

All of these elements must be regularly reviewed. It must be remembered that well-trained, well-supervised security personnel may be the best possible protection available, but badly selected, badly supervised guards are not an asset at all—and, worse, could themselves be a danger to security. They can, after all, succumb to temptation like any other individual. The opportunities for theft are far greater than those offered to an average employee.

Issuance and use of keys to stockrooms, security storage and other repositories of valuable merchandise or materials must be limited and their use strictly accounted for. Fire protection, not otherwise covered by sensors or sprinkler systems, is not often a major problem in such spaces, and in the event a fire were to threaten, the door could be broken or a guardhouse key could be used to enter the endangered area.

Although all security personnel have been subjected to a thorough background investigation before assignment, and they have been closely observed during the early period of employment, they may still yield to the heavy pressure of temptation and opportunity presented to them when they have free and unlimited access to all of the firm's goods. Although it is not good, sound personnel practice to show distrust or lack of faith in the sincerity of security personnel, neither is it fair or reasonable business practice to subject them to such a test of probity nor to expose company property to such risks.

The Image of Security

The next element with which the security manager must deal is the image or representation of the security function.

In order for security to be effective in any organization, it must have the implied approval and confidence of that organization. Every time a guard acts the stern father figure, and every time a system is installed which is cumbersome and inefficient, the image of security suffers.

It is the task of the security manager to undertake a regular program of indoctrination to clearly define the role of security and of security personnel within the organization. Since employee participation and cooperation are essential to the success of any security effort, it is of extreme importance that a thorough indoctrination program eliminate any tendency to alienate these important allies by overbearing or bullying attitudes on the part of security personnel. Such a program must also impress upon all security people the importance of their role in public relations, both as employees of the company and as members of the security department.

No matter how well the department is organized, it cannot be effective without the full support of the people in the organization it serves. To achieve this support, the department must be educated in attitudes, duties and demeanor—and proper supervision must ensure that these attitudes are maintained. The very fact that security personnel are controlling the movement and conduct of other members of their community suggests that they must themselves be carefully controlled to avoid giving rise to feelings of resentment and hostility. The entire organization can suffer great harm as a result of general animosity directed at only one member of the security department who has conducted himself improperly or unwisely.

Departmental Evaluation

Regular departmental evaluations should try to determine whether security policies and procedures are being properly followed, and whether such existing policies and procedures are still desirable in their present form or should be modified to better achieve predetermined goals. These evaluations should also review all manpower and equipment needs and the efficiency of their current use in the conduct of the security program.

Since security concerns itself with prevention of damage, disruption or loss, its effectiveness is never easy to evaluate. The absence of events is not in itself revealing, unless there are accurate accounts of actual crimes against the organization during some prior, analogous period to provide a standard of measurement.

Even such a comparison is of dubious value as the basis for a thorough, ongoing, objective analysis of security effectiveness. Circumstances change, personnel terminate, motivating elements alter or disappear. And there is always the haunting uneasiness created by the possibility that security procedures might be so ineffective that crimes against the company have gone virtually undetected—in which case the reduction or absence of detected crime presents a totally misleading picture of the company's position.

Personnel Review. In reviewing departmental performance, all records of individual personnel performance should be examined. The degree of familiarity of each man with his duties, and the extent of his authority, departmental and organizational goals should be examined. Corrective indoctrination should be required as indicated. The health, appearance, and general morale of each man should be noted.

Equipment Review. The state of all security equipment should be reviewed regularly with an eye to its current condition and the possible need for replacement, repair or substitution. This review should cover all space assigned for use by security personnel on or off duty, uniforms, arms, if any, communication and surveillance equipment; vehicles, keys, and report forms.

Carelessness or inadequate maintenance of such gear should be corrected immediately.

Procedures Review. A review of security department procedures is essential to the continued efficacy of the department. All personnel should be examined periodically for their compliance with directives governing their area of responsibility.

Familiarity with departmental policies should be evaluated and corrective action taken where necessary. At the same time, the very usefulness of prescribed procedures should be reviewed, and changes should be initiated where they are deemed advisable.

REVIEW QUESTIONS

1. Discuss the statement: "In general, the security manager can be considered to be serving a staff function."
2. Why is functional authority important to the security manager?
3. Describe the duties of the security officer on patrol.
4. Why did the Rand Report criticize the training of the 'typical private guard'? Was their criticism justified?
5. What categories of concern should be reviewed during a security department evaluation?

Chapter 4

GOVERNMENT SECURITY AND ORGANIZATION

The security of government at all levels is of prime concern among sovereign entities throughout the world. The United States is no exception, and over the years has evolved into a complex and sophisticated security apparatus for its protection against attack from within or without. Some aspects of the federal security organization are replicated at state, county and municipal levels. Each level, though related in some ways and frequently in cooperation, is autonomous and functions in its own behalf.

For a clear understanding of the total security picture, it is essential that the various components of security in the public sector be examined in a general way. The responsibilities, limitations, and relationships between these elements have frequently proven to be a source of confusion and misunderstanding. A broad overview of this picture will serve to dispel the misapprehensions which so frequently surround this area of government.

FEDERAL AGENCIES

At the federal level, security from aggression is ultimately provided by the armed services. These forces play a role analogous to the police in terms of enforcement. The analogy does not bear close scrutiny in that, unlike police agencies, national armies act unilaterally and without clearly defined authority or restriction in the international arena. Nonetheless, the ultimate security reaction at the federal level is the deployment of armed forces.

Short of such a solution, there are many federal agencies engaged in security functions which are directed toward information accumulation or toward action against internal threats. All federal agencies, by their very nature, accumulate information which may prove useful in security applications. Historically, this information was largely wasted in that no central bureau coordinated its evaluation and dissemination in such a way that it could be put to any effective use by the administration or the executive branch. To overcome this weakness in the processing of information, the National Security Council (NSC) was created as the ultimate supervisory body over the coordinated efforts of a network of many federal agencies which are charged, in varying degrees of responsibility, with the collection and transmission of specified data. It is important to recognize that the national security is largely based on accurate and timely information collected from thousands of sources. This information is then coordinated, evaluated, analyzed and presented to the decision-making entity for action. The structure for this vital role makes up the intelligence system which is ultimately supervised by the NSC.

National Security Council

The National Security Council was established by the National Security Act of 1947 (as amended 1949). Its function is to advise the President on foreign and domestic matters which affect the national security. This involves a careful appraisal of all circumstances involving security with respect to foreign powers and assessments of data collected by appropriate agencies with respect to domestic security. The Council actively directs the acquisition of information as the need arises, or as the President commands, and coordinates any related activities by departments or agencies or the government as the President may deem necessary in the interests of national security. The NSC membership is an exclusive one, consisting of the President, Vice President, Secretary of State, and the Secretary of Defense.

The National Foreign Intelligence Board (NFIB)

The NFIB (formerly known as the United States Intelligence Board) is a coordinating body which further refines intelligence data submitted by agencies actively engaged in intelligence gathering activities. The Board is at a level just below the NSC and feeds evaluated information up to it. Its membership is comprised of top echelon members of the principal agencies of the intelligence system such as the Director of the CIA; Director of Intelligence and Research of the Department of State; the Director of Defense Intelligence Agency (DIA); the Assistant Chief of Staff Intelligence, U.S. Army (observer); the Assistant Chief of Staff for Naval Operations Intelligence, Navy (observer); the Assistant Chief of Staff Intelligence, Air Force (observer); the Assistant General Manager Atomic Energy Commission (AEC) (now the Energy Research and Development Administration—ERDA); the Assistant to the Director, Federal Bureau of Investigation (FBI); and the Director of the National Security Agency (NSA). The high ranking nature of its membership allows the NFIB to maintain the closest possible relationship with the other agencies.

Both the National Security Council and the National Foreign Intelligence Board have supervisory and coordinating responsibilities. The major tasks of the remaining agencies is the collection, evaluation, and dissemination of gathered information. The relationships within this structure are shown in Figure 4-1.

Central Intelligence Agency (CIA)

The Central Intelligence Agency was established by the National Security Act of 1947, the same act that created the NSC which was to supervise the activities of the CIA. Under the direction of the National Security Council, the CIA is responsible for:

1. advising the NSC in all matters concerning those intelligence activities of departments and agencies affecting national security;
2. making recommendations to the NSC for coordinating intelligence activities of departments and agencies relative to national security;
3. evaluating intelligence with respect to national secuirty;
4. providing for the dissemination of such intelligence within the government; and
5. performing such functions and duties as the NSC may direct, relative to intelligence which affects national security.[32]

Figure 4-1. Structure of the National Security System.

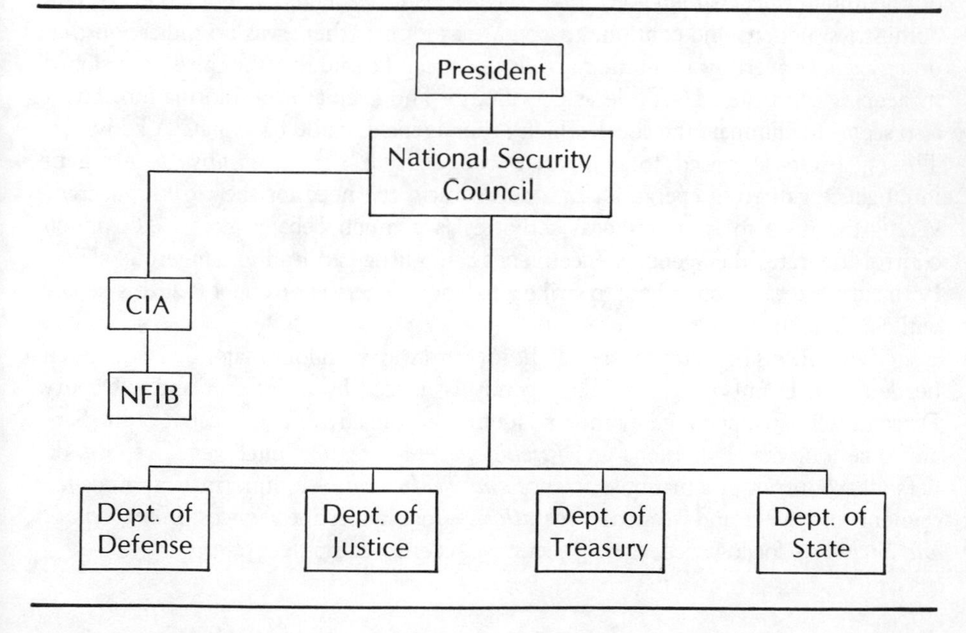

The dilemma facing the Congress was obvious, and subsequent events have shown that the matter was not fully and satisfactorily resolved at the time of the establishment of the CIA. The need for secrecy of operations of any successful intelligence gathering body is beyond debate. The dangers inherent in a covert agency operating outside of normal administration, supervision, and inspection are equally indisputable. Such an agency could, under some circumstances, be a far greater threat to national security than those perils from which it was chartered to protect the nation. The legislation creating the agency was carefully drawn to allow it the widest possible scope in performing its mission, while, at the same time, limiting its arena to protect the country from its possible excesses. The agency was restricted in several ways, such as: "(1) The CIA shall have no police subpoena or law enforcement powers and no internal security functions; (2) The CIA will not exceed most departmental intelligence functions."[33] The National Security Act goes on to state that every department "will continue to collect, correlate, evaluate, and disseminate departmental intelligence; (3) The Director of Central Intelligence, with certain approvals, has a right to inspect the intelligence production of all government security agencies . . . This regulation does not apply to the Federal Bureau of Investigation."[34]

The last provision is interesting in that it could open up an area of abuse by the CIA. The Congress was very careful in the first two provisions to limit the CIA's domestic activities. The intelligence gathering of the other agencies could be seen as a check on the CIA in that their intelligence would alert them to any possible unauthorized domestic operations of the CIA. If, however, the CIA has access to all of the intelligence gathered by these other agencies, could it not enter certain domestic situations and continue to operate as long as there was no indication that other agencies were aware of their activities? It can be said that the third provision is in keeping with the CIA's role as coordinator and evaluator of information, but it also seems to eliminate the check which other agencies could have on CIA activities. This question is raised to emphasize the delicate nature of any government intelligence gathering operation. In a democracy, the need for the cloak of secrecy versus the need to control illegal activities is a much debated point. Too much control frustrates the agency's effectiveness; too little can lead to dangerous abuse. Both sides are valid, but where to strike a balance? There is no doubt that this debate will continue for some time.

The CIA is organized in a straight forward way with four major divisions, each headed by a Deputy Director. The agency is headed by a Director and a Deputy Director who are appointed by the President with the advice and consent of the Senate. The four major divisions are: *Intelligence*—for regular intelligence reports on persons, countries or situations; *Science and Technology*—for information on developments in science and weapons; *Operations*—for clandestine services; *Management and Services*—for logistics, agency security, records, equipment, etc.

The Department of Defense (DOD)

The Department of Defense has a vital role in the security organization beyond that of the last resort security of armed conflict. Intelligence data is generated from an incredible number of units within the DOD, the most important being the Joint Chiefs of Staff (JCS). It is through this unit that such data is channeled up to the Secretary of Defense, the main source of intelligence input to the NSC and, thence to the President. The JCS is composed of a chairman, the Chiefs of Staff of the Army and the Air Force, and the Chief of Naval Operations. The Commandant of the Marine Corps attends meetings when matters within his sphere are under discussion. The Defense Intelligence Agency (DIA), also within the DOD, is administered by the Joint Chiefs. This agency covers a vast field of military intelligence and could be thought of as a key security area either in war or in maintaining the peace. Directorates whose functions are not far removed from the assigned missions of the four divisions of the CIA comprise the departmental headings: Collection and Surveillance Intelligence; Scientific and Technical Intelligence; and Command and Defense Intelligence School. The Attache Affairs Directorate and the Directorate for Support, which handles counterintelligence and security complete the list.

The DIA is responsible for:

1. the organization, direction, management, and control of Department of Defense intelligence resources assigned to or included within DIA;
2. the review and coordination of those Department of Defense intelligence functions retained by or assigned to the military departments;
3. the development of guidance for the conduct and management of such functions for review, approval, and promulgation by the Secretary of Defense; and
4. the supervision of the execution of all plans, programs, policies, and procedures for those Department of Defense general activities and functions for which DIA has management responsibility.[35]

The National Security Agency (NSA), which was established in 1952 by Presidential directive in the Department, reports directly to the Secretary and, unlike the DIA, is not under the direct supervision of the Joint Chiefs of Staff. Its role is highly technical and is essentially involved with technical surveillance of non-domestic military and industrial activity.

The Department of Justice

The Department of Justice is the principle federal security agency for combatting a wide range of internal threats. There are seven major divisions, each of which specializes in some area of law enforcement and data collection.

1. *The Internal Security Division*—This unit was formed in 1954 in response to a perceived threat to internal security. Its function is prosecution of cases involving treason, espionage, sedition, and crimes committed under definition by the Atomic Energy Act.
2. *The Federal Bureau of Investigation*—The FBI is the principal law enforcement arm of the Department. Its jurisdiction is strictly domestic. In this role, it is a fact gathering and reporting agency. It draws no conclusions, and makes no recommendations to report recipients. It is active in the investigation of federal crimes, such as kidnapping, bank robbery, interstate crimes affecting commerce, etc. It is deeply involved in the accumulation of data on the activities of subversive organizations operating in the United States and is involved in the investigation of crimes also covered by the Internal Security Division. The FBI operates a Computer Systems Division which administers the work of the National Crime Information Center (NCIC) which serves a vast network of police activities across the country and has shown itself capable of highly efficient information transmission.
3. *Immigration and Naturalization Service (INS)*—The INS is responsible for enforcing immigration and naturalization laws, including investigation of violations. It is also responsible to patrol the land borders of the United States and to protect them against smugglers, drug dealers, and illegal immigrants.

   The INS does have an intelligence accumulation capability through agents stationed in foreign countries.
4. *Drug Enforcement Administration (DEA)*—The DEA is responsible for the enforcement of laws relating to the possession, use and sale of illegal, mind-altering drugs. It is also responsible for the regulation of the legal trade in narcotics and dangerous drugs.

   Agents on foreign assignment provide valuable intelligence input.
5. *Criminal Division*—This division is essentially concerned with prosecutions, extraditions, and criminal cases before the Supreme Court.
6. *Law Enforcement Assistance Administration (LEAA)*—LEAA was established by the so-called Safe Streets Act of 1968. Its purpose is to assist state, county, and municipal governments to handle crime. To this end, the program makes grants and provides technical assistance in solving specific or general problems.
7. *U.S. Marshalls*—The U.S. Marshalls Service has been a bureau in the Department of Justice since 1973. Marshalls have certain responsibil-

ities in handling federal prisoners, provide security and service in federal courts and may be assigned protection of federal property during times of anticipated trouble.

The Department of the Treasury

The Department of the Treasury, besides being the government's banker, has within its jurisdiction the management of a wide variety of law enforcement and intelligence gathering functions. Five of the major functions are discussed below.

1. *The Secret Service*—This service was established to prevent counterfeiting—a role it still occupies—but its duties were broadened to include the protection of the President and his family, the Vice President, designated dignitaries, visiting heads of state and, in presidential election years, the major contending candidates. The Secret Service is totally responsible for planning, briefing, and protecting those designated. The Office of Protective Intelligence plays a significant role in preparing for these duties.
2. *U.S. Customs Service*—The Customs Service is charged with:
    1. enforcing collection of duties on goods entering the United States;
    2. protecting U.S. aircraft and passengers from hijacking;
    3. identifying and suppressing smuggling; and
    4. controlling carriers and merchandise imported/exported to or from the U.S.[36]

    An intelligence system within the Fraud Investigation Division is effective in the control of import/export crime.
3. *Office of Law Enforcement*—This office is responsible for formulating departmental policy on its law enforcement activities. It is also the expediting entity at which all interactions within and without the department take place.
4. *Internal Revenue Service (IRS)*—The prime goals of the IRS is the collection of taxes. In order to do this effectively, an Inspection Division was created to enforce its regulations or to bring action against offenders. The division is empowered to conduct investigations, serve warrants, and to serve federal subpoenas and summonses. This is aimed at internal security and at maintaining the integrity of the IRS.

    The Compliance Division is involved with investigating and dealing with tax fraud. An Intelligence Division in Compliance handles investigations of suspected tax frauds.
5. *Bureau of Alcohol, Tobacco and Firearms*—This bureau carries on investigations and prevention programs relating to violations of laws relating to liquor, tobacco, firearms, and explosives.

The Department of State

Ever since Tallyrand made an art out of the collection of intelligence in international diplomacy, foreign offices have become increasingly dependent on the assimilation of data touching on virtually everything within their sphere of activity. Our Department of State is no different in this regard. Since this department is responsible for the development of U.S. foreign policy one of its major tasks is the collection, evaluation and assimulation of foreign intelligence. In this, it has the resources of the rest of the intelligence community to call upon as well as its own Bureau of Intelligence and Research whose primary responsibility is the collection of intelligence information. The director of this bureau sits on the USIB and supervises the department's collection of foreign intelligence which will be used for the development and guidance of foreign policy. The Office of External Research is under the supervision of the Bureau of Intelligence and Research and develops political and economic studies.

Within the Department of State is the Office of Security which provides for a total security program for personnel, files and installations. It may also, upon request, conduct investigations into passport or visa violations.

LOCAL LAW ENFORCEMENT AGENCIES

The agencies discussed above have as their major purpose the national security of the country. There isn't room here to discuss the security functions of the hundreds of other federal agencies. Almost any federal agency has certain security requirements peculiar to its specific operation. The Department of Agriculture, Federal Emergency Management Agency, Department of Commerce, and Federal Trade Commission are just some of the very many federal agencies which must routinely deal with security matters.

Law enforcement functions largely at the state, county, or municipal level. It is with these local agencies that the average citizen comes into closest contact. Whereas there are slightly over 50 federal law enforcement agencies, there are more than 40,000 such entities at levels below that of the Federal Government. At the state level, state policies, crime laboratories, the office of the state attorney, general crime commissions, state liquor control, state vehicle control, state licensing, state bureaus of criminal records and identification all serve as part of the security system. County efforts center largely on either the Sheriff or the District Attorney and certain licensing or inspection units, but it is the municipal level that provides the greatest volume of security in terms of manpower and the number of autonomous units. Most of the security at the municipal level is in the form of police units. This is not to minimize the significant role in health and safety of the many boards and commissions charged with inspections, licensing and investigations, but the day to day security of the community rests largely with the police. The effectiveness of that service will be determined by many factors, but not until we have examined the organizational position of the police function and the climate in which it is administered will we be able to fully understand this most important element of government security located at the opposite end from the federal effort.

Points of Police Administration

Any security function must occupy some position in the organizational structure of the community it serves. The police are no different from the department of public works in this respect. The police department fits into the framework and is responsible to the controlling entity in the administrative line. This control will differ with the level at which police duties are performed.

Presumably, a national police force could be operated at a high level of efficiency which could effect enormous savings to every municipality and, hence, to every taxpayer in the country. As noted above, however, the dangers to any society of a national police force are sufficient enough to create concern in an informed citizenry. History is replete with examples of nations subverted by a strong national police. It was to prevent this possibility that the Constitution left these powers to the various states. In the United States, the principle of local autonomy takes precedence in virtually every area of government. Education, most aspects of public health administration, as well as police functions operate without federal interference. The municipalities and the counties are created under state constitutions from which they derive their authority to make ordinances and to enforce them. In spite of this relationship, there is little or no interference in the affairs of the local jurisdiction from the state capital.

Strangely enough, this was not always the case. In the early nineteenth century when urban police departments were being established along the lines of Sir Robert Peel's London Metropolitan police, state governments created police boards in many states to take control of the municipal police establishments. The reason given in most cases was that city authorities were unable to maintain proper standards of police conduct, though it soon became evident that state politicians saw an opportunity to pick a rich patronage plum for their own benefit. State controlled police departments were found in many cities, including New York, Chicago, Boston, Detroit, Cleveland, Baltimore, Cincinnati, St. Louis, New Orleans, and Kansas City. Friction between urban and state leaders grew with the inefficiency in departmental administration. State control of urban police was gradually abolished. Boston regained autonomy in 1962 and today only twelve cities remain under state control (Baltimore, Kansas City, St. Louis and St. Joseph, Missouri, and eight New Hampshire cities). With these exceptions, all police functions are now administered locally. Historically, the administration of the police has been competent or incompetent, corrupt or honest, all depending on many factors. Many of these factors came down to the administrative atmosphere in which the police were required to function.

This climate was fostered by the effectiveness of the municipal government itself, since it is unlikely that any police department can operate honestly and effectively in a dishonest and incompetent municipal organization, or conversely, will be an efficient and effective force if the structure within which it operates is equally sound.

How this works may best be seen within the framework of the three most common forms of municipal government—the Mayor-Council plan, the Commission plan, or the Council-Manager plan.

The Mayor-Council Plan. The Mayor-Council form of municipal government is the most common and familiar pattern of municipal organization. Under this plan, both the mayor and the individual members of the Council are elected to office. The Council constitutes the principal legislative and, to some degree, policy-making body. As an aspect of its legislative prerogatives and responsibilities, the Council holds the bank account and controls all expenditures. The Mayor acts as chief executive, appointing department heads and administering the routine of government. If the Mayor is a person of vision, energy, integrity, and talent, his competence will invigorate and encourage the municipality to heights of achievement. If he lacks these qualities, the administration will suffer accordingly. If political pressures are so intense that either he or the Council is forced into disabling compromises, the city will suffer and may eventually give way to graft and incompetence. In such a climate, the security delivered by a demoralized or corrupt police will be minimal or worse.

The Commission Plan. The Commission Plan came into being early in the twentieth century in response to municipal corruption. It simply combined the legislative and administrative functions of government in a powerful body of five members chosen by popular vote. Collectively, the Commission constituted the totality of municipal government; individually, each Commissioner was in charge in one or several departments of the government.

In its earliest days, the plan was hailed as a great success, as bringing efficiency while stamping out corruption in local government. By 1917, five hundred cities had adopted the plan.

Clearly, the system has flaws. Without the checks and balances between the legislative and administrative branches, corruption could, and often did, fluorish. Popular figures rather than trained professionals were running complex city functions. The security systems sometimes suffered with amateurs running fire and police protection from a position of great authority but without a clearly defined concommittant responsibility. It invited the Commissioner of Public Safety to take the bows on popular issues and to disclaim any responsibility in unfavorable situations.

The Council-Manager Plan. The Council-Manager Plan was first adopted by a large city in 1914 when Dayton, Ohio, found itself confronted with a crisis, both natural (from flood waters) and man-made (from bad administration and debt). Essentially, this arrangement is a combination of the previous two systems except that in the case of the City Manager the city is run by an experienced professional instead of a successful politician who may or may not know anything about administration. The plan borrows from management techniques of modern business. The City Council functions primarily as a board of directors. It determines policy and hires the city manager. The manager is a technician in the field of administration and is placed in charge of all municipal management. He is given wide authority and control and is supervised only in a general way with respect to policy and audit procedures. The system has been highly successful in bringing professionalism to city government and today there are nearly three thousand cities and counties operating under the Manager plan of local government.

Alternatives for the Future

The position of the police in existing form of municipal government varies with the community and its charter. By and large, the police, under the City Manager, have been treated in a more professional manner and have been less subject to political pressures unrelated to their job in the community. Yet, there are still many ways in which this security function can be improved. In our dynamic changing society it is necessary to continue to examine the alternatives.

Almost 70% of America's population lives in metropolitan areas covering less than 10% of its land mass and these areas are growing into great megalopolises in which millions of people live in contiguous, overlapping jurisdictions, each with its own autonomous government and services. The duplication in money and effort is staggering to contemplate.[37]

Efforts are being made in various areas to overcome the confusion and complexity created by this growth and several alternatives have appeared that have created great interest among city officials and police professionals around the country.

Federated Police System. In this system, a metropolitan police department would form a cooperative federation with local departments. This operation would eliminate certain duplication of effort but still maintain the autonomy of the local departments. This would include centralized communication and training and laboratory facilities and a central record system.

Contract Law Enforcement. Under this arrangement, a municipality contracts with the county or state for delivery of police service. The Los Angeles County Sheriff's Department has over twenty contracts with incorporated cities within county limits to provide specified contracted police services.

Consolidation of City and County. The trend toward consolidation of city and county services has effected dramatic savings in areas where that has been accomplished.

At a time of high inflation, high taxes, and shifting population trends, the need for professional public security services delivered in the most efficient way needs the closest possible examination.

REVIEW QUESTIONS

1. What is the function of the National Security Council and what is its relationship to other agencies?
2. Explain how the National Security Act of 1947 limits the arena of the CIA. In what way has this Act opened up a possible area of abuse within the CIA?
3. What are the seven major divisions of the Department of Justice?
4. Discuss the five major functions of the Department of the Treasury.
5. How does municipal government affect police administration? Discuss the various forms of municipal government, describing both the strengths and weaknesses of each.

Chapter 5

SECURITY AND THE LAW

Whereas public police and protection services derive their authority to act from a variety of statutes, ordinances, and orders enacted at various levels of government, private police function essentially as private citizens. Their authority to so function is no more than the exercise of the right of every citizen to protect his own property. Every citizen has common law and statutory powers that include arrest, search, and seizure. The security officer has these same rights, both as a citizen and as an extension of an employee's right to protect his employer's property. Similarly, this common law recognition of the right of defense of self and of property is the legal underpinning for the right of every citizen to employ the services of others to protect his property against any kind of incursion by others.

The broad statement of such rights, however, in no way suggests the full legal complexities that surround the question. In common law, case law, state statutes, as well as in the basic authority of the Constitution, the rights, privileges, and restrictions further defining these rights abound. The body of law covering the complex question of individual rights of defense of person and of property contains many apparent contradictions and much ambiguity. In their efforts to create a perfect balance between the rights of individuals and the needs of society, the courts and the legislatures have had to walk a narrow path. As the perception of society's needs changed or as the need for the protection of the individual became more prominent, a swing in the attitudes of the courts or the legislatures was apparent. This led to some confusion—especially among those with little or no knowledge of the law.

It is of enormous value, therefore, for everyone engaged in security to pursue the study of criminal law. Such studies are not aimed at acquiring a law degree, and certainly not at developing the skills to practice law. It is directed toward developing a background in those principles and rules of law that will be useful in the performance of the complex job of security.

Without some knowledge of the law, a security officer frequently cannot serve his client's interests. He may subject himself or his employer to ruinous lawsuits through well-meaning but misguided conduct. In cases which must eventually go to court, his handling of evidence, reports, and interrogations may be critical to the case; without an understanding of legal processes and how they operate, the case could be lost.

In short, the pursuit of security itself involves contact with others. In each such contact, there is the delicate consideration of conflicting rights. Without an appreciation of the elements involved, the security officer cannot perform properly.

Although for the purposes of this book, we are primarily concerned with criminal law, it might be useful to distinguish between criminal and civil law. *Criminal law* deals with those offenses against society. Every state has its criminal code which classifies and defines criminal offenses. Serious crimes like murder, rape, arson, armed robbery, and aggravated assault are felonies. Misdemeanors include charges such as disorderly conduct and criminal damage to property. When these offenses are brought into court, the state takes an active part considering itself to be the offended party.

Civil law, on the other hand, has more to do with the personal relations and conflicts between individuals. Broken agreements, sales that leave a customer dissatisfied, outstanding debts, disputes with a government agency, accidental injuries, and marital breakup all fall under the purview of civil law. In these cases, private citizens are the offended parties, and the party found at fault is required to compensate directly the victim.

This chapter is intended as a guide to some of the intricacies of criminal and tort law. It is aimed at those subjects with which a security officer would most likely be confronted. It will deal with *substantive law* or that portion of the law which concerns the rights, duties, and penalties of individuals in their relationships with each other. *Procedural law* is the other of the two divisions of the law. It deals with the rules of procedure—the mechanics of the legal machine.

SOURCES OF LAW

Common Law

The principal source of law in the United States is the English common law. This is law which had been established over generations by common agreement among reasonable men as to what constituted acceptable and unacceptable conduct. Certain acts have been considered criminal as far back as history can trace man in society. It is hard to imagine any social structure which would condone such conduct. So it was in England's legal system which simply accepted the accumulated common understanding of the ages in agreeing that certain acts were criminal. They are still referred to as common law crimes. Treason, murder, battery, robbery, arson, larceny, burglary, kidnapping, and rape are the major crimes on this list.

The original common law in England was augmented by Parliamentary enactments which added new crimes to this list of offenses. These enactments were further refined by court decisions and interpretations which, in turn, became part of the body of the common law.

When the first Englishmen came to these shores, they brought with them their law and their court procedures. Even after the colonies became independent, the common law continued in effect. As new states joined the Union, they, too, adopted the English system. Currently, we are in the position where 19 states have preserved the status of common law offenses by case law (the procedure of court rulings on a series of cases); an almost equal number (18) have abolished the common law as such and have written into statutes most of the principles of the common law; eight states have formally adopted the common law by ratification; the remaining seven are unclear as to the status of the common law, per se (see Table 5-1).

Case Law

When a case goes to court, it is usually preceded by numerous cases of similar nature. Those preceding cases have usually been resolved in such a way as to put to rest any doubts as to the meaning of the governing statutes, as well as to clarify the attitude of the courts regarding the violation involved. Each of these cases has established a precedent which will guide the court in subsequent cases based on the same essential facts. Since the facts in any two cases are rarely precisely the same, opposing attorneys cite prior cases whose facts more readily conform to their own theory or argument in the case at hand. They, too, build their case on precedents or case law already established. It is up to the court to choose one of the two sides, or to establish its own theory. This is a very significant source of our law, in addition to the common law.

Since society is in a constant state of change, it is essential that the law adapt to these changes. At the same time, there must be a stability in the law if it is to guide behavior. People must know that the law as it appears today will be the same tomorrow; that they will not be punished tomorrow for behavior that was permitted today. They need to know that each decision represents a settled statement of the law and that they can conduct their affairs accordingly. So the decisions of the courts become guides to the meaning of the law and, in effect, become the law itself. This is not to say that the courts create laws or that they occupy a position superior to the laws they interpret. Rather, their judgments flesh out legislative enactments to give them a clear outline. Such interpretation, based on precedents in this way, are never regarded lightly.

This does not mean that each decided case locks the courts forever into automatic compliance. Conditions which created the climate of the earlier decision may have changed, rendering the precedent invalid. And there are cases decided in such a narrow way that they cannot be applied beyond that case. Further, there is nothing

Table 5-1
ATTITUDES OF STATES REGARDING COMMON LAW

	Common Law Adhered to	Common Law Preserved by Statutory Retention	Statutory Law Only (Common Law Abolished)	Attitude Unclear
Alabama	X			
Alaska		X		
Arizona			X	
Arkansas	X			
California			X	
Colorado				X
Connecticut	X			
Delaware	X			
District of Columbia	X			
Florida		X		
Georgia			X	
Hawaii			X	
Idaho			X	
Illinois			X	
Indiana			X	
Iowa			X	
Kansas			X	
Kentucky	X			
Louisiana			X	
Maine	X			
Maryland	X			
Massachusetts	X			
Michigan	X			
Minnesota			X	
Misissippi		X		
Missouri	X			
Montana				X
Nebraska			X	
Nevada		X		
New Hampshire	X			
New Jersey		X		
New Mexico		X		
New York			X	
North Carolina	X			
North Dakota				X
Ohio			X	
Oklahoma			X	
Oregon				X
Pennsylvania		X		
Puerto Rico			X	
Rhode Island		X		
South Carolina	X			
South Dakota				X

Tennessee	X		
Texas		X	
Utah			X
Vermont	X		
Virginia	X		
Washington	X		
West Virginia	X		
Wisconsin		X	
Wyoming			X

(Table 5-1 continued)

that prevents review of a decision at the time of a later case. If the court agrees that the earlier case was in error, it will not be influenced by a faulty precedent.

So, it can be seen that case law is an important source of the law; it provides a climate of legal stability without closing the law to responsiveness to changing needs.

Law From Legislation

Congress and state legislatures are empowered to enact laws that describe additional crimes. The authority to do so emanates from the U.S. Constitution and from the individual state constitutions. These constitutions do not specifically establish a body of criminal law. In general, they are more concerned with setting forth the limitations governmental power over the rights of individuals. But they do provide both for the authority of legislative action in establishing criminal law and for a court system to handle these as well as civil matters.

The criminal law is, in fact, the creation of the legislatures. However the legislatures define the criminal law that is what it is. It is the legislatures that are exclusively responsible for making laws. The courts may find some laws are unconstitutional, or vague, and thus set them aside, but they may not create laws. Only the legislatures are thus empowered.

THE COURTS: THE PROCEDURE

In the event of an arrest, the accused must, by law, be taken without unnecessary delay before the nearest judge or magistrate. The court may proceed with the trial, in the case of a misdemeanor charge, unless the accused demands a jury trial or requests a continuance and such is ordered by the court. This trial, whether held right away or later, is conducted in *trial court* or a *court of primary jurisdiction.*

If the charge is a felony, the judge or magistrate conducts a preliminary hearing, an informal process designed to determine if reasonable grounds exist for believing the accused committed the offense as charged. If such grounds do not appear, the accused will be discharged.

If the judge finds that there are reasonable grounds for believing that the accused *may* have committed the offense as charged, he will "bind him over" for the action of the grand jury. He will be held in jail in the interim unless, if the offense is bailable, he makes bond.

The grand jury is required by many states to consider the evidence in any felonious matter. Grand juries, usually consisting of 23 citizens of whom 16 constitute a quorum, do not conduct a trial. They hear only the state's evidence. The accused may not be accompanied by his attorney in the hearing room and may not, in most cases, offer evidence in his own behalf. Misdemeanors are not handled by grand jury action but are usually prosecuted on an "information," a document filed by the prosecuting attorney upon receipt of a sworn complaint of the victim or a witness or other person who is personally informed as to the circumstances of the incident alleged.

For a felony charge, the jury proceedings must result in a vote of at least 12 members for the accused to be indicted, and an indictment must be obtained in those jurisdictions which require it, even if it was determined that there was reasonable grounds for prosecution at the preliminary hearing. This procedure was instituted as a constitutional guarantee in federal cases as a safeguard against abitrary prosecutorial action. Those states that have the same requirement are also motivated to provide protection at the state and municipal level.

If an indictment, also known as a "true bill," is voted, the next step is the appearance of the accused before a judge who is empowered to try felony cases. At this time, the accused is confronted with the charges against him and he is asked to plead. If he pleads "guilty," he may be sentenced without further court action; if he pleads "not guilty," his trial is set for some future date.

THE TRIAL

The Sixth Amendment to the U.S. Constitution states explicitly: "In all criminal prosecutions, the accused shall enjoy the right to a speedy and public trial by an impartial jury of the State and district wherein the crime shall have been committed, which district shall have been previously ascertained by law, and to be informed of the nature and cause of the accusation; to be confronted with the witnesses against him; to have compulsory process for obtaining witnesses in his favor, and to have assistance of counsel for his defence" (sic).

Most states have statutes specifying what constitutes "a speedy . . . trial" and the accused must be released and be thereafter immune from prosecution for that offense if he is not prosecuted within the time and provisions of the statutes. The accused, though entitled to be tried by an impartial jury, may waive that right and be tried by a judge alone. In that case, the judge is solely responsible for matters of fact, usually determined by a jury, as well as matters of law which he would adjudicate at all trials. If the case is heard by a jury, it will determine the validity of the facts presented and ultimately determine whether the accused is guilty as charged or not.

The jury is usually composed of 12 citizens who have, as a rule, been impanelled from the voter rolls. In determining the 12 jurors who will hear any given case, the prosecuting attorney and the defense attorney will individually question a larger panel of prospective jurors. Each attorney has a fixed number of peremptory challenges by which he may unseat any juror he feels may be unsympathetic to his case. After his peremptory challenges are exhausted, either attorney may challenge with cause, in which case the judge must agree that the juror in question would not render a fair and impartial judgment before he can be dismissed. The purpose of this sometimes lengthy process of selection is to obtain a truly impartial jury that will weigh without prejudice the cases of both the accused and the people.

After the attorneys for both sides have presented their opening statements, in which they outline the case they are about to present, the prosecuting attorney presents the case "For the People." He presents evidence, witnesses, and testimony. It is his job to present the case proven beyond any reasonable doubt. During this phase of the trial, the witnesses called are summoned for the purpose of presenting the state's case. The defense may cross-examine but may not, at this time, initiate a new line of questioning.

After the prosecution has presented all of the elements of its case, the defense may ask the court for a directed verdict of "not guilty" on the grounds that the prosecution has failed to present a case against the accused that would persuade reasonable jurors of his guilt. If the court is not so persuaded, the defense may present evidence refuting the prosecution's case. The roles are reversed when the defense presents its case, in that the prosecution may now cross-examine defense witnesses but it may not present its own witnesses nor may its questions be other than relative to that line of questioning already undertaken by the defense. In cross-examination, opposing attorneys may seek to discredit the witness or they may try to shake the witness's testimony by showing confusion, uncertainty or bias, but they may not launch into new areas of investigation in their cross examinations.

The accused is not obliged to testify at any point in the trail pursuant to the protection against being "compelled in any criminal case to be a witness against himself" as guaranteed by the Fifth Amendment. If he chooses not to appear, the prosecutor may not comment on that fact to the jury nor in any way suggest that by availing himself of his constitutional rights, the accused has in some way acknowledged his guilt in the matter charged.

After the defense has rested, or completed the presentation of its case, the prosecution may rebut with additional evidence. Usually, evidence ends after the prosecution rebuttal. At this point, the defense may again ask for a directed verdict of "not guilty" based on a number of factors, but chiefly on the failure of the prosecution to present a substantial case against the accused.

It is important to stress that the accused is not obliged to defend himself in any way. It is the burden of the prosecution to present what is known as a *prima facie* case against the defendant. A *prima facie* case is one which is proved "on the face of it". It means that the case in itself establishes the probability of guilt beyond a reasonable doubt. Such a case, at this point, is proved "on the face of it". The defense may then destroy that case by testimony explaining away otherwise damaging evidence. The defense may elect to present no case at all, confident that the jury will not convict on the case the prosecution has presented. In most cases, however, the prosecution has a case and the defense feels obliged to defend against it.

After both prosecution and defense have rested their cases, they make closing arguments in which they analyze the weakness of the opposition and the strength of their own case in order to sway the jurors to their point of view.

Following the closing arguments, the judge instructs the jury on what points of law may be applicable. It is the jury's function to determine what it believes the facts to be, and does so in private in the jury room. In most states, the jury must reach a unanimous decision as to the guilt of the accused. If the jury is unable to arrive at a unanimous decision as to the guilt or innocence of the defendant, and if the judge is satisfied that its deliberations have explored every possibility of arriving at such a decision, and that further deliberation would be fruitless, he will dismiss what is referred to as the "hung jury" and declare a mistrial. The prosecution will then determine whether it wishes to bring the case to court again or whether it will drop the charges.

If the jury determines the defendant is not guilty, he will be freed immediately and the matter is forever dismissed. If, on the other hand, he is found guilty, he is then subject to the sentence. In most states, the sentence is determined by the judge, with the jury functioning to determine the guilt or innocence, except in cases dealing with murder or rape. In such cases, most states place the sentencing responsibility on the jury. The jury decides whether the penalty is to be death or imprisonment, and even how many years are to be served in prison, if that is the penalty.

APPEAL

If the accused is convicted in the trial court, he may appeal the case to a higher court. The higher or appellate courts usually consist of several judges, together or "en banc," who will consider the appeal from conviction in a lower court. In such an appeal, no new evidence or testimony is presented to the court. There is no jury to determine guilt. The court examines the stenographic record of the trial court and considers the arguments, both written and oral, of defense and prosecution. The court then considers, as a matter of law, whether the trial was properly conducted, whether matters prejudicial to the rights of the accused were introduced in the trial proceedings, whether the evidence provided to the jury was legally obtained and properly presented and was sufficient for a jury to be entitled to return a guilty verdict. Any of a number of errors that might appear in the trial proceedings could be sufficient grounds for overturning the original judgment. Upon consideration of the matters presented, the reviewing court will render a written decision which either affirms or reverses the trial court's conviction. If the original judgment is affirmed, it means that the decision of the trial court stands; if it is reversed, the conviction is set aside and the accused is freed. If the appellate court rules that the original judgment is "reversed and remanded," it means that the conviction is nullified because of some error in the proceedings, but that the defendant may be tried again by the trial court.

Much of the law, or more accurately, its interpretation and, hence, its application, comes from the continuing judgments of the courts in cases all over the country. Most of these hundreds of cases heard daily are routine and, however significant they are to the participants, represent no particularly startling legal principle nor any significant upheaval in the day to day conduct either of the courts or of the average person. However, patterns in jurisprudence emerge and landmark cases which may have routine beginnings do come into view.

It is essential for a lawyer to keep up with this flood of information because the very practice of his profession is being constantly reshaped by events in court rooms around the country. Even the casual student, with only a sporadic interest in the dynamics of the legal world, should have some way to research areas of his immediate concern.

Legal Research. There are many sources of information available to the researcher: legal encyclopedias, dictionaries, legal periodicals, and code books setting forth the statutes. The encyclopedias focus on legal principles and theories, along with cases in which such principles predominate. There are also digests which index these cases. Perhaps the most useful sources are the bound volumes of reported cases called "Reporters" which list the decisions of the appellate court. These decisions establish the precedents which are the cornerstone of the judicial system.

In legal writing, cases are frequently cited to show how a legal principle was applied by a court. Each such citation is followed by certain figures and abbreviations which are simply a convenient way to indicate the location of a description of the elements of the case in a Reporter. Cases are arranged with volumes for cases in each state. In addition to State Reporters, a private publisher has established what is termed the National Reporter System in which blocks of states by geographical area are combined in various volumes. For example, appellate decisions from courts in Illinois, Massachusetts, Indiana, Ohio, and New York are contained in the *Northeastern Reporter* (N.E.)

For a certain 1966 wiretapping case in Illinois, the citation is: *People v. Kurth*. 34 Ill. 2d 387, 216 N.E. 2d 154. Translated, this means that the decision of the appellate court in the case of *People v.* (versus or against) *Kurth* can be found in the *Illinois Reporter*, second series, Volume 34 on page 387; this same decision can be found in Volume 216 of the *Northeastern Reporter*, second series, on page 154.

The reported decision indicates the contending parties, a synopsis of the case up to the time it appeared before the reviewing court, the decision of the court, the relevant points of law considered and decided by the court (in the opinion of the legal experts employed by the publisher), the majority opinion and the minority opinion, if there is one included. Other information includes dates, names of justices and contending attorneys, even citations of the case if it has passed through prior appeals before the current one under consideration. For this wealth of information, the Reporters are invaluable aids to any research of points of law, and the cases in which they are found.

CLASSES OF CRIMES

The California Penal Code specifies: ''A crime or public offense is an act committed or omitted in violation of a law forbidding or commanding it, and to which is annexed upon conviction, either of the following punishments: 1) death; 2) imprisonment; 3) fine; 4) removal from office; and 5) disqualification to hold and enjoy any office of honor, trust, or profit in this state.''

Since such a definition encompasses violations from the most trivial to the most disruptive and repugnant, efforts have long been made to classify crimes in some way. In the common law, crimes are classified according to seriousness from treason (the most serious) to misdemeanors (the least serious). Crimes in most states do not list treason separately and deal with felonies as the most serious crimes; misdemeanors as the next in seriousness; with different approaches to the least serious crimes which are known as *infractions* in some jurisdictions, *less than misdemeanors* in others, and *petty offenses* in still others. It will become apparent why any security specialist should understand the nature of a given crime and its classification, since such considerations will be important in determining his right to arrest, his right to use force in making the arrest, his right to search, and various other considerations that must be determined under possibly difficult circumstances, and without delay.

Felonies

From the time of Henry II of England there has been general understanding that felonies comprised the more serious crimes. This is true in modern American law as far as it goes, but clearly the definition of *felony* must be pinned down more precisely than that if it is to be used as a classification of crime and if courts are to respond differently to felons than they would to another type of law breaker. The definition of a felony is by no means standard throughout the United States. In some jurisdictions there is no distinction between felonies and misdemeanors.

The federal definition of a felony is an offense punishable by death or by imprisonment for a term exceeding one year. The test, then, for a felony is the length of time that punishment is imposed on the convicted person.

A number of states follow the federal definition. In those states, felony is a crime punishable by more than a year's imprisonment. The act remains a felony whatever the ultimate sentence may actually be. The crime is classified as a felony because it *could* be punished by a sentence of more than a year.

Other states provide that ''A felony is a crime . . . punishable with death or by imprisonment in the state prison.'' This definition hinges on the *place* of confinement rather than, as in the federal description, the length of confinement.

Some states bestow broad discretionary powers on the judge by providing that certain acts may be considered either a felony or a misdemeanor, depending on the sentence. The penalty clauses in the statutes thus involved specificially state that if the judge should sentence the defendant to a state prison, the act for which he was convicted shall be a felony (under the state definition of a felony) but if the sentence be less than such confinement, the crime shall be a misdemeanor.

The distinction can be very important in that in most states where arrest by private citizens (eg. security personnel) are covered by statute, it is clear that arrests for crimes less than a felony may be made only where the offense is committed in the presence of the arrestor. In the case of arrest for a felony, the felony must, in fact, have been committed (though not necessarily in the presence of the arrestor), and there must be reasonable grounds to believe the person arrested committed it. In other words, the security employee, unlike a police officer, acts at his own peril. "A police officer has the right to arrest without a warrant where he reasonably believes that a felony has been committed and that the person arrested is guilty, even if, in fact, no felony has occurred. A private citizen, on the other hand, is privileged to make an arrest only when he has reasonable grounds for believing in the guilt of the person arrested and a felony has *in fact* been committed." *U.S. v. Hillsman*, 522 F. 2d 454, 461 (7 Cir. 1975). The making of a citizen's arrest, which must be recognized as the only kind of arrest made by a security officer, is a privilege, not a right, and as such, is carefully limited by law. Such limitation is enforced by the ever-present potential for either criminal prosecution or tort action against the unwise or uninformed action of a security professional.

Tort Law: Source of Power and Limit

Tort law is the law that defines the general duties of citizens toward each other and allows lawsuits to recover damages for injury caused by a citizen's breach of such duty.

Tort law is the primary source of both the power and restrictions on the activities of security personnel. Its effect is basically remedial. It has the effect of restraining careless conduct because of the possibility of lawsuits brought against anyone who causes injury.

The following are the primary torts relevant to security officers:

1. *Battery*—intentional harmful or otherwise touching of another.
2. *Assault*—intentional causing of fear of a harmful or offensive touching.
3. *Intentional Infliction of Emotional Distress*—intentionally causing mental or emotional distress in another.
4. *False Imprisonment*—intentionally confining or restricting the movement or freedom of another.
5. *Malicious Prosecution*—groundlessly instituting criminal proceedings against another.
6. *Trespass to Land*—unauthorized entering upon the property of another.
7. *Trespass to Personal Property*—unauthorized taking or damaging of another's goods.

8. *Negligence*—causing injury to persons or property by taking an unreasonable risk or by failing to use reasonable care.
9. *Defamation* (Slander and Libel)—injuring the reputation of another by publicly making untrue statements.
10. *Invasion of Privacy*—intruding upon another's physical solitude, disclosing private information about another, or publicly placing another in a false light.[38]

Since security personnel can be expected to be active in detaining, questioning, and even searching suspected shoplifters; directing traffic, both vehicular and pedestrian; handling disturbances and unruly persons; removing people from premises; inspecting packages and vehicles; and many other direct confrontations, such activity clearly places many security officers directly in the path of a tort action. This threat of suit should serve to deter security officers from illegal activities in these areas, although to what degree it is effective as a deterrent is difficult to evaluate.

The security officer is empowered by tort law with various privileges which tend to shelter him from incurring such liabilities, but for each privilege or immunity which he may enjoy, there are restrictions that allow of little or no latitude.

Arrest

As indicated above, every citizen has some privilege to arrest a person who is committing or has committed a crime, with the intention of turning that person over to authorities. However, there are variations, from state to state on the application of this privilege. In common law states, the security officer, like the public police officer, is required to have reasonable grounds to believe the arrestee is guilty of the crime, whether or not the arrestee is, in fact, guilty. Thus, whether the arrestee is guilty or not does not affect the validity of the arrest, but does affect the matter of the arrestor's civil or criminal liability. New York, however, does not require reasonable grounds for believing the arrestee has, in fact, committed a felony. If it cannot be shown that the arrestee did, in fact, commit a felony, the arrestor would be exposed to criminal and civil responsibility, no matter how reasonable his actions may appear to have been. Thus, in states where common law still controls, a security officer can arrest for a felony committed either in his presence or out of his presence if he has reasonable grounds to believe the arrested person committed the felony, but he can arrest for a misdemeanor only when it involves a breach of the peace *and* when it is committed in his presence. (It is clear that, without so-called "shoplifting statutes" to extend his arrest privilege, a security officer would rarely be able to arrest for shoplifting under such restrictions in the shoplifting, per se, is hardly a breach of the peace.) In other states, the breach of peace requirement is not present and a private citizen may arrest for any misdemeanor committed in his presence.

Recognizing the restrictive nature of the civil and criminal penalties which serve to discourage or virtually eliminate private citizen involvement in handling shoplifters and, at the same time recognizing the enormous losses suffered by merchants, 45 states have enacted statutes making shoplifting a crime, separate and apart from theft or larceny (California, New Hampshire, New York, Vermont, and Texas are the five states which have no legislation dealing with shoplifting). Shoplifting must be formulated as a separate crime in that theft and larceny require proof of the mental element of "intent to permanently deprive the true owner of the property," an intent which is often difficult to prove in episodes in a store open to the public for business. Consequently, an arrestor is significantly exposed to civil and criminal liability for failure to meet the requirements of a valid citizen's arrest.

Forty-four of the forty-five states with shoplifter statutes provide that a merchant or his employee or agent (security people are covered under the extension "agent" or "employee") has the authority on his premises to stop a person suspected of shoplifting and question, search, and detain a person in a reasonable manner and for a reasonable period until the arrival of the public police. It can be seen that, whereas under citizen's arrest laws a merchant who stopped, questioned, searched, and detained a person would be in many ways subject to liabilities, the shoplifting statutes provide a substantial immunity from such liability provided it is conducted "reasonably." None of the statutes attempts to define what is meant by "reasonable."

The important element to keep in mind is that the power of citizen's arrest is of questionable value to the security professional. Any mistake in the arrest can eradicate the privilege. As we have seen, in the case of an arrest for a felony, there can be no mistake about the fact of the felony itself even though there may be some reasonable error in the identity of the perpetrator. No mistakes are allowed in the case of a misdemeanor. The arrested person must be guilty of the misdemeanor and it must have been committed in the presence of the arrestor. The arrest is further affected by the classification of the crime for which it is made. This requires a knowledge of the penal code of the jurisdiction in which the arrest takes place, as well as the ability to recognize whether the act is a felony or a misdemeanor according to that code.

It is interesting to note that citizen's arrest is the only authorization for all federal agents with arrest powers such as the F.B.I., the Drug Enforcement Agency, the Secret Service, Postal Inspectors, Customs Inspectors, Treasury Agents, and others. In the absence of specific federal legislation giving them the authority to act pursuant to federal jurisdiction, they operate under the citizen's arrest law of the state where it takes place.

Detention

"Detention" is a concept which has grown up largely in response to the difficulties faced by merchants in protecting their property from shoplifters and the problems and dangers they faced when making an arrest. Generally, this privilege differs from arrest in that it permits a merchant to briefly detain a suspected shoplifter without turning over to the police. An arrest requires that the arrestee be turned over to the authorities as soon as practicable and, in any event, without unreasonable delay.

All the shoplifting statutes refer to "detain" not to "arrest," a terminology probably derived from the thought that a distinction could be made between the two. The distinction was based on the fact that an arrest is for the purpose of delivering the suspect to the authorities and exercising strict physical control over that person until the authorities arrive. A detention, or temporary delay, would not be termed an arrest, as defined. The distinction is clearly frivolous but the statutes (with the exception of Rhode Island, which permits detention and search only by a peace officer) are clear. In Florida, for example:

> "A peace officer, or a merchant, or a merchant's employee who has probable cause for believing that goods held for sale by the merchant have been unlawfully taken by a person and that he can recover them by taking the person into custody, may, for the purpose of attempting to effect such recovery, take the person into custody and detain him in a reasonable manner for a reasonable length of time. Such taking into custody and detention by a peace officer, merchant, or merchant's employee shall not render such police officer, merchant, or merchant's employee criminally or civilly liable for false arrest, false imprisonment, or unlawful detention."[39]

California was one of the first with such merchant immunity established in a 1936 Supreme Court decision. In *Collyer v. S.H. Kress & Co.*, 5 Cal. 2d 175, 54 p. 2d 20 (1936), the court upheld the right of a department store official to detain a suspected shoplifter for 20 minutes.

Most statutes include the merchant, employee, agent, private police, and peace officer as authorized to detain suspects, but they do not include citizens at large, such as another shopper. Most of the statutes also describe the purpose of the detention and the manner in which it may be conducted. These purposes are: search, interrogate, investigate suspicious behavior, recover goods, and await a peace officer.

The manner in which the detention is to be conducted is generally described as "reasonable" and for "a reasonable period of time." Only five states describe the time of the detention: West Virginia, Maine and Montana provide for a maximum of 30 minutes; Indiana and Louisiana provide for a maximum of one hour. No state describes the reasonableness of the manner in which the detention is to be performed.

The privilege of detention is, however, subject to some problems. There must be probable cause to believe larceny has or is about to take place before a merchant may detain anyone. Probable cause is an elusive concept and one which has undergone many different interpretations by the courts. It is frequently difficult to predict how the court will rule on a given set of circumstances which may, at the time, clearly indicate probable cause to detain. Secondly, reasonableness must exist both in time and manner of the detention or the privilege will be lost.

USE OF FORCE

All of the privileges dealt with here sanction some use of force in order to effect the privilege. Here, as elsewhere, reasonableness is the key. If excessive or unreasonable force is used, not only is there liability for the torts (such as battery) which may be involved, but the original privilege which justified the use of reasonable force is lost.

Generalizations about the amount of force permitted in making an arrest, detaining a shoplifter, preventing a crime, protecting property, or maintaining order are not easy to make but certain guidelines usually apply.

As a general rule, the use of force in the performance of a security officer's duty is not a license but a limited privilege to be used sparingly by the arrestor. Where property rights are involved, a request for voluntary cooperation must usually precede the use of any force and, in any instance involving property, the use of deadly force is never justified unless the incident involving property entails a threat to life.

Deadly force cannot be used to prevent, or arrest for, a misdemeanor but it may be used to prevent or arrest for a felony which threatens the life or safety of a human being. Though many states allow the use of deadly force in an arrest for any felony, there is some question about the use of such force in other than life threatening felonies. In *Commonwealth v. Chermanski,* 430 Pa. 170, 242 A.2d 237 (1968), the Supreme Court of Pennsylvania changed its law regarding the use of deadly force in a felony arrest. In this case, the court declared that although, heretofore, it was permissible to use deadly force when the arrestor was in fresh pursuit of a felon fleeing from the commission of a felony, and if, but for the use of deadly force, the felon could not otherwise be taken, hereafter deadly force would be permissible only to prevent or arrest for felonies which generally cause or threaten death or great bodily harm. This ruling observed that the line between felonies and misdemeanors was no longer clear. A similar ruling appears in a California case, *People v. Piorkowski,* 115 Cal. Rptr. 833 (1974) in which the court held that though the evidence indicated the commission of a burglary which gave the defendant cause to make a citizen's arrest, the use of deadly force was not warranted. The circumstances of the crime were not such as to constitute any threat to life or limb. The court stated: "At common law, one could use deadly force to prevent the commission of a felony. Statutory expansion of the class of crimes punishable as felonies has made the common law rule manifestly too broad. It appears that the principle that deadly force may be directed toward the arrest of a felon is a correct statement of the law only where the felony committed is one which threatens death or great bodily harm."

INTERROGATION

A person who is legally detained (as outlined under "Detention") may be interrogated. Many of the statutes covering detention specifically authorize interrogation as a purpose of the detention. This is not to say that the suspect is obliged to answer any questions. He may remain silent, and threats or physical force to press a reply are prohibited. Any confessions, releases or agreements signed under duress or coercion are invalid and public questioning of a suspect may lead to liability for slander.

The acceptability of any interrogation turns on the legality of the detention and the manner in which the interrogation is conducted. The guiding principle here, as in so many of these questions, is reasonableness.

Search

Here again, as in the matter of interrogation, the legality of the search is dependent on the legality of the detention in which the search originates. If the detention is illegal, any search is prohibited. Even when the detention is perfectly proper, there is substantial question surrounding the issue of search. Many of the so-called shoplifting statutes, in allowing the recovery of property as a reason for detaining a suspect, do not go beyond that, thus leaving unanswered the question as to how that property is to be located and recovered. The Indiana statute, for example, states that: "(a) security agent of a mercantile establishment . . . who has probable cause . . . may detain that person . . . to determine whether such person has in his possession unpurchased merchandise taken from such mercantile establishment."[40] Oklahoma is specific in stating that a purpose of the detention is "performing a reasonable search of the detained person and his belongings when it appears that the merchandise or money may otherwise be lost."[41] Or in the case of Minnesota: "A merchant or merchant's employee . . . may detain such person for the sole purpose of delivering him to a peace officer without unnecessary delay . . . The person detained shall be informed promptly of the purpose of the detention and shall not be subjected to unnecessary or unreasonable force, nor to interrogation against his will."[42] Finally, the Montana statute states: "Unless evidence of concealment is obvious and apparent to the merchant this section shall not authorize a search of the detained person other than a search of his coat or other outer garments and any package, briefcase, or other container unless the search is done by a peace officer under proper legal authority."[43] It can be seen, in these examples, that there is a substantial difference in approach by different jurisdiction. Indiana implies the right to search, Oklahoma affirms the right to a "reasonable search," Minnesota forbids a search, and Montana permits a limited search.

The above discussion concerns itself only with the recovery of property by a merchant who is given certain authorizations and immunities under anti-shoplifting legislation. There are other areas of search in which security personnel may be involved, and these may be even more confusing.

The search of employee lockers, packages, or automobiles on the employer's premises is usually a matter dictated by company policy and is part of the employment contract, either specifically or by implication. In any event, in such a situation, searches are undertaken with the consent of the searched. Such consent will always validate a search if the search is questioned, unless it is conducted in a totally unreasonable manner. In that case, the search, itself, would not come into question, although some attendant circumstances might.

The law is not clear as to the rights of citizens in searching for weapons, and even less clear with respect to searching for incriminating evidence. The common law provides that every citizen has the right of self-defense, and such right may be invoked in searching for offensive weapons. The California Penal Code states that: "Any person making an arrest may take from the person arrested all offensive weapons which he may have about his person . . . ."[44] Both the common law and the California provision deal with weapons and with persons already arrested. No such right of search for weapons is given in laws governing detention.

Rights of search by private citizens are not clear except in certain state statutes limited to dealing with shoplifting, and even those differ widely. Again, generalizations are difficult, but search is valid in cases where consent is given, where an anti-shoplifting statute authorizes recapture of merchandise and a search attendant thereto, and in arrest situations where weapons are the object of the search. In all cases, the search must be conducted in a reasonable manner.

GENERAL POWER OF SECURITY PERSONNEL

From the discussion above, it can be seen that security personnel are limited to the exercise of powers possessed by every citizen. There is no legal area where the position of a security officer, per se, confers any greater rights, powers, or privileges than those possessed by every other citizen. As a practical matter, if the officer is uniformed, he will very likely find that, in most cases, people will comply with his requests. Thus, he will obtain consent to directives that may, in fact, be if not illegal, beyond his power to command. Many people are not aware of their own rights nor of the limitations of the powers of a security officer. Therefore, they comply. This acquiescence is usually harmless enough, but in those cases where a security officer has taken liberties with his authority, he and his employer will be subject to the penalties of a tort action. The litigation involved in suing a security officer and his employer for tortious conduct is slow and expensive, factors which may take such recourse out of the reach of the poor and those unfamiliar with their rights. But the judgements which have been awarded have had a sobering effect on security professionals, generally, and have probably served to reduce the number of such incidents.

Criminal law also serves to regulate security activities. Major crimes such as battery, manslaughter, kidnapping, and breaking and entering—any one of which might be confronted in the course of security activities—are substantially deterred by the criminal sanctions in effect.

Further limitations may be imposed upon the authority of a security force by licensing laws, administrative regulations, and specific statutes directed at security activities. Operating contracts between employers and security firms may also specify limits on the activities of the contracted personnel.

SECURITY, PUBLIC POLICE AND THE CONSTITUTION

The framers of the U.S. Constitution, with their grievances against England uppermost in mind when creating a new government, were primarily concerned with the manner in which the powerless citizen was, or could be, abused by the enormous power of government. The document they drew up was concerned, therefore, not with the rights of citizens as against each other, but with those rights with respect to federal or state action. Breaking and entering by one citizen upon another may be criminal and subject to tort action as well, but it is not a violation of any constitutional right. Similar action by public police is a clear violation of Fourth Amendment rights and, as such, is expressly forbidden by the Constitution.

The public police have substantially greater powers than security personnel in their powers of arrest, detention, search, and interrogation. Where security people are, as a rule, limited to the premises of their employer, public police operate through a much wider jurisdiction.

At the same time, the public police are limited by various restrictions imposed by the Constitution. Although the issue is not entirely clear, private police are not, as a rule, touched by these same restrictions.

Public police are limited by federal statutes which make it a crime for officials to deny others their constitutional rights. The Fourth and Fourteenth Amendments are most frequently invoked as the cornerstones in citizen protection against arbitrary police action.

The Fourth Amendment guarantees:

> "The right of the people to be secure in their persons, houses, papers, and effects, against unreasonable searches and seizures, shall not be violated, and no Warrants shall issue, but upon probable cause, supported by Oath or affirmation, and particularly describing the place to be searched, and the persons or things to be seized"(sic).

In an historic decision, the Supreme Court rules that any and all evidence uncovered in violation of the Fourth Amendment will be excluded from consideration. That means *all* evidence, no matter how trustworthy or probative, will be inadmissible if it is illegally obtained. This case (*Mapp v. Ohio,* 367 U.S. 643 (1961)) remains a landmark in leading to the development of the "exclusionary rule," which restates the principle of *Mapp* and others by rendering evidence illegally seized (and its fruits) inadmissable in any state or federal proceeding. *Mapp* had, however, been preceded in 1914 by *Weeks v. United States,* 232 U.S. 383 (1914), which had set the stage for the later all-inclusive decision in *Mapp* by holding that evidence acquired by officials of the *federal government* in violation of the Fourth Amendment must be excluded in a *federal* prosecution.

Mapp is clear in its application of the exclusionary rule to state and federal prosecutions. The question is, does the exclusionary rule apply to private parties? The determining case in this area is *Burdeau v. McDowell*, 256 U.S. 465 (1921), in which the Supreme Court held that evidence wrongfully obtained by a private individual is legally admissible in evidence. In its decision, the Court declared: "It is manifest that there was no invasion of the security afforded by the Fourth Amendment against unreasonable search and seizure, as whatever wrong was done was the act of individuals in taking the property of another." There is considerable controversy over *Burdeau* in that there are those who feel that constitutional guarantees are threatened by the acceptance of the evidence illegally obtained by private security personnel. Such evidence cannot, of course, be used in cases either directly or indirectly. Any involvement by government officials constitutes "state action" or an action "under color of law" and is limited by any constitutional restrictions which are applicable to public police actions. In *State v. Scrotsky,* 39 NJ 410, 416, 189 A.2d 23 (1963), the court excluded evidence obtained when a detective merely accompanied a theft victim to the defendant's apartment to identify and recover stolen goods. The court held: "The search and seizure by one served the purpose of both and must be deemed to have been participated in by both." The exclusionary rule is applied here, as in many others, to discourage government officials from conducting improper searches and from using private individuals to conduct them.

In cases where private parties act totally independent of government involvement, the courts have not been so clear. In a significant case, *People v. Randazzo,* 220 Cal. App. 2d 268, 34 Cal. Rptr. 65 (1963), the court admitted evidence obtained by a merchant in a shoplifting case. The court did not even deal with any questions of Fourth Amendment violation since there was no state action involved. The court held that the remedy for the victim of an unreasonable search conducted by a private individual not under color of law is a tort action and not the application of the exclusionary rule. In *Thacker v. Commonwealth,* 310 Ky. 702, 221 SW 2d 682 (1949), the court held that a private party acts for the state when he makes an arrest in accordance with the state's arrest statute. On the other hand, a federal district court found no state action in a case where the plaintiff alleged she was wrongfully detained, slapped, beaten, harassed, and searched by the manager and an employee of the store. *Weyandt v. Mason's Stores, Inc.*, 279 F. Supp. 283, 287 (W.D. Pa. 1968). Plaintiff sued, alleging, among other things, that the employee, a security officer, was acting "under color of law" because he was licensed under the Pennsylvania Private Detective Act. The court rejected this argument and found that the Pennsylvania law "invests the licensee with no authority of state law."

Generally, although public police are clearly limited by constitutional restructions, private police are not. Provided they act as private parties, in no way involved with public officials, they are limited by criminal and civil sanctions, but are not, at this time, bound by most constitutional restrictions.

CONTRACT SECURITY

The principle of *respondeat superior* ("let the master respond") is well established in common law. It is not, in itself, the subject of any substantial dispute, and at those times when it becomes an issue in a dispute, the area of contention is factual rather than the doctrine itself. There is no question that an employer (master) is liable for injuries caused by an employee (servant) who is acting within the scope of his employment. This is not to say that the employee is relieved of all liability. He is, in fact, the principal in any action, but since he rarely has the capability to satisfy a third party suit, an injured person will look beyond the employee to the employer for compensations for damages suffered.

In the doctrine of *respondeat superior,* the master is responsible for the actions of his servant while his servant is acting in his master's behalf. Clearly, the relationship needs definition. "A servant is a person employed by a master to perform service in his affairs, whose physical conduct in the performance of the service is controlled or is subject to the right of control by the master. This court has stated that the right of control and not necessarily the exercise of that right is the test of the relation of master and servant. Basically, it is distinction between a person who is subject to orders as to how he does his work and one who agrees only to do the work in his own way." *Graalum v. Radisson Ramp,* 245 Minn. 54, 71 NW 2d 904, 908 (1955). Under this definition, in-house security officers are servants, whereas contract security personnel may not be. In the latter case, contract personnel are employees of the supplying agency and, in most cases, the hiring company will not be held liable for his acts. The relationship is a complex one, however.

If the security officer is acting within the scope of his employment and commits a wrongful act, then his employer is liable for his acts. The matter then turns on the scope of his employment and his employer/employee relationship.

One court described the scope of his employment as depending on:

1. The act of being of the kind the offender is employed to perform;
2. It occurring substantially within the authorized time and space limits of the employment, and
3. The offender being motivated, at least in part, by a purpose to serve the master. *Fournier v. Churchill Downs - Latonia* 292 Ky. 215, 166 SW 2d 38 (1942).

This is further refined in another case in which the court stated that, with respect to an officer: "If he acts maliciously or in pursuit of some purpose of his own, the defendant is not bound by his conduct, but if, while acting within the general scope of his employment, he simply disregards his master's orders or exceeds his powers, the master will be reasonable for his conduct." *Hayes v. Sears, Roebuck Co.,* 209 P. 2d 468, 478 (1949).

Liability, then, is a function of the control exercised or permitted, in the relationship between the security officer and the hiring company. If the hiring company maintains a totally hands-off posture with respect to personnel supplied by the agency, it may well avoid liability for wrongful acts performed by such personnel. On the other hand, there is some precedent for considering the hiring company as sharing some liability, simply by virtue of its underlying rights of control over its own premises, no matter how it wishes to exericse that control. Many hiring companies are, however, motivated to contractually reject any control of security personnel on their premises in order to avoid liabilities. This, as pointed out in *Private Police* works to discourage hiring companies from regulating the activity of security employees and "the company that exercises controls, e.g. carefully examines the credentials of the guard, carefully determines the procedures the guard will follow, and pays close attention to all his activities, may still be substantially increasing its risk of liability to any third persons who are, in fact, injured by an act of the guard."[45]

It is further suggested in this excellent study that there may be an expansion of certain "non-delegable duty" rules into consideration of the responsibilities for the actions of security personnel. This concept of the "non-delegable duty" provides that there are certain duties and responsibilities which are imposed on an individual and for which he remains responsible even though he hires an independent contractor to implement them. Such duties currently encompass keeping a safe place to work and keeping the premises reasonably safe for business visitors. It is also possible that the courts may find negligence in cases where hiring companies, in an effort to avoid liability, have neglected to exercise any control over the selection and training of personnel and that such negligence on the part of the hiring company has led to injury to third party victims.

The legal road is filled with bumps and potholes. There is no easy way with it. The alert security manager will familiarize himself with the climate in his jurisdiction and keep abreast of the latest developments. There are rewards for the knowledgeable professional in an acquaintance with the law and its changes in his field. For the unwary or uncaring, the road can be troublesome, indeed.

REVIEW QUESTIONS

1. Why is a practical knowledge of the law important to the security officer and the security manager?
2. What impact does tort law have on the private security industry?
3. What makes an 'arrest' different from a 'detention'?
4. What are the major legal differences between public police and private security officers?
5. Why is the legal term 'Respondeat Superior' important to the contract security industry?

[illegible]

[illegible]

[illegible]

QUESTIONS

1. Why [illegible]

2. What [illegible]

3. What [illegible]

4. [illegible]

5. [illegible]

Chapter 6

RISK ANALYSIS AND THE SECURITY SURVEY

Once security goals and responsibilities have been defined and an organization created to carry them out, the ongoing task of security management is to identify potential areas of loss and to develop and install appropriate security countermeasures. Implicit in this approach is the concept of security as a comprehensive, integrated function of the organization, as discussed in a previous chapter.

This view of the loss prevention function might be contrasted with more limited security responses, such as the following:

- *One-dimensional* security, which relies on a single deterrent, such as guards.
- *Piecemeal* security, in which ingredients are added to the loss-preventive function piece by piece as a need arises, without a comprehensive plan.
- *Reactive* security, which responds only to specific loss events.
- *Packaged* security, which installs standard security systems (equipment or personnel, or both), without relation to specific threats, either because "everybody's doing it" or on the theory that packaged systems will take care of any problems that might arise. This is akin to prescribing a remedy without diagnosing the illness, like a broad-spectrum antibiotic that will "knock out any bug you might have."

As we have indicated in a preceding chapter, an integrated or systems approach to security is not *always* the desired solution. A small business, particularly one with minimal loss potential or relative ease of defense, might adequately be served (as many are) by a good lock on the door and an alarm system, or by a contract guard patrol. But, as the areas of loss increase and become more complex, and as the ability to protect a growing company against those losses with one-dimensional responses decreases, then it becomes increasingly necessary to adopt a more comprehensive security program.

If security is not to be one-dimensional, piecemeal, reactive or prepackaged, it must be based upon analysis of the total risk potential. In order to set up defenses against losses from crime, accident or natural disasters, in other words, there must first be a means of identification and evaluation of the risks.

Threat Assessment

The first stage in risk analysis involves identifying the threat. This means identifying, in every area of the company, the *vulnerability* to loss. Where could a loss occur? What assets of the company would attract a thief, either external or internal? How would it be possible for this "enemy" to carry out the theft? Similarly, what is the frequency and potential of accidents? Of natural disasters, such as flood or windstorm? Of fire or explosion?

Obviously, the answers to such questions will vary with every type of business or facility. The potential for shoplifting is great in a retail store; the potential for credit-card fraud is a continuing problem for hotels and airlines; the potential for explosion is high in particular types of manufacturing and storage. And a bewildering gamut of crimes threaten *every* business.

A viable threat assessment involves going through every department of a company, every operation and every function to find and identify these vulnerabilities. This search will be aided by past experience, including specific crimes or other loss incidents as well as the recent pattern of losses. The physical and social environment will also contribute to this assessment of risks. A facility located in a high-crime area must expect great vulnerability to particular types of crimes, including robbery, burglary, vandalism and assault. Some types of businesses also become "target risks" during periods of social upheaval or labor strife.

In addition to identifying vulnerability, threat assessment must also consider the *probability* of loss. The existence of valuable assets and/or the presence of a security weakness do not necessarily mean that a loss is probable. Once vulnerabilities are identified, it then becomes necessary to analyze the factors favoring loss. A piece of jewelry in a department store display might have a value equal to that of a television set in another display, but the ease with which the jewelry could be stolen far exceeds that of the TV set, and the protective measures that should be adopted will be different for each. Again, past history can serve as a guideline—although it must always be remembered that the fact that a loss has not occurred before (or been discovered before) does not mean that it will not occur in the future.

Threat Evaluation

To the recognition of vulnerability and probability of loss must be added a third dimension: *criticality* of loss. How serious would be the effect upon the company if a specific loss occurred? The criticality of a given loss may range from zero (no measurable loss) through minor, serious and critical to catastrophic (a loss so damaging as to threaten the continued survival of the business).

An important aspect in measuring the criticality of loss is the potential *frequency* of that loss. A minor theft from the shipping dock may become serious or even critical if such thefts continue to occur frequently.

For example, if several pallets of merchandise are exposed on the loading dock prior to shipment each day—the potential loss in the course of a year can be considerable. If one case is valued at $50, then the estimated loss ("frequently" means three cases per day at five days per week), is $50 X 3 X 5 X 52 weeks per year, or $39,000. Such dollar estimates for each vulnerability become essential in management's determination of the necessity to install countermeasures, whose cost can generally be more precisely estimated.

Cost Effectiveness of Security

It is unlikely that any evaluation will ever absolutely determine the cost-effectiveness of any security operation. A low rate of crime—whether compared to past experience, to like concerns, or to neighboring businesses—is an indication that the security department is performing effectively. But how much is being protected that would otherwise be damaged, stolen or destroyed? This can be any figure from the total exposure of the entire organization to some more refined estimate based upon the incidence of criminal attack locally or nationally, the average losses suffered by the industry in general, or the reduction in losses by the organization over a given period.

An estimate based on such figures might well serve as a practical guide to the usefulness of the security function. On the other hand, if a security operation costing $400,000 annually were estimated, by some formula using a mix of the data referred to, to have saved a potential in theft and vandalism of $300,000, would it be deemed advisable to reduce the department's operating budget by $100,000 or more? Obviously not. This would be roughly analogous to reducing or canceling insurance because damage or loss and subsequent insurance recovery for a specific period or incident were less than the cost of the premium. Security can be considered as "insurance" against unacceptable risks.

Cost-effectiveness studies must be made, however, as part of a periodic review of protection systems, even though such studies cannot be used, as a general rule, in devising a magic formula for computing the cost-per-$1,000 actually saved in cash or goods that would otherwise have been lost. Such a review would consider, for example, the savings that could result from the substitution of functionally equivalent electronic or other gear for manpower (the most expensive deterrent), and the feasibility of taking such a step.

THE SECURITY SURVEY

In the process of risk analysis which proceeds from threat assessment (identifying risk) to threat evaluation (determining the criticality and dollar cost of that risk) to the selection of security countermeasures designed to contain or prevent that risk, one of management's most valuable tools is the security survey.

A security survey is essentially an exhaustive physical examination of the premises and a thorough inspection of all operational systems and procedures. Such an examination or survey has as its overall objective the analysis of a facility to determine the existing state of its security, to locate weaknesses in its defenses, to determine the degree of protection required, and, ultimately, to lead to recommendations establishing a total security program.

Motivation setting the survey in motion should come from top management to insure that adequate funds for the undertaking are available and to guarantee the cooperation of all personnel in the facility. Since a thorough survey will require an examination of procedures and routines in regular operation as well as an inspection of the physical plant and its environs, management's interest in the project is of the highest priority.

The survey may be conducted by staff security personnel or by qualified security specialists employed for this purpose. Some experts suggest that outside security people could approach the job with more objectivity and would have less of a tendency to take certain areas or practices for granted, thus providing a more complete appraisal of existing conditions.

Whoever undertakes the survey should have training in the field and have achieved a high level of ability. It is also important that at least some members of the survey team be totally familiar with the facility and its operation. Without such familiarity, it would be difficult to formulate the survey plan, and the survey itself must be planned in advance in order to make the best use of personnel and in order to study the operation in every phase.

Part of the plan may come from previous studies and recommendations. These should be studied for any information they may offer. Another part of the survey plan will include a check list made up by the survey team in preparation for the actual inspection. This list will serve as a guide and reminder of areas that must be examined and, once drawn, should be followed systematically. In the event some area or procedure has been omitted in the preparation of the original check list, it should be included in the inspection and its disposition noted in the evaluation and recommendation.

Since no two facilities are alike—not even those in the same business—no check list exists that could universally apply for survey purposes. The following discussion is intended only to indicate those areas where a risk may exist. It should be considered as merely a guide to the kinds of questions or specific problems that might be dealt with.

The Facility

- Consider the *perimeter* as a security problem. Check fencing, gates, culverts, drains. Check lighting, including standby lights and power. Check overhangs and concealing areas. Can vehicles drive up to the fence?
- Consider the *parking lot* as a problem. Are employee automobiles adequately protected from theft or vandalism? How? Is the lot sufficiently isolated from plant or office to prevent unsupervised back and forth traffic? Are there gates or turnstiles for the inspection of traffic, if that is necessary? Are these inspection points properly lighted? Can packages be thrown over or pushed through the fence into or out of the parking lot?
- Consider all *adjacent building* windows and roof tops as security problems. Are spaces near these adjacencies accessible to them? Are they properly secured? How?
- Consider all *doors and windows* less than 18 feet above ground level as security problems. How are these openings secured?
- Consider the *roof* as a security problem. What means are employed to prevent access of the roof?
- Consider the issuance of *main entrance keys* to all tenants in a building a security problem. How often are entrance locks changed? What is building procedure when keys are lost or not returned? How many tenants are in the building? What business are they in?
- Coinsider any *shared occupancy*—as in office buildings—a security problem. Does the building have a properly supervised sign-in log for off-hours? Do elevators switch to manual, or can floors be locked against access outside of business hours? When are they so switched? By whom can they then be operated? Who collects the trash, and how and when is it removed from the building? Are lobbies and hallways adequately lighted? What guard protection does the building have? How can they be reached? Are washrooms open to the public? Are equipment rooms locked? Is a master key system in use? How are keys controlled and secured? Is there a receptionist or guard in the lobby? Can the building be accessed by stair or elevator from basement parking facilities?
- Consider all areas containing *valuables* to be a security problem. Do safes, vaults, or rooms containing valuables have adequate alarms? What alarms are in place to protect against burglary, fire, robbery or surreptitious entry?
- Consider the *off-hours* when the facility is not in operation or all night-time hours to be a security problem. How many guards are on duty at various times of day? Are guards alert and efficient? How are guards

equipped? How many patrols are there and how often do they make their rounds? What is their tour? What is the guard communication system?

- Consider the control and supervision of *entry into the facility* a security problem. What method is used to identify employees? How are applicants screened before they are employed? How are visitors (including salesmen, vendors and customers) controlled? How are privately owned vehicles controlled? Who delivers the morning mail and when? How are empty mail sacks handled? Do you authorize salesmen or solicitors for charity in the facility? How are they controlled? Are their credentials checked? Who does the cleaning? Do they have keys? Who is responsible for these keys? Are they bonded? Who does maintenance or service work? Are their tool boxes inspected when they leave? Are their credentials checked? By whom? Are alarm and telephone company men allowed unlimited access? Is the call for their service verified? By whom? How is furniture or equipment moved in or out? What security is provided when this takes place at night or on weekends? Are messengers permitted to deliver directly to the addressee? How are they controlled? Which areas have the heaviest traffic? Are visitors claiming official status, such as building or fire inspectors, permitted free access? Are their credentials checked? By whom?
- Consider *keys and key control* a security problem. Are keys properly secured when not in use? Are locks replaced or recored when a key is lost? Are locks and locking devices adequate for their purpose? Are all keys accounted for and logged? What system is used for the control of master and sub-master keys?
- Consider *fire* a basic security problem. Are there sufficient fire boxes throughout the facility? Are they properly located? Is the type and number of fire extinguishers adequate? Are they frequently inspected? How far is the nearest public fire department? Have they ever been invited to inspect the facility? Does the building have automatic sprinklers and automatic fire alarms? Are there adequate fire barriers in the building? Is there an employee fire brigade? Are fire doors adequate? Are "No Smoking" signs enforced? Are flammable substances properly stored? Is there a program of fire prevention education? Are fire drills conducted on a regular basis?

For all its seeming length, this is in reality only a sample of the kinds of questions that must be asked in conducting a survey of any facility. This list covers only a general overview of some of the aspects to be covered.

General Departmental Evaluations

Each department in the organization should be evaluated separately in terms of its potential for loss. These department evaluations will eventually be consolidated into the master survey for final recommendation and action. Basic questions might be as follows:

- Is the department function such that it is vulnerable to embezzlement?
- Does the department have cash funds or negotiable instruments on hand?
- Does the department house confidential records?
- What equipment, tools, supplies or merchandise can be stolen from the department?
- Does the department have heavy external traffic? Internal traffic?
- Does the department have "target" items in it such as drugs, jewelry, furs?
- What is the special fire hazard in the department or from adjacencies?

These are questions that may serve to guide the survey in focusing on particular areas of risk in each department to be examined. Where particular risks predominate, special attention must be paid to provide some counter-action to remove them.

Personnel Department

- Can the department area be locked off from the rest of the floor or building after hours?
- How are door and file keys secured?
- Are files kept locked during the day when not in use?
- What system is followed with regard to the payroll department when employees are hired or terminated?
- What are the relationships between persons in personnel and payroll?
- What are the employment procedures? How are applicants screened?
- How closely does personnel work with security on personnel employment procedures?

Security of personnel files is of extreme importance. Normally these files will contain information on every employee, past and present, from the president on down. This information is highly confidential and must be handled that way. There can be no exceptions to this firm policy.

Accounting

The accounting department has total supervision over the firm's money and will generally be the area most vulnerable to major loss due to crime. Certainly protective systems have been in operation in this area from the company's founding, but these systems must be re-evaluated regularly, in light of ongoing experience, to find ways of improving both their efficiency and security.

Cashier.

- How accessible is the cash operation to hallways, stairs and elevators?
- Do posted signs clearly announce the location and operating hours of the cashier?
- Is there generally sufficient cash on hand to invite an employee to abscond with it? To attract an attempted burglary or armed robbery?
- What are the present systems of audit and controls? What forms are used?
- What are the controls put on cashier embezzlement?
- What are the opportunities for collusion in this operation?
- Is the security adequate to the risk?

Accounts Receivable. In evaluating the frauds to which accounts receivable is vulnerable, it will be necessary to consider, from experience, all the possibilities, and in this light to examine every step of the procedures currently in operation, from billing advice through billing to the credit of the account. All flow of information and action documents must be studied minutely to find if any flaw or weakness could be exploited for criminal purposes.

- Consider the billing procedure with particular attention to the forms used and the authorizations required.
- Try to determine how difficult it might be to cash a check payable to the company.
- What are the opportunities to destroy billing records and to keep and cash a check?
- What are the possibilities of altering invoices to show a lesser amount payable?

Accounts Payable. Accounts payable as a disbursing entity invites more attention from thieves than most other areas. It is particularly susceptible to internal attack in a number of ways. The most common is the dummy invoice by which forged authorizations permit payment to a non-existent account. This is relatively easy to overcome, however, by an alert staff working within a system that provides reasonable security.

- Examine all forms and systems on a step-by-step basis.
- How is the authenticity of new accounts established?

Payroll.

- What system is used to introduce a new employee into the payroll?
- Do the records in personnel and in payroll correspond? Are these cross checked? How? By whom?
- Could an employee in personnel conspire with an employee in payroll to introduce fictitious employees into the records? What would prevent this?

Company Bank Accounts.

- Can one person transfer unlimited funds?
- Is there a ceiling limiting withdrawal of company funds?
- What instructions have been given to the bank? By whom? Who can change these instructions?
- Who audits company bank accounts? How often?
- Could the transfer or controller leave with the company bank account?

It is well to remember here, especially when it appears that questions concerning the probity of company officers are posed, that the job of the security survey is not to make judgments on whether a criminal act is *likely* to occur, but whether it *could*. A survey of this nature in no way implies that the treasurer is apt or in any way inclined to abscond with company funds. It simply asks whether or not the procedures are such that he could. If this were the case, the recommendation would surely suggest that safeguards such as countersignatures be set up so that the current or future treasurers could not perform what they very likely would not perform in any event.

Data Processing

- Are adequate auditing procedures in effect on all programs?
- How are printouts of confidential information handled?
- What is the off-site storage procedure? How are such tapes updated?
- What is the system governing program access?
- How is computer use logged? How is the accuracy of this record verified?
- Who has keys to the computer spaces? How often is the list of authorized key holders evaluated?
- What controls are exercised over access? How often is the list of those authorized to enter updated?

- What fire prevention and fire protection procedures are in effect? What training is given employees in fire prevention and protection? What is the number, location and condition of fire extinguishers and the basic extinguishing system?

Purchasing

This is an area subject to many temptations. Graft is often freely offered in the form of cash, expensive gifts, lavish entertainment and luxurious vacations—all in the name of seeking the good will of the purchasing agent. Generally speaking, this is not a security matter unless he succumbs to the extent that he pays for goods never delivered or perhaps pays an invoice twice. If his judgment is distorted by all the attention from vendors and he buys unwisely, that is a concern of management in which security plays no role.

There are, however, some areas in the purchasing function in which security might be involved:

- What are the procedures preventing double payment of invoices? Fraudulent invoices? Invoices for goods never received? How often is this area audited?
- Are competitive bids invited for all purchases? Must the lowest bid be awarded the contract?
- What forms are used for ordering, authorizing payment? How are they routed?
- Since purchasing is frequently responsible for the sale of scrap, waste paper and other recoverable items, who verifies the amount actually trucked away? Who negotiates the sale of waste or scrap material? Are several prospective buyers invited to bid? How is old equipment or furniture sold? What records are kept of such sales? Is the system audited? What controls are placed on the authority of the seller?

Shipping and Receiving Areas

Freight and merchandise handling areas are particularly troublesome. There is a great potential for theft in these areas. Close attention must be paid to current operations and efforts must continually be directed to their improvement.

- What inspections are made of employees entering or leaving such areas?
- How is traffic in and to such areas controlled? Are these areas separated from the rest of the facility by a fence or barrier?

- Where is merchandise stored after receipt or before shipment? What is the security of such area? What is the nature of supervision in these areas?
- What is the system for accountability of shipments and receipts?
- Is the area guarded?
- What losses are being experienced in these areas? What is the profile of such loss (type, average amount, time of day)?
- Is merchandise left unattended in these areas?
- Are truck drivers provided with rest room facilities separate from dock personnel? Are they isolated from them at all times to prevent collusion?
- How many people and who are authorized in security storage areas.

Miscellaneous

- What records are kept of postage meter usage? What controls are established over meter usage?
- How is the use of supplies and materials controlled?
- How are forms controlled?

Report of the Survey

After the survey has documented the full scope of its examination, a report should be prepared indicating those areas which are weak in security and recommending those measures which might reasonably bring the security of the facility up to acceptable standards. On the basis of the status as described in the survey, and considering the recommendations made, the security plan can now be drawn up.

In some cases compromises may have to be made. The siting of the facility or the area involved, for example, may make the ideal security program of full coverage of all contingencies too costly to be practical. In such cases, the plan must be restudied to find the best approach to achieve acceptable security standards within these limitations.

It must be understood that the security director will rarely get all of what he wants to do the job. As in every department, he must work within the framework of the possible. Where he is denied extra personnel, he must find hardware that will help to replace them. Where his request for more coverage by closed circuit television is turned down, he must develop inspection procedures or barriers that may serve a similar purpose. If, at any point, he feels that he has been cut to a point where the stated objective cannot be achieved, he is obliged to communicate that opinion to management, who will then determine whether to diminish their original objective or authorize more money. It is important, however, that the security director exhaust every alternative method of coverage before he goes to management with an opinion that requries this kind of decision.

Periodic Review

Even after the security plan is formulated, it is essential that the survey process be continued. A security plan, to be effective, must be dynamic. It must change regularly in various details to accommodate changing circumstances in a given facility. Only regular inspections can provide a basis for the ongoing evaluation of the security status of the company. Exposures and vulnerability change constantly. What may appear to be a minor alteration in operational routines may have a profound effect on the security of the entire facility.

Security Files

The survey and its resultant report are also valuable in the building of security files. From this evaluation emerges a detailed current profile of the firm's regular activities. With such a file, the security department can operate with increased effectiveness, but it should, by inspections and additional surveys, be kept current.

Such a data base could be augmented by texts, periodicals, official papers and relevant articles in the general press related to security matters. Special attention should be paid to subjects of local significance. Although national crime statistics are significant and help to build familiarity with a complex subject, local conditions have a more immediate import to the security of the company.

As these files are broadened, they will become increasingly useful to the security operation. Patterns may emerge, seasons may become significant, economic conditions may predict events to be alert to.

For example:

- Certain days or seasons may emerge as those on which problems occur.
- Targets for crime may become evident as more data is amassed. This may enable the security director to reassign priorities.
- A profile of the type and incidence of crimes—possibly even the criminal himself—may emerge.
- Patterns of crime and its *modus operandi* on payday or holiday weekends may become evident.
- Criminal assaults on company property may take a definable or predictable shape or description, again enabling the security director to better shape his countermeasures.

The careful collection and use of data concerning crime in a given facility can be an invaluable tool for the conscientious security man. It can add an important dimension to his regular re-examination of the status of crime in his company.

REVIEW QUESTIONS

1. What is the difference between *vulnerability* to loss and loss *probability*?
2. What is meant by *criticality* of loss?
3. If a security counter-measure costs as much or more than the loss being protected against for a given period, does it follow that the security measure should be discontinued because it is not cost-effective?
4. Why should accounting procedures be a part of a security survey?
5. Why are security files significant in protection planning?

PART III
BASICS OF DEFENSE

Chapter 7

The Outer Defenses: Building and Perimeter Protection

The cause of security can be furthered simply by making it more difficult (or to be more accurate—*less easy)* for criminals to get into the premises being protected. And these premises should then be further protected from criminal attack by denying ready access to interior spaces in the event that exterior barriers are surmounted by a determined intruder.

This must be the first concern in security planning.

True, every security program must be an integrated whole—and each element must grow out of the specific needs dictated by the circumstances affecting the facility to be protected. But the first and basic defense is still the physical protection of the facility. Planning this defense is neither difficult nor complicated, but it requires meticulous attention to detail.

Whereas the development of anti-embezzlement systems, or even the establishing of shipping and receiving safeguards, requires particularized sophistication and expertise, the implementation of an effective program of physical security is the product of common sense and a lot of leg work expended in the inspection of the area.

Physical security concerns itself with those means by which a given facility protects itself against theft, vandalism, sabotage, unauthorized entry, fires, accidents, and natural disaster. And in this context, a facility is a plant, building, office, institution, or any commercial or industrial structure or complex with all the attendant structures and functions that are part of an integrated operation. An international manufacturing operation, for example, might have many *facilities* within its total organization.

Physical security planning includes protection of the facility's perimeter and the building itself (discussed in this chapter) and of the building's interior and its contents (discussed in the following chapter).

BARRIERS, FENCES AND WALLS

A facility's perimeter will usually be determined by the function and location of the facility itself. An urban office building or retail enterprise will frequently occupy all the real estate where it is located. In such a case, the perimeter may well be the walls of the building itself. Most industrial operations, however, require yard space and warehousing, even in urban areas. In that case, the perimeter is the boundary of the property owned by the company. But, in *either* case, the defense begins at the perimeter—the first line which must be crossed by an intruder.

Barriers

Natural and structural barriers are the elements by which boundaries are defined and penetration is deterred.

Natural barriers comprise those topographical features that assist in impeding or denying access to an area. They may consist of rivers, cliffs, canyons, dense growth, or any other terrain or feature that is difficult to overcome. Structural barriers are permanent or temporary devices such as fences, walls, grilles, doors, roadblocks, screens, or any other construction that will serve as a deterrent to unauthorized entry.

It is important to remember that structural barriers rarely, if ever, prevent the penetration. Fences can be climbed, walls can be scaled, and locked doors and grilled windows can eventually be bypassed by a resolute assault.

The same is generally true of natural barriers. They almost never constitute a positive prevention of intrusion. Ultimately, all such barriers must be supported by additional security, and most natural barriers should be further strengthened by structural barriers of some kind. It is a mistake to suppose that a high, steep cliff, for example, is by itself protection against unauthorized entry.

Fencing

The most common type of fencing normally used for the protection of a facility is chain link, although barbed wire is useful in certain permanent applications, and concertina barbed wire is occasionally used in temporary or emergency situations.

Chain link. Chain link fencing should meet certain specifications developed by the Defense Department in order to be fully effective.

It should be constructed of a wire of a number 11 or heavier gauge with twisted and barbed selvage top and bottom. The fence itself should be at least seven feet tall and should begin no more than two inches from the ground. If the soil is sandy or subject to erosion the bottom edge of the fence should be installed below ground level. The fence should be stretched and fastened to rigid metal posts set in concrete, with such additional bracing as may be necessary at corners and gate openings. Mesh openings should be no more than two inches square. The fence should, additionally, be augmented by a top guard or overhang of three strands of stretched barbed wire, angled at 45 degrees away from the protected property. (See Figure 7-1). This overhang should extend out and up far enough to increase the height of the fence by at least one foot, to an overall height of eight feet or more.

To protect the fence from washouts or channeling under it, culverts or troughs should be provided at natural drainage points. If any of these drainage openings are larger than 96 square inches, they, too, should be provided with physical barriers that will protect the perimeter—without, however, impeding the drainage.

If buildings, trees, hillocks, or other vertical features are within 10 feet of the fence, it should be heightened or protected with a Y-shaped top guard.

Barbed Wire. When a fence consists of barbed wire, a 12 gauge, twisted double strand with four–point barbs four inches apart is generally used. These fences, like

Figure 7-1. A standard chain link fence with a 'top guard' for added security. (Courtesy of American Security Fence Corporation.)

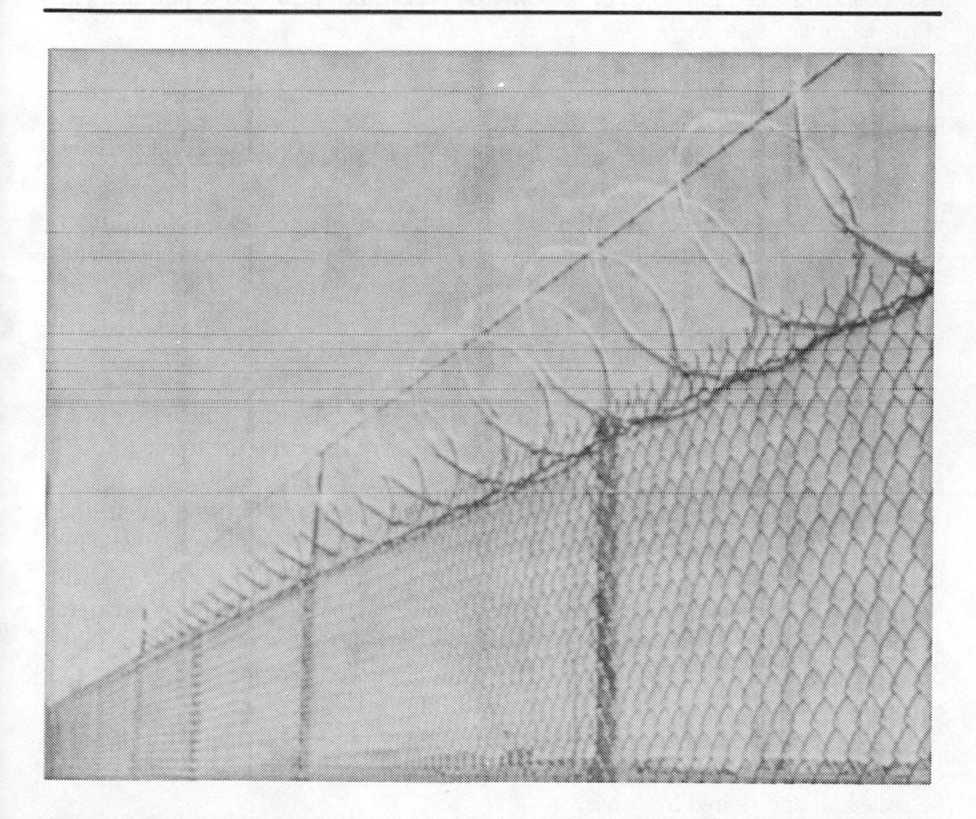

chain link, should also be at least seven feet high and they should, additionally, carry a top guard. Posts should be metal and spaced no more than six feet apart. Vertical distance between strands should be no more than six inches, and preferably less.

Concertina Wire. Concertina wire is a coil of steel wire clipped together at intervals to form a cylinder. When it is opened it forms a barrier 50 feet long and three feet high. Developed by the military for rapid laying, it can be used in multiple coils. It can be used either with one roll atop another, or in a pyramid with two rolls along the bottom and one on the top. Ends should be fastened together, and the base wires should be staked to the ground.

Concertina wire is probably the most difficult fence to penetrate, but it is unsightly and is rarely used except in a temporary application.

Walls

In some instances, masonry walls may be used to form all or part of the perimeter barrier. Such walls may be constructed for aesthetic reasons to replace the less decorative chain link look, or possibly to conceal the operations within that part of the facility.

In those areas where a masonry wall is used, it should be at least seven feet high with a top guard of three or four strands of barbed wire, as in the case of a chain link fence.

Since concealment of the inside activity must be paid for by also cutting off the view of any activity outside the wall, extra efforts must be made to prevent scaling the wall. Ideally, the perimeter line should also be staggered in a way that permits observation of the area in front of the wall from a position or positions inside the perimeter.

GATES AND OTHER BARRIER BREACHES

Every opening in the perimeter barrier is a potential security hazard. The more gates, the more security personnel must be deployed to supervise the traffic through them. Obviously, these openings must be kept to a minimum.

During shift changes, there may be more gates open and in use than at other times of the day when the smallest practical number of such gates are in operation. Certainly there must be enough gates in use at any time to facilitate the efficient movement of necessary traffic. The number can only be determined by a careful analysis of the needs at various times of the day—but every effort must be made to reduce the number of operating gates to the minimum consistent with safety and efficiency.

If it develops that some gates are not necessary to the operation of the facility, or that they can be dispensed with by changing traffic patterns, the openings should be sealed off and retired from use.

Padlocking

Gates used only at peak periods or emergency gates should be padlocked and secured. If possible, the lock used should be distinctive and immediately identifiable. It is common for thieves to cut off the plant padlock and substitute their own, so they can work at collecting their loot without an alarm being given by a passing patrol that has spotted an otherwise missing lock. A lock of distinctive color or design could compromise this ploy.

It is important that all locked gates be checked frequently. This is especially important where, as is usually the case, these gates are out of the current traffic pattern and are remote from the general activity of the facility.

Personnel and Vehicle Gates

Personnel gates are usually from four to seven feet wide to permit single line entrance or exit. It is important that they not be so wide that control of personnel is lost. Vehicular gates, on the other hand, must be wide enough to handle the type of traffic typical of the facility. They may handle two-way traffic, or if the need for control is particularly pressing, they may be limited to one-way traffic at any given time.

A drop or railroad crossing type of barrier is normally used to cut off traffic in either direction when the need arises. The gate itself might be single or double swing, rolling or overhead. It could be a manual or an electrical operation. Railroad gates should be secured in the same manner as other gates on the perimeter except during those times when cars are being hauled through it. At these times, the operation should be under inspection by a security guard.

Miscellaneous Openings

Virtually every facility has a number of miscellaneous openings that penetrate the perimeter. All too frequently these are overlooked in security planning, but they *must* be taken into account, because they are frequently the most effective ways of gaining an entrance into the facility without being observed.

These openings or barrier breaches consist of sewers, culverts, drainpipes, utility tunnels, exhaust conduits, air intake pipes, manhole covers, coal chutes and sidewalk elevators. All must be accounted for in the security plan.

Any one of these openings having a cross section area of 96 square inches or more must be protected by bars, grillwork, barbed wire, or doors with adequate locking devices. Sidewalk elevators and manhole covers must be secured from below to prevent their unauthorized use. Storm sewers must be fitted with deterrents which can be removed for inspection of the sewer after a rain.

BUILDINGS ON THE PERIMETER

When the building forms part of the perimeter barrier, or when, in some urban situations, the building walls are the entire perimeter of the facility, it should be viewed in the same light as the rest of the barrier or it should be evaluated in the same way as the outer structural barrier. It must be evaluated in terms of its strength and all openings must be properly secured.

In cases where a fence joins the building as a continuation of the perimeter, there should be no more than two inches between these two structures. Depending on the placement of windows, ledges or setbacks, it might be wise to gradually double the fence height to the point where it joins the building. In such a case the higher section of the fence should extend six to eight feet out from the building.

Windows and Doors

Windows and other openings larger than 96 square inches should be protected by grilles, metal bars, or heavy screening when they are less than 18 feet from the ground or when they are less than 14 feet from structures outside the barrier.

Doors which penetrate the perimeter walls must be of heavy construction and fitted with a strong lock.

Since both the law and good sense require that there be adequate emergency exits in the event of fire or other danger, provision must be made for such eventualities. Doors that have been created for emergency purposes only should have exterior hardware removed so that they cannot be opened from the outside. They can be secured by a remotely operated electromagnetic holding device, or they can be alarmed so that their use from inside will be substantially reduced or eliminated.

Roofs and Common Walls

An important, though often overlooked, part of the perimeter is the roof of the building. In urban shopping centers or even in small, freestanding commercial situations where the building walls are the perimeter, entry through the roof is common. Entry can be made through skylights or by chopping through the roof—an activity rarely detected by passersby or even by patrols.

Buildings sharing a common wall have also frequently been entered by breaking through the wall from a poorly secured neighboring occupancy. All of these means of entry circumvent normal perimeter alarm systems and can, therefore, be particularly damaging.

BARRIER PROTECTION

In order for the barrier to be most effective in preventing intrusion, it must be patrolled and inspected regularly. Any fence or wall can be scaled, and unless these barriers are kept under observation, they will be neutralizing the security effectiveness of the structure.

Clear Zone

A clear zone should be maintained on both sides of the barrier to make immediately visible any approach to the barrier from the outside or any movement from the barrier to areas inside the perimeter. Anything outside the barrier, such as refuse piles, weed patches, heavy undergrowth or anything else that might conceal a man's approach, should be eliminated. Inside the perimeter, everything should be cleared away from the barrier to create as wide a clear zone as possible.

Unfortunately, it is frequently impossible to achieve an uninterrupted clear zone. Most perimeter barriers are indeed on the perimeter of the property line, which means that there is no opportunity to control the area outside the barrier. The size of the facility and the amount of space needed for its operation will determine how much space can be given up to the creation of a clear zone inside the barrier. It is important, however, to create some kind of a clear zone, however small it must be.

In situations where the clear zone is necessarily so small as to endanger the effectiveness of the barrier, thought should be given to increasing the barrier height in critical areas or to the installation of an intrustion detection device to give due and timely warning of an intrusion to an alert guard force.

Inspection

Having established the perimeter defense and the clear zones to the maximum possible and practical, it is essential that a regular inspection routine be set up. Gates should be examined carefully to determine whether locks or hinges have been tampered with, fence lines should be observed for any signs of forced entry or tunneling, walls should be checked for marks that might indicate they have been scaled or that such an attempt has been made, top guards must be examined for their effectiveness, miscellaneous service penetrations must be examined for any signs of attack, brush and weeds must be cleared away, erosion areas must be filled in, any potential scaling devices such as ladders, ropes, oildrums or stacks of pallets must be cleared out of the area. Any condition which could, even in the smallest degree, compromise the integrity of the perimeter must be both reported and corrected.

Such an inspection should be undertaken no less than weekly, and possibly more often if the conditions indicate.

INSIDE THE PERIMETER

Unroofed or outside areas within the perimeter must be considered a second line of defense, since these areas can usually be observed from the outside and targets selected before an assault is made.

In an area where materials and equipment are stored in a helter-skelter manner, it is difficult for guards to determine if they have been disturbed in any way. On the other hand, storage which is neat, uniform and symmetrical can be readily observed and any disarray can be detected at a glance.

Discarded machinery, scrap lumber and junk of all kinds haphazardly thrown about the area create safety hazards, as well as provide cover for any intruder. Such conditions must never be permitted to develop. Efficient housekeeping is basic security.

Parking

The parking of privately owned vehicles within the perimeter barrier should never be allowed. There should be no exceptions to this rule. Facilities which can't or won't establish parking lots outside the perimeter barrier are almost invariably plagued by a high incidence of pilferage, due to the ease with which employees can conceal goods in their cars at any point during the day.

In those cases where the perimeter barrier encompasses the employee and visitor parking areas, additional fencing should be constructed to create new barriers which excludes the parking areas. Appropriate guarded pedestrian gates must, of course, be installed to accommodate the movement of employees to and from their cars.

The parking lot, itself, should be fenced and patrolled to protect against car thieves and vandals. Few things are more damaging to morale than the insecurity which an unprotected parking area in a crime-ridden neighborhood can create.

Company cars and trucks—especially loaded or partially loaded vehicles—should be parked within the perimeter for added security. This inside parking area should be well lighted, and it should be regularly patrolled or kept under constant surveillance.

Loaded or partially loaded trucks and trailers should be sealed or padlocked and should, further, be parked close enough together and close enough to a building wall, or even back to back, so that neither their side doors nor rear doors can be opened without actually moving the vehicles.

Surveillance

The entire outside area within the security barrier must be kept under surveillance at all times, particularly at night. Since goods stored in this area are particularly vulnerable to theft or pilferage, this is the area most likely to attract the thief's first attention. With planning and study in cooperation with production personnel, it will undoubtedly be possible to lay out this yard area so that there are long, uninterrupted sight lines which will permit inspection of the entire area with a minimum of movement.

LIGHTING

Depending on the nature of the facility, protective lighting will be designed either to emphasize the illumination of the perimeter barrier and the outside approaches to it, or to concentrate on the area and the buildings within the perimeter. In either case, it must produce sufficient light to create a psychological deterrent to intrusion as well as to make detection virtually certain in the event an entry is made.

It must avoid glare that would reduce the visibility of security personnel, while creating glare to deter intruders. It must also avoid casting annoying or dangerous lights into neighboring areas. This is particularly important where the facility abuts streets, highways, or navigable waterways.

The system must be reliable and designed with overlapping illumination to avoid creating unprotected areas in the event of individual light failures. It must be easy to maintain and service, and it must itself be secured against attack. Poles should be within the barrier, power lines should be buried and the switch box or boxes must be secure.

There should be a back-up power supply in the event of power failure. Supplementary lighting, including searchlights and portable lights, should also be a part of the system. These lights are provided for special or emergency situations and, although they may not be used with any regularity, they must be available to the security force.

The system could be operated automatically by a photoelectric cell which responds to the amount of light to which it is subjected. Such an arrangement allows for lights to be turned on at darkness and extinguished at daylight. This can be set up to activate individual lamps or to turn on the entire system at one time.

Other controls are timed, which simply means that lights are switched on and off by the clock. Such a system must be adjusted regularly to coincide with the changing hours of sunset and sunrise. The lights may also be operated manually.

Types of Lighting

Lamps used in protective lighting are either incandescent, gaseous discharge or quartz. Each type has special characteristics suitable for specific assignments.

Incandescent. These are common light bulbs of the type found in the home. They have the advantage of providing instant illumination when the switch is thrown, and are thus the most commonly used in protective lighting systems. Some incandescents are manufactured with interior coatings which reflect the light, and with a built-in lens to focus or diffuse the light. Regular high-wattage incandescents can be enclosed in a fixture, giving much the same result.

Gaseous Discharge Lamps. Mercury vapor lamps give out a strong light with a bluish cast. They are more efficient than incandescents because of a considerably longer lamp life. Sodium vapor lights give out a soft yellow light and are even more efficient than mercury vapor. They are widely used in areas where fog is a frequent problem, since the yellow penetrates the mist more readily than a white light. They are frequently found on highways and bridges.

The use of gaseous discharge lamps in protective lighting is somewhat limited, since they require a period of from two to five minutes to light when they are cold and an even longer period to relight, when hot, after a power interruption.

Quartz lamps. These lamps emit a very bright white light and snap on almost as rapidly as the incandescent bulb. They are frequently used at very high wattage—1,500 to 2,000 watts is not uncommon in protective systems—and they are excellent for use along the perimeter barrier and in troublesome areas.

Types of Equipment

No one type of lighting unit is applicable to every need in a protective lighting system, although manufacturers are continually working to develop just such a fixture.

Amid the great profusion of equipment in the market, there are four basic types which are in general use in security applictions: floodlights, searchlights, Fresnels, and street lights. (The first three of these might, in the strictest sense, be considered as a single type, since they are all, basically, reflection units in which a parabolic mirror directs the light in various ways. We will, however, deal with them separately.)

Street lights are pendant lighting units which are built as either symmetrical or asymmetrical. The symmetrical units distribute light evenly. Those units are used where a large area is to be lighted without the need for highlighting particular spots. They are normally centrally located in the area to be illuminated.

Asymmetrical units direct the light by reflection in the direction where the light is required. They are used in situations where the lamp must be placed some distance from the target area. Since these are not highly focused units, they do not create a glare problem.

Street lights are rated by wattage or, even more frequently, by lumens, and in protective lighting applications may vary from 4,000 to 10,000 lumens, depending on their use.

Floodlights are fabricated to form a beam so that light can be concentrated and directed to specific areas. They can create considerable glare.

Although many floods specify beam width in degrees, they are generally referred to as wide, medium, or narrow, and the lamp is described in wattage. Lamps may run from 300 to 1,000 watts in most protective applications, but there is a wide latitude in this and the choice of one will depend upon a study of its mission.

Fresnel lights are wide beam units, primarily used to extend the illumination in long, horizontal strips to protect the approaches to the perimeter barrier. Unlike floodlights and searchlights, which project a focused round beam, Fresnels project a narrow, horizontal beam which is approximately 180 degrees in the horizontal and from 15 to 30 degrees in the vertical plane.

These units are especially good for creating a glare for the intruder while the facility remains in comparative darkness. They are normally equipped with a 300 to 500 watts lamp.

Searchlights are highly focused incandescent lamps which are used to pinpoint potential trouble spots. They can be directed to any location within or without the property and, although they can be automated, they are normally controlled manually.

They are rated according to wattage, which may range from 250 to 3,000 watts, and the diameter of the reflector, which may range from six inches to two feet (the average is around 18 inches). The beam width is from three to 10 degrees, although this may vary in adjustable or focusing models.

Maintenance

As with every other element of a security system, electrical circuits and fixtures must be inspected regularly to replace worn parts, verify connections, repair worn insulation, check for corrosion in weatherproof fixtures, and clean reflecting surfaces and lenses. Lamps should be logged as to their operational hours and replaced at between 80 and 90 percent of their rated life.

Perimeter Lighting

Every effort should be made to locate lighting units far enough inside the fence and high enough to illuminate areas both inside and outside the boundary. The farther outside the boundary the lighted areas extend, the more readily guards will be able to detect the approach of an intruder.

This light should be directed down and away from the protected area. The location of light units should be such that they avoid throwing a glare in the eyes of the guard, do not create shadow areas, and do create a glare problem for anyone approaching the boundary.

Fixtures used in barrier and approach lighting should be located inside the barrier. As a rule of thumb, they should be around 30 feet within the perimeter, spaced 150 feet apart, and about 30 feet high. These figures are, of course, approximations and will not apply to every installation. Local conditions will always dictate placement.

Floodlights or Fresnel units are indicated in illuminating isolated or semi-isolated fence boundaries where some glare is called for. In either case, it is important to light from 20 feet inside the fence to as far into the approach as is practical. In the case of the isolated fence, this could be as much as 250 feet. Semi-isolated and non-isolated fence lines cannot be lighted as far into the approach, since such lighting is restricted by streets, highways, and other occupancies.

Since glare cannot be employed in illuminating a non-isolated fence line, street lights are recommended.

Where a building of the facility is near the perimeter or is, itself, part of the perimeter, lights can be mounted directly on it. Doorways of such buildings should be individually lighted to eliminate shadows cast by other illumination.

In areas where the property line is on a body of water, lighting should be designed to eliminate shaded areas on or near the water or along the shore line. This is especially true for piers and docks, where both land and water approaches must be lighted or capable of being lighted on demand. Before finalizing any plans for protective lighting in the vicinity of navigable waters, however, the United States Coast Guard must be consulted.

Gates and Thoroughfares

It is important that the lighting at all gates and along all interior thoroughfares be sufficient for the operation of the facility.

Since both pedestrian and vehicular gates are normally manned by guards inspecting credentials, as well as checking for contraband or stolen property, it is critical that the areas be lighted to at least one footcandle. Pedestrian gates should be lighted to about 25 feet on either side of the gate, if possible, and the range for vehicular gates should be twice that. Streetlighting is recommended in these applications, but floodlights can also be used if glare is strictly controlled.

Thoroughfares used for pedestrians, vehicles, or forklifts should be lighted to 0.10 footcandles for security purposes. Much more light may be required for operational efficiency, but this level should be maintained as a minimum, no matter what the conditions of traffic may be.

Other Areas

Open or unroofed areas within the perimeter, but not directly connected to it, require an overall intensity of illumination of about 0.05 footcandles (up to 0.10 footcandles in areas of higher sensitivity). These areas, when non-operational, are usually used for material storage or for parking. Any particularly vulnerable installations in the area should not be lighted at all, but the approaches to them should be well lighted for at least 20 feet to aid in the observation of any movement.

Searchlights may be indicated in some facilities, especially in remote mountainous areas or in waterfront locations where small boats could readily approach the facility.

General

In sum, a well thought-out plan of lighting along the security barrier and the approaches to it, an adequate overall level of light in storage, parking, and other non-operational areas within the perimeter, and reasonable lighting along all thoroughfares are essential to any basic security program. The lighting required in operational areas will usually be much higher than the minimums required for security and will, therefore, serve a security purpose as well.

PLANNING SECURITY

No business exists without a security problem of some kind, and no building housing a business is without security risk. Yet few such buildings are ever designed with any thought given to the steps that must eventually be taken to protect them from criminal assault.

A building must be many things in order for it to satisfy its occupant. It must be functional and efficient, it must achieve certain aesthetic standards, it must be properly located and accessible to the markets served by the occupant, and it must provide security from interference, interruption, and attack. Most of these elements are provided by the architect—but all too frequently the important element of security is overlooked.

Good security requires thought and planning—a carefully integrated system. Most security problems arise simply because no one has thought about them.

This is especially true where a company building is concerned. Since few architects have any training or knowledge in security matters, they design buildings that assist the burglar or the vandal by doing nothing to deter him. The clients themselves seldom consider security in the planning stage, the buildings are erected which provide needless opportunities for crime.

Old Construction

Older buildings, particularly, though certainly not exclusively, office buildings, present a host of different and difficult problems to the security staff. Exterior fire escapes, old and frequently badly worn locks, common walls, roof access from neighboring buildings, unused and forgotten connecting doors—all increase the exposure to burglary.

It is vital that all such openings be surveyed and plans made for securing them. Those windows not designated as emergency exits must be barred or screened. Where windows lead to a fire escape or are accessible to adjacent fire escapes, their essential security must be accomplished within the regulations of the local fire codes. Fire safety must be a primary consideration. In cases where prudence or the law (or both) dictate that locks would be a hazard to safety, windows should be alarmed and the interior areas to which these windows provide access must be further secured. Here, security can be likened to any army retreating to secondary or tertiary lines of defense to establish a strong and defensible position.

It is also well to consider the danger of attack from neighboring occupancies in shared space where entry might be made from a low risk, badly secured premise into a higher risk area that might otherwise be well protected against a more direct attack.

New Construction

Modern urban buildings, though security conscious in varying degrees, present their own problems. Most interior construction is standardized. Fire and building codes are such that corridor doors can resist most attack if the hardware is adequate. Corridor ceilings are fixed, and entrances to individual offices usually offer a fairly high degree of security.

On the other hand, modern construction creates offices which are essentially open-top boxes. It has solid exterior walls (though interior walls are frequently plaster board) and a concrete floor. But nothing of any security value protects the top. The ceiling is simply a layer of acoustical tiles lying loose on runners suspended between partition walls. In the space above these tiles—between them and the concrete slab above—are vital air conditioning ducts, and wiring for power and telephones.

In effect, any given floor of a building has a crawlspace that runs from exterior wall to exterior wall. This may not be literally so in every case, but the net result stands. It means that virtually every room and every office is accessible through this space. Once this crawlspace is reached from any occupancy, the remaining offices on that floor are accessible.

Extending dividing walls up to the next floor will not solve the problem, since this "drywall" construction is easily broken through and, in any case, it must be breached to allow passage of all utilities. Alarms of various kinds, which are discussed later, are recommended to overcome this problem.

Security at the Building Design Stage

Once a building has been constructed, the damage has been done. Security weaknesses begin to manifest themselves, but it is far too expensive to make basic structural changes to correct them. Guard services and protective devices that might not otherwise have been necessary must be instituted. In any event, there will be some considerable expense for protection that could easily have been incorporated into the design of the building before construction began. This kind of oversight can be very expensive, indeed.

Unfortunately, we have not yet arrived at the point where the need for security from criminal acts is as automatic a consideration as the need for efficiency or of profits. The architect's interest in design for protection of buildings and grounds is usually minimal. He traditionally leaves such demands to the client, who usually is unaware of the availability of protective hardware and is rarely competent to deal in the problems of protective design.

The rising crime rate and the growing awareness of the problem has, however, directed more attention toward the important role that building design can play in security. There now appear to be some efforts on the part of the federal government to accentuate the architect's role in security.

Under the umbrella of environment security, concepts of crime prevention through environment design (known as CPTED) have received encouragement from the LEAA. Early work in this field concentrated on residential security, particularly in public housing, with Oscar Newman's major study of "defensible space" being a pioneering work. The LEAA also funded a study by Westinghouse on "Elements of CPTED," and later provided the funds for CPTED projects in residential, commercial and school modes.

This emerging approach to crime prevention through environmental design has important implications for private security. It seeks to bring together many disciplines—among them urban planning, architectural design, public law enforcement and private security—to create an improved quality of urban life through crime prevention. And, in particular, it encourages awareness of crime prevention techniques through physical design.

Security Principles in Design

There are certain principles which should always be considered in planning any building. Without them, it can be dangerously vulnerable. Some considerations are as follows:

1. The number of perimeter and building openings should be kept to a minimum consistent with safety codes.
2. Perimeter protection should be planned as part of the overall design.

3. Exterior windows, if they are less than 14 feet above ground level, should be constructed of glass brick, laminated glass or plastic materials, or shielded with heavy screening or steel grilles.
4. Points of possible access or escape which breach the exterior of the building, or the perimeter protection should be protected. Points to be considered are skylights, air conditioning vents, sewer ducts, manholes, or any opening larger than 96 square inches.
5. High quality locks with readily changeable cylinders should be employed on all exterior and restricted area doors for protection and quick-change capability in the event of key loss.
6. Protective lighting should be installed.
7. Shipping and receiving areas should be widely separated.
8. Exterior doors intended for emergency use only should be alarmed.
9. Exterior service doors should lead directly into the service area so that non-employee traffic is restricted in its movement.
10. Dock areas should be designed so drivers can report to shipping or receiving clerks without moving through storage areas.
11. Employment offices should be located so that applicants either enter directly from outside, or move through as little of the building as possible.
12. Employee entrances should be located directly off the gate to the parking lot.
13. Employee locker rooms should be located by employee entrance and exit doors.
14. Doors in remote areas should be alarmed.

REVIEW QUESTIONS

1. How can the various openings in a perimeter be effectively protected and secured?
2. Why should parking not be allowed inside the controlled perimeter?
3. Discuss the different security applications for the various types of lighting equipment.
4. What considerations must be taken into account when installing a security lighting system?
5. Discuss the principles of security design in construction.

Chapter 8

THE INNER DEFENSES: INTRUSION AND ACCESS CONTROL

Once the facility's perimeter is secured, the next step in physical security planning is to minimize or control access to the building's interior. The extent of this control will depend upon the nature and function of the facility; the controls must not interfere with the facility's operation. It is theoretically possible to *completely* seal off access to a given operation, but it would be difficult to imagine how useful the operation would be in such an atmosphere.

Certainly no commercial establishment can be open for business while it is closed to the public. A steady stream of outsiders, from customers to service personnel, is essential to its economic health. In such cases, the security problem is to control this traffic without interfering with the function of the business being protected. Isolated manufacturing facilities must also provide for the traffic created by the delivery of raw materials, the shipping of fabricated goods, services and, of course, the labor force—which may be operating in several shifts.

All such traffic tends to compromise the physical security of the facility. But security must be provided, and it must be provided appropriately for the operation of the facility to be served.

Within any building, whether or not it is located inside a perimeter barrier other than the walls of the building itself, it is necessary to consider the need to protect against the internal thief as well as the intruder. Whereas the boundary fence is primarily designed to keep out unwanted visitors (not altogether forgetting its function in the control of movement of authorized personnel), interior security must provide some protection against the free movement of employees bent on pilferage, as well as establishing a second line of defense against the intruder.

Since every building is used differently and has its own unique traffic composition and flow, each building presents a different security problem. Each must be examined and analyzed in great detail before a really effective security program can be developed.

It cannot be overemphasized that such a program must be implemented without, in any way, interfering with the orderly and efficient operation of the facility to be protected. It must not be obtrusive, and yet, it must provide a predetermined level of protection against criminal attack from without or within.

The first points of examination must be the doors and windows. These must be considered in terms of effectiveness whether the building walls form a part of or constitute, in themselves, the perimeter barrier, as we have already discussed, or whether they are a true second defense line where the building under examination is completely within the protection of a barrier.

WINDOWS

It is axiomatic that windows should be protected. Since the ease with which most windows can be entered makes them a ready target for the intruder, they must be viewed as a potential weak spot in any building's defense. Over 50 percent of all break-ins are through window glass—whether such glass is installed in doors or windows.

In most industrial facilities, windows should be protected with grillwork, heavy screening, or chain link fencing. In cases where caution dictates, they may be needed as emergency exits beyond strict requirements of fire laws. Or where they might be needed to lead in fire hoses, consideration should be given to hinging and padlocking with protective coverings.

Burglary-Resistant Glass

In applications such as prominent administration buildings, office buildings, and the like, where architectural considerations preclude the use of such relatively clumsy installations as mesh or industrial screen, the windows can be immeasurably strengthened by the use of either UL-listed (a term which means that the material or item so designated has met certain standards of Underwriter's Laboratories for burglary resistance) burglary-resistant glass, or one of the brands of UL-listed polycarbonate glazing material. Both of these products are considerably more expensive than plate glass, and are generally used only in those areas where attack can be expected or where a reduction in insurance premium would justify the added expense.

As opposed to tempered glass, which is designed to protect people from the danger of flying shards in the event of breakage, UL-listed burglary-resistant glass (frequently referred to as "safety glass"), resists heat, flame, cold, picks, rocks, and most other paraphernalia from the intruders arsenal. It is a useful security glazing material because it is durable, weathers well, and is noncombustible. On the other hand, it is heavy, difficult to install, and expensive.

The plastic glazing which is sold under the trade name of either Tuffak (Rohm and Haas) or Lexan (General Electric) is optically clear, thin, and easy to install.

Acrylic glazing material which appears as Plexiglass (Rohm and Haas) does not meet UL standards for burglar-resistant material, however, it is much stronger than ordinary glass and has many useful applications in window security applications. It is also lighter in weight and cheaper than either safety glass, Lexan or Tuffak, and, at 1¼ inches thick, is UL approved as a bulletresistant barrier.

All these materials have the appearance of ordinary glass. Obviously, any window so hardened against entry must be securely locked from the inside to protect against intrusion from the outside.

"Smash and Grab" Attacks

Burglary-resistant glass is used to a considerable degree in banks and retail stores where there has been a very real need to prevent the "smash and grab" raid on window displays and showcases.

It should be noted that UL-listed burglary-resistant glass is a laminate of two sheets of flat glass (usually 3/16 inches thick) held together by a 1/16 inch layer of polyvinyl butyral, a soft transparent material. In this thickness, laminated glass is virtually indistinguishable from ordinary glass, hence the burglar tries with a hammer or iron bar to break what he supposes to be a plate glass window. It is only after he has made a few unsuccessful tries that he realizes he is up against a material he is not apt to penetrate, certainly not in the time he has between first attack and police or guard response.

Even though the attacker may flee empty-handed, the owner in such situations is left with a window with a web of cracks over the surface of the outer layer of glass, making replacement of the entire pane necessary in applications where appearance is important. Insurers, in many cases, require that laminated glass be clearly identified so as to discourage what would be a futile but damaging assault by the "smash and grab" attacker.

Screening

It can also be important to screen windows to protect their use as a means by which employees can temporarily dispose of goods for later recovery. The smaller the goods being manufactured or available on the premises, the smaller the mesh in the screen must be to protect against this kind of pilferage.

Generally speaking, any windows less than 18 feet from the ground or less than 14 feet from trees, poles, or adjoining buildings should receive some protective treatment unless they are well within the perimeter barrier and open directly onto an area outside the building which is particularly well secured.

DOORS

Every door within the building must be carefully examined to determine the degree of security required of it. Such an examination will also determine the type of construction, as well as the locking system to be used on each door.

Whatever security measures may be required at any specific door will be determined by the operations in progress or by the value of the assets stored or available in the various areas. The need for adequate security cannot be overemphasized, but it must be provided as part of an overall plan for the safe and efficient conduct of the business at hand.

When this balance is lost, the business must suffer. Either the security function will be downgraded in favor of a more immediate convenience, or the smooth flow of business will be impeded to conform to obtrusive security standards. Either of these conditions is intolerable in any business, and it is a management responsibility to determine the balance required in establishing systems that will recognize and accommodate all such company needs.

Door Construction and Hardware

Doors are frequently much weaker than the surface into which they are positioned. Panels may be thin, easily broken wood, or glass. Locks may be old and ineffective. The door frame may be so constructed that a lever or a plastic card can be inserted between the door and jamb to disengage the bolt in the lock. Even with a properly hung door, if the jamb is of soft, unreinforced aluminum it can be peeled or ripped away from a long-throw bolt. Heavy wood or metal doors with reinforced jambs can go a long way to overcome these problems.

In some cases, doors are entered by ''pulling''—a technique whereby the lock cylinder is ripped from the door and the lock is operated through the opening left in its face. The installation of a special, hardened-steel cylinder guard can overcome this kind of assault.

Door hinges may also contribute to a door's weakness. Surface mounted hinges with mounting screws or hinge pins exposed on the exterior side of the door can be removed, and entrance gained on the hinge side. To complicate the matter, the door can be replaced on its hinges after the intruder has finished his business and, in most cases, the intrusion would never be detected. Without any visible sign of forced entry, very few insurance policies would pay off on the stolen merchandise.

To prevent this unhappy chain of events, hinges should be installed with the screws concealed and with the hinge pins either welded or flanged to prevent removal.

Traffic Patterns

Doors must be analyzed for their function in laying out the security plan. In some cases, they may serve a dual purpose, as, for example, fire doors which are designed to close automatically in the event of a fire. These doors, which may remain open at the discretion of management, must be fitted to form an effective and automatic barrier to the spread of fire. They may be desirable when fire doors separate a production area from a warehouse or storage area. During those times when the production area is in operation but the warehouse is not, such fire doors can perform a security function by remaining closed.

In other cases, doors must be examined in an effort to establish a schedule for their use. Employee entrances that are the authorized points of passage for all employees may be manned by security personnel, depending on whether the control point is established there or farther out on the perimeter. These doors could be secured once the employees have entered, thus denying entrance to unauthorized visitors, as well as preventing any employees from wandering out to the fence or the parking lot or any other location where they might cache contraband for later pick-up or transport.

It is axiomatic, however, that any door used as an entrance will, in a time of emergency, be used as an exit by some employees. This is true in apartments and office buildings, and even in industrial facilities where the employees are thoroughly familiar with the premises. No matter what or how many designated emergency exits or procedures there may be, some individuals in a time of tension or near-panic will seek out the door with which they are most familiar.

The entrance, then, must always be considered an emergency exit, and it should be equipped with panic hardware. To protect against surreptitious use it should also be fitted with a local alarm.

The same, of course, is true of the designated emergency exits. These doors should, additionally, be stripped of all exterior hardware from the outside, since they are not intended for operational use at any time.

Personnel doors leading to and from the dock area must be carefully controlled and supervised at all times when the dock is in the use. These and dock doors must be secured when the area is no longer operational.

Fire doors in office buildings should be alarmed to prevent surreptitious use, and access to interior, public stairwells should be prohibited or discouraged unless doors from them open out into reception areas.

Doors to Sensitive Areas

Doors to telephone equipment rooms, computer installations, research and development, and other sensitive areas should be equipped with automatic door closing devices and fitted with a strong dead bolt and a heavy latch.

In cases where an area is under heavy security, but has any degree of traffic, it might be well to consider the installation of an electric strike to secure the operation and control the traffic. This kind of unit is a locking device controlled remotely by a security person, permitting entry of a recognized, authorized person only when a button is pressed to release the lock. Since it requires someone on hand at all times for its operation, this system can be expensive. It must be examined with the cost-versus-security-cost equation in mind.

Supply rooms and toolroom doors should be secured whenever those rooms are not actually in operation. Even when they are, entrance into these areas must be restricted. The usual construction of such restraints consists of either a dutch door in which the bottom half is secured, or a counter which can be closed off by heavy screening, chain link fencing material, or reinforced shutters.

Special care should be taken in the storage of small items of value. Such merchandise or material is highly pilferable by virtue of its value for resale or personal use, combined with the ease with which it can be stolen. Although such items may be stored in a facility of any construction capable of providing security, it has been the experience of many firms that uniformly stacked rows, piles, or pallets of such items within a cage-type construction that provides instant eyeball inventory is the best protection. Such precautions will vary from business to business, but they must be carefully systematized to control this potentially troublesome area of loss.

Office Area Doors

Doors between production and office areas, or heavily trafficked areas and office spaces, must be examined for the likelihood of their use for criminal purposes. Their construction and locking hardware will be determined by such a survey. In most cases, these passages would be minimum security areas during regular working hours, since there is usually a need for movement between these areas. When there is little or no use of the office area, these doors should be secured.

Locking Schedules

Door-locking schedules and responsibilities must be established and supervised vigorously. The system must be set up in such a way that a procedure for altering the routine to fit immediate needs is possible, but in all respects the schedule, whether the master plan or the temporary, must be adhered to in every detail. A breakdown in such a system, especially in large offices institutions, or industrial facilities, could represent just the opportunity that an alert criminal is waiting for.

LOCKS AND KEYS

Attacks Against Locks

Although direct forcible assault is the method generally used to gain entry, more highly skilled burglars may concentrate on the lock. This may be their only practical means of ingress if the door and jamb are well-designed in security terms and essentially impervious to forcible attack.

Picking the lock or making a key by impression are the methods generally used. Both require a degree of expertise. In the former method, metal picks are used to align the levers or tumblers as an authorized key would, thus enabling the lock to operate. Making a key by taking impressions is a technique requiring even greater skill, since it is a delicate, painstaking operation requiring repeated trials.

Because both of these techniques are apt to take time, they are customarily used to attack those doors where the intruder feels he may work undisturbed and unobserved for adequate periods of time. The picked lock rarely shows any signs of illegal entry, and often the insurance is uncollectible.

Locks As Delaying Devices

The best defense against lock-picking and the making of keys by impression is the installation of special pick-resistant, impression-resistant lock cylinders. They are more expensive than standard cylinders but, in many applications, may well be worth the added cost. Generally speaking, in fact, locks are the cheapest security investment that can be made. Cost-cutting in their purchase is usually a poor economy, since a lock of poor quality is virtually useless and—effectively—no lock at all.

The elementary but often overlooked fact of locking devices is that, in the first place, they are simply mechanisms that extend the door or window into the wall that holds them. If, therefore, the wall or the door itself is weak or easily destroyed, the lock cannot be effective.

In the second place, it must be recognized that any lock will eventually yield to an attack. They must be thought of only as delaying devices. But this delay is of primary importance. The longer an intruder is stalled in an exposed position while he works at gaining entry, the greater are the chances of discovery. Since many types of locks in general use today provide no appreciable delay to even the unskilled prowler, they have no place in security applications.

Even the highest quality locking devices are only one part of door and entrance security. Locks, cylinders, door and frame construction, and key control are inseparable elements; all must be equally effective. If any one element is weak, the system breaks down.

Kinds of Locks

A brief review of the types of locks in general use, with notations of their characteristic, may serve to familiarize those as yet unacquainted with them with the variety available.

Warded Locks are those found generally in pre-war construction, in which the keyway is open and can be seen through. These are also recognized by the single plate which includes the doorknob and the keyway. The security value of these locks is nil.

Disc Tumbler Locks were designed for the use of the automobile industry and are in general use in car doors today. Because this lock is easy and cheap to manufacture, its use has expanded to other areas such as desks, files and padlocks. The life of these locks is limited because of their soft metal construction. Although these locks provide more security than warded locks, they cannot be considered as very effective. The delay afforded is, approximately, three minutes.

Pin Tumbler Locks are in wide use in industry as well as in residences (see figures 8-1, 8-2 and 8-3). They can be recognized by the keyway, which is irregular in shape, and the key, which is grooved on both sides. Such locks can be master keyed in a number of ways, a feature which recommends them to a wide variety of industrial applications, although the delay factor is ten minutes or less.

Lever Locks are difficult to define in terms of security, since they vary greatly in their effectiveness. The best lever locks are used in safe deposit boxes and are, for all practical purposes, pick-proof. The least of these locks are used in desks, lockers and cabinets and are generally less secure than pin tumbler locks. The best of this variety are rarely used in common applications, such as doors, because they are bulky and expensive.

Combination Locks are difficult to defeat since they cannot be picked and few experts can so manipulate the device as to discover the combination. Most of these locks have three dials which must be aligned in the proper order before the lock will open. Some such locks may have four dials for greater security. Many also have the capability of changing the combination quickly.

Code-Operated Locks are combination-type locks in that no keys are used. They are opened by pressing a series of numbered buttons in the proper sequence. Some of them are equipped to alarm if the wrong sequence should be pressed. The combination of these locks can be changed readily. These are high security locking devices. Because this type of lock can be compromised by ''tailgating'' (more than one person entering on an authorized opening), it should never be used as a substitute for a guard or receptionist.

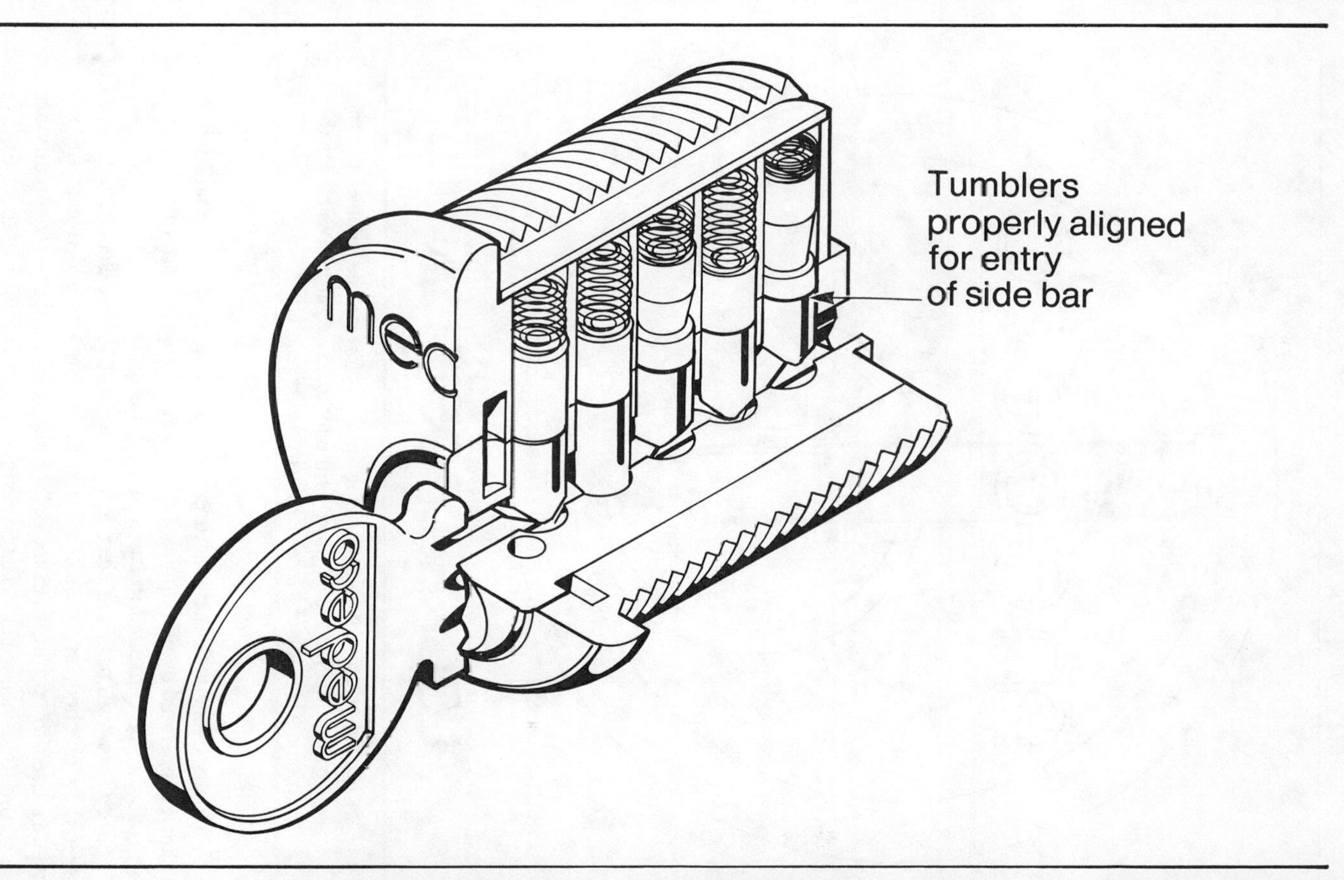

Figure 8-1. A cut-away of a pin tumbler lock showing the springs and tumblers. When the correct key is inserted into the lock, it will align all of the tumblers in a straight line to allow the plug to turn and operate the locking mechanism. (Courtesy of Medeco Security Locks, Inc.)

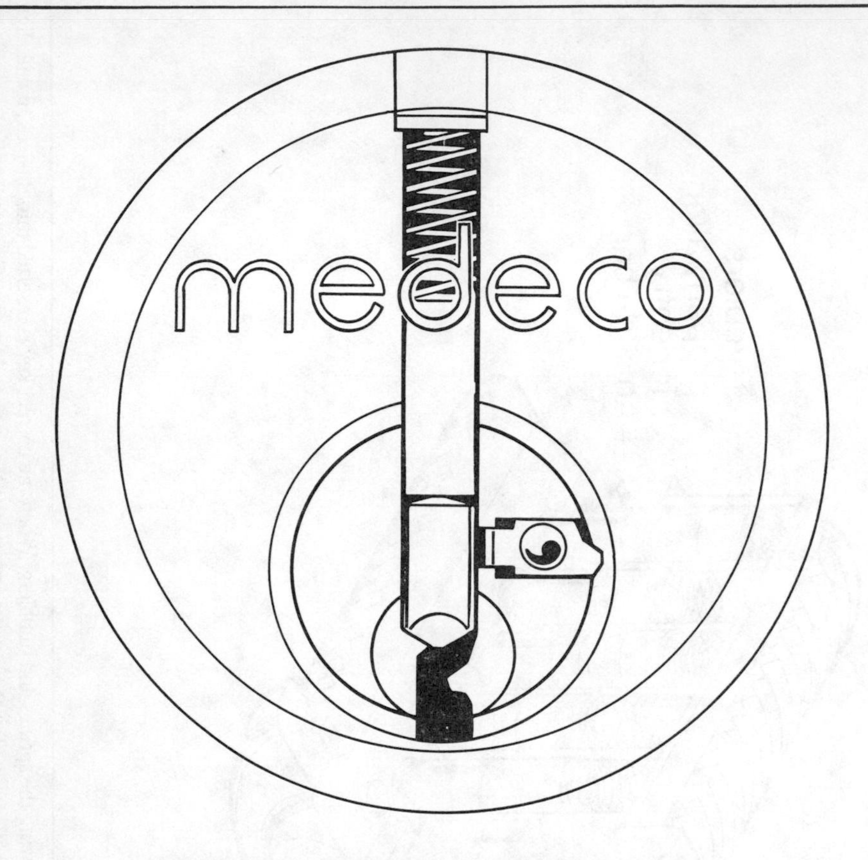

Figure 8-2. Locked position. Notice how the spring is forcing the tumbler to project part way into the inner core (plug) of the lock, making it impossible for the plug to rotate. (Courtesy of Medeco Security Locks, Inc.)

Virtually every lock manufacturer makes some kind of special high security lock which is operated by non-duplicable keys. A reliable locksmith or various manufacturers should be consulted in cases of such needs.

Card Operated Locks are electrical or, more usually, electromagnetic. Coded cards—either notched, embossed, or containing an embedded pattern of copper flecks—are used to operate such locks (See figure 8-5). These frequently are fitted with a recording device which registers time of use and identity of the user. The cards serving as keys may also serve as company ID cards. As with code-operated locks, tailgating can occur with this lock as well.

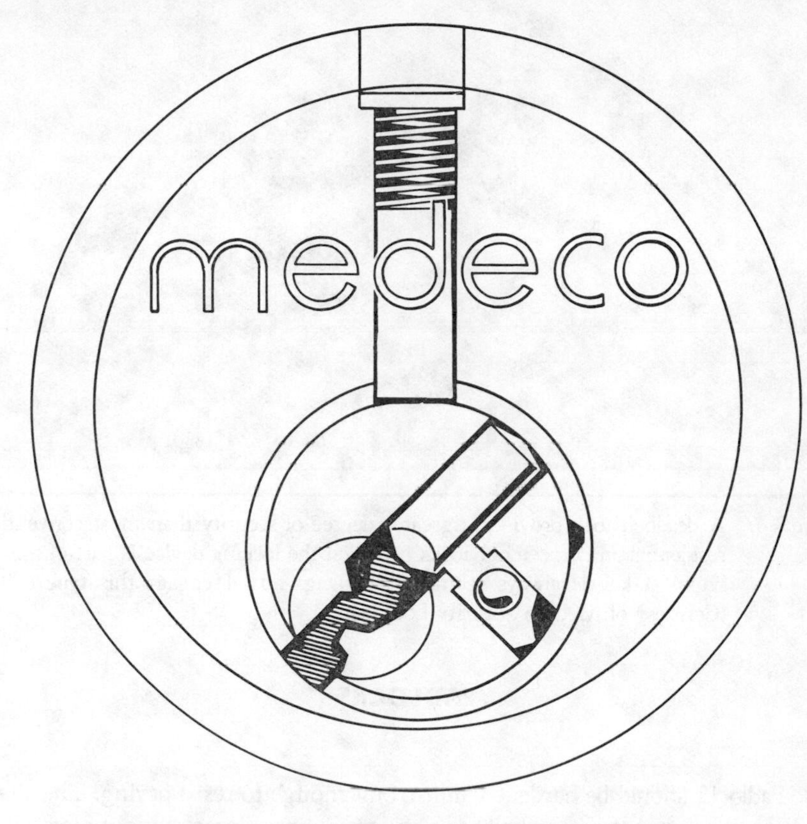

UNLOCKED POSITION

Figure 8-3. Unlocked position. Tumbler is now outside of the plug, allowing it to be rotated. (Courtesy of Medeco Security Locks, Inc.)

Electromagnetic Locks are devices holding a door closed by magnetism. These are electrical units consisting of the electromagnet and a metal holding plate. When the power is on and the door secured, they will resist a pressure of up to 1,000 pounds. A high frequency of mechanical failures with this type of lock can create problems. Inconvenienced employees will often block the door open or jam the door bolting mechanism so that the lock no longer operates. Quality equipment, preventive maintenance, frequent inspections, and quick response to problems will minimize these problems.

Figure 8-4. A deadbolt lock provides a greater degree of security than most conventional residential and office door locks in which the locking device is part of the door knob. A key is always required to engage and disengage this type of lock. (Courtesy of Medeco Security Locks, Inc.)

PADLOCKS

Padlocks should be hardened and strong enough to resist prying. The shackle should be close enough to the body to prevent the insertion of a tool to force it. No lock which will be used for security purposes should have less than five pins in the cylinder.

It is important to establish a procedure requiring that all padlocks be locked at all times even when they are not securing an area. This will prevent the possibility of the lock being replaced by another to which a thief has the key.

The hardware used in conjunction with the padlock is as important as the lock itself. It should be of hardened steel, without accessible screws or rivets, and should be bolted through the door to the inside, preferably through a backing plate. The bolt ends should be burred.

Locking Devices

In the previous list we have considered the types of locks that are generally available. It must be remembered, however, that locks must work in conjunction with other hardware which effect the actual closure. These devices may be fitted with

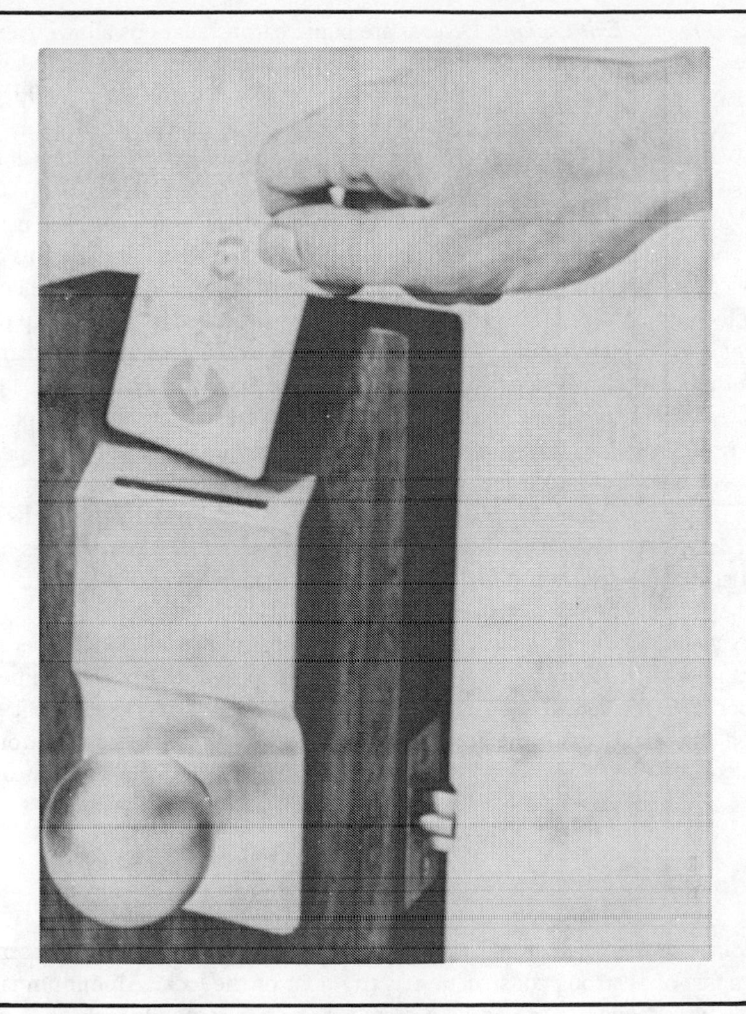

Figure 8-5. Card operated lock. (Courtesy of Cardkey Systems.)

locks of varying degrees of security, and themselves provide security to various levels. In a security locking system, both of these factors must be taken into consideration before determining which system will be most effective for specific needs.

Double Cylinder Locking Devices are installed in doors that must be secured from both sides. They require a key to open them from either side. Their most common application is in doors with glass panels which might otherwise be broken to allow an intruder to reach in and open the door from the other side. Such devices cannot be used in interior fire stairwell doors, since firemen break the glass to unlock the door from the inside in this case.

Emergency Exit Locking Devices are panic-bar installations allowing exit without use of a key. This device locks the door against entrance. Since such devices frequently provide an alarm feature that sounds when exit is made, they are fitted with a lock which allows exit without alarming when a key is used.

Recording Devices provide for a printout of door use by time of day and by the key used.

Vertical Throw Devices lock into the jamb vertically instead of the usual horizontal bolt. Some versions lock into both jamb and lintel. A variation of this device is the Police Lock, which consists of a bar angled to a well in the floor. The end of the bar contacting the door is curved so that when it is unlocked it will slide up the door, allowing it to open. When it is locked, it is secured to the door at one end and set in the floor at the other. A door locked in this manner is virtually impossible to force.

Electric Locking Devices are installed in the same manner as other locks. They are activated remotely by an electric current which releases the strike, permitting entrance. Many of these devices provide minimal security, since the engaging mechanisms frequently offers no security feature not offered by standard hardware. The electric feature provides a convenient method of opening the door. It does not, in itself, offer locking security. Since such doors are usually intended for remote operation, they should be fitted with a closing device.

Sequence Locking Devices are designed to insure that all doors covered by the system are locked. The doors must be closed and locked in a predetermined order. No door can be locked until its designated predecessor has been. Exit is made through the final door in the sequence, and entry can be made only through that same door.

Door Jambs

Since doors, when closed, are as a rule attached to the jamb, it is essential that the jamb be of as strong construction as the door or the lock. Aluminum jambs, for example, are frequently spread by a crowbar or an automobile jack. If they cannot resist such attack, the door can be opened easily. The locking bolt must be at least an inch into the jamb for security and to help prevent spreading. Cylinders should be flush or inset to prevent their being wrenched out or "popped."

Removable Cores

In facilities requiring that a number of keys be issued, the loss or theft of keys is an ever-present possibility. In such situations, it might be well to consider removable cores on all locks. These devices are made to be removed if necessary with a core key, allowing a new core to be inserted. Since the core is the lock, this has the effect of rekeying without the necessity of changing the entire device, as would be the case with fixed cylinder mechanisms.

Keying Systems

Keys are generally divided into change keys, submaster keys, master keys, and occasionally grand master keys.

1. *The change key*—a key to a single lock within a master-keyed system.
2. *The sub-master key*—a key that will open all the locks within a particular area or grouping in a given facility. In an office, a submaster might open all doors in the Accounting Department; in an industrial facility, it might open all locks in the loading dock area. Typically, such groupings concern themselves with a common function, or they may simply be located in the same area, even if they are not otherwise related.
3. *The master key*—where two or more sub-master systems exist, a master key system is established. Such a key would open any of the systems.
4. *The grand master key*—a key that will open everything in a system involving two or more master key groups. This system is relatively rare, but might be used by a multi-premise operation in which each location was master keyed while the grand master would function on any premise.

Obviously, master and sub-master keys must be treated with the greatest care. If a master key is lost, the entire system is threatened. Rekeying is the only really secure thing that can be considered, but the cost of such an effort can be enormous.

Master and grand master keys are normally machined to be very thin so that the use of each one is very limited. This is deemed to be a security measure, guarding against their extensive use in the event of loss or theft. This is a dubious proposition at best, since the loss of one of these keys effectively compromises the system and substantially reduces its security value. Even one or two ventures through the facility with such keys could do serious harm and, thin as they are, they might well stand up for that many uses. Unfortunately, when keys of such sensitivity are lost, rekeying is the only answer.

Any master key system is vulnerable. Beyond the danger of loss of the master itself and the subsequent staggering cost of rekeying—or, even more unfortunate, the use of such a key by enterprising criminals to loot the facility—there is the problem that it necessarily serves a lesser lock. Locks in such a system are neither pick-resistant nor resistant to making a key by impression.

On the other hand, relative security coupled with convenience may make such a system preferable in some applications where it would not be in others. Only the most careful evaluation of the particular circumstances of a given facility will determine the most efficient and most effective keying system.

Rekeying

In any sizable facility, rekeying can be very expensive, but there are methods to lessen the disruption and staggering cost that can be involved in rekeying. Outer or perimeter locks can be changed first, and the old locks moved to interior spaces requiring a lower level of security. After an evaluation, a determination of priorities can be made and rekeying can be accomplished over a period of time, rather than requiring one huge capital outlay at once.

Of prime importance is the securing of keys so that such problems do not arise.

Key Control

Every effort should be exerted to develop ways whereby keys remain in the hands of security personnel or management personnel. In those cases where this is not possible or practical, there must be a system of inventory and accountability. In any event, keys should be issued only to those demonstrably responsible persons who have compelling need for them. Though possession of keys is frequently a status symbol in many companies, management must never issue them on that basis.

Keys should never be issued on a long-term basis to outside janitorial personnel. The high employee turnover rate in this field would suggest that this could be a dangerous practice. Employees of this service should be admitted by guards or other building employees and issued interior keys which they must return before leaving the building.

By the same token, it is bad practice to issue entrance keys to tenants of an office building. If such is done, control of this vital security point is lost. A guard or building employee should control entry and exit before and after regular building hours. If keys must be issued to tenants, however, the lock cylinder in the entrance should be changed every few months and new keys issued to authorize tenants.

A careful, strictly supervised record of all keys issued must be maintained by the security department (See Figure 8-6). This record should indicate the name and department of the person to whom the key was issued, as well as the date of issue.

A key depository for securing keys during non-working hours should be centrally located, locked, and kept under the supervision of security personnel. Keys issued on a daily basis, or those issued for a specific one-time purpose, should be acccounted for daily. Keys should be counted and signed for by the security supervisor at the beginning of each working day.

When a key is lost, the circumstances should be investigated and set forth in writing. In some instances, if the lost key provides access to sensitive areas, locks should be changed. All keys issued should be physically inspected periodically to ensure that they have not been lost, though unreported as such.

Master keys should be kept to a minimum. If possible, sub-masters should be used, and they should be issued only to a limited list of personnel especially selected

SECURITY DEPARTMENT KEY LOG

DATE	TIME OUT	NAME	DEPARTMENT	KEY SET NO.	SIGNATURE	RELEASING OFFICER	TIME IN	DATE	ACCEPTING OFFICER

Figure 8-6. Security department key log. A typical form used to record the issuance of keys.

by management. Careful records should be kept of such issuance. The list should be reviewed periodically to determine whether all those authorized should continue to hold such keys.

Before a decision can be reached with respect to the master and submaster key systems and how such keys should be issued, there must be a careful survey of existing and proposed security plans, along with a study of current and planned locking devices. Where security plans have been developed with operational needs of the facility in mind, the composition of the various keying systems can be readily developed.

FILES, SAFES, AND VAULTS

The final line of defense at any facility is in the high security storage areas where papers, records, plans or cashable instruments, precious metals, or other especially valuable assets are protected. These security containers will be of a size and quantity which the nature of the business dictates.

Every facility will have its own particular needs, but certain general observations apply. The choice of the proper security container for specific applications is influenced largely by the value and the vulnerability of the items to be stored in them. Irreplaceable papers or original documents may not have any intrinsic or marketable value, so they may not be a likely target for a thief, but since they do have great value to the owners, they must be protected against fire. On the other hand, uncut precious stones, or even recorded negotiable papers which can be replaced, may not be in danger from fire, but they would surely be attractive to a thief. They must therefore be protected from him.

In protecting property, it is essential to recognize that, generally speaking, protective containers are designed to secure against burglary *or* fire. Each type of equipment has a specialized function, and each type provides only minimal protection against the other risk. There are containers designed with a burglary-resistant chest within a fire-resistant container which are useful in many instances, but these, too, must be evaluated in terms of the mission.

Whatever the equipment, the staff must be educated and reminded of the different roles played by the two types of containers. It is all too common for company personnel to assume that the fire-resistant safe is also burglary-resistant, and vice versa.

Files

Burglary-resistant files are secure against most surreptitious attacks. On the other hand, they can be pried open in less than half an hour if the burglar is permitted to work undisturbed and is not concerned with the noise created in the operation.

Such files are suitable for non-negotiable papers or even proprietary information, since these items are normally only targeted by surreptitious assault.

Filing cabinets with a fire-rating of one hour, and further fitted with a combination lock, would probably be suited for all uses but the storage of government classified documents.[46]

Safes

Safes are expensive, but if they are selected wisely, they can be one of the most important investments in security. It is to be emphasized that safes are not simply safes. They are each designed to perform a particular job to a particular level of protection. Two types of safes of most interest to the security professional are the record safe (fire–resistant) and the money safe (burglary-resistant). See Figure 8-7. To use fire-resistant safes for the storage of valuables—an all too common practice—is to invite disaster. At that same time, it would be equally careless to use a burglary-resistant safe for the storage of valuable papers or records, since, if a fire were to occur, the contents of such a safe would be reduced to ashes.

Safes are rated to describe the degree of protection they afford. Naturally, the more protection provided, the more expensive the safe will be. In selecting the best one for the requirements of the facility, a number of questions must be considered. How great is the threat of fire or burglary? What is the value of the safe's contents? How much protection time is required in the event of a fire or burglary attempt? Only then can a reasonable, permissible capital outlay for their protection be arrived at.

Record Safes. Fire resistant containers are classified according to the maximum internal temperature permitted after exposure to heat for varying periods. A record safe with a UL rating of 350-4 (formerly designated "A") can withstand exterior temperatures building to 2000°F for four hours without permitting the interior temperature to rise above 350°F (See Figure 8-8).

The UL tests which result in the various classifications (See Table 8-1) are conducted in such a way as to simulate a major fire with its gradual build-up of heat to 2000°F including circumstances where the safe might fall several stories through the fire damaged building. Additionally, an explosion test simulates a cold safe dropping into a fire which has already reached 2000°F.

The actual procedure for the 350-4 rating involves the safe staying four hours in a furnace temperature that reaches 2000°F. The furnace is turned off after four hours but the safe remains inside until it is cool. The interior temperature must remain below 350°F during heating and cooling-out period. This interior temperature is determined by sensors sealed inside the safe in six specified locations to provide a continuous record of the temperatures during the test. Papers are also placed in the safe to simulate records. The explosion impact test is conducted with

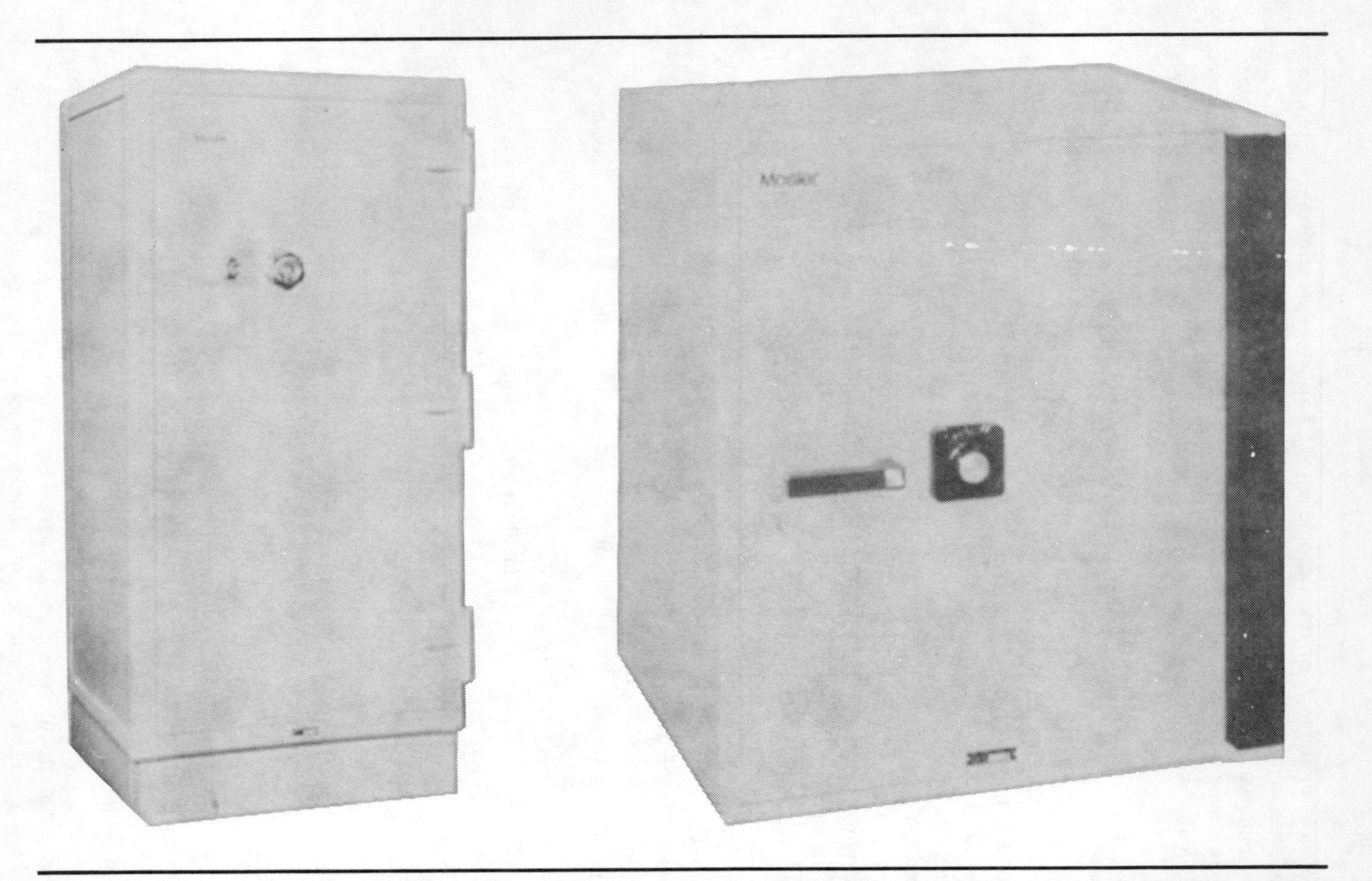

Figure 8-7. UL rated Money Safes and Record Safes can be distinguished by noting two factors: 1) Wheels—Money safes never have wheels; 2) Weight—Money safes are much heavier. They must weigh at least 750 pounds or be anchored to the floor. The safe on the left is a modern record safe. A dust skirt at the bottom hides the wheels. The safe on the right is a TL-15 money safe. Courtesy of

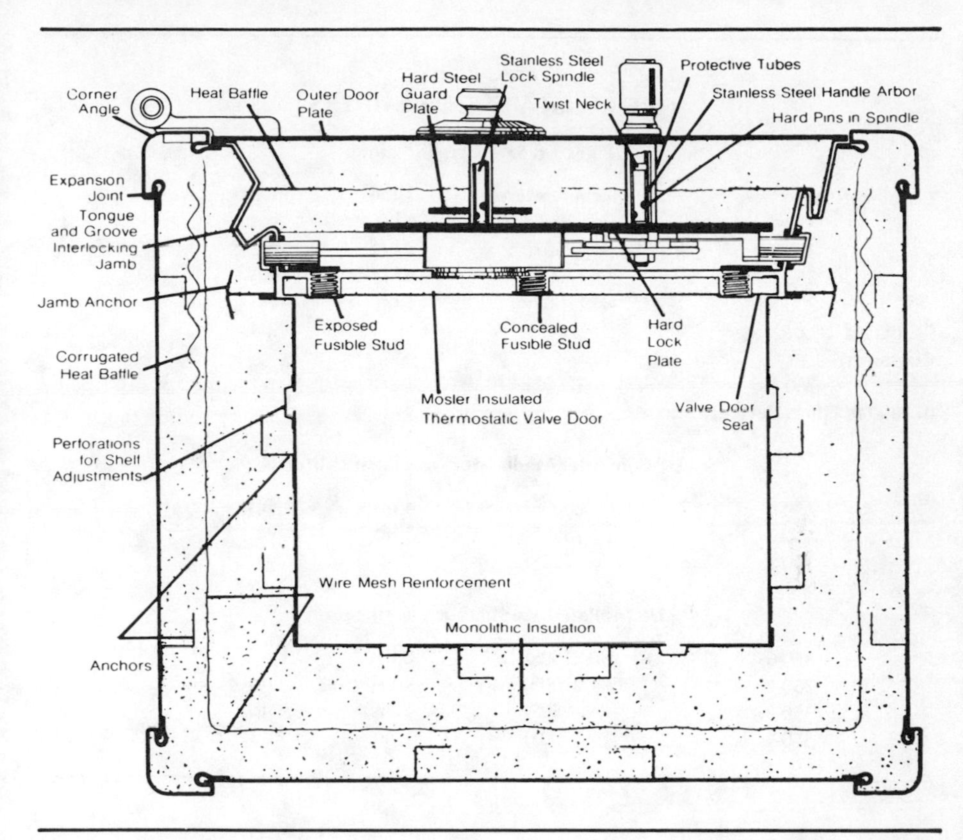

Figure 8-8. Cut-away view shows horizontal section through a typical record safe. Basic design is two shells with insulation poured in between. Other design details enhance the safe's resistance to heat penetration. (Courtesy of the Mosler Safe Company.)

another safe on the same model which is placed for one-half hour in a furnace preheated to 2000°F. If no explosion occurs, the furnace is set at 1550°F and raised to 1700°F over a half-hour period. After this hour in the explosion test, the safe is removed and dropped 30 feet onto rubble. The safe is then returned to the furnace and reheated for one hour at 1700°F. The furnace and safe are allowed to cool after which the papers inside must be legible and uncharred.

350-2 record safes protect against exposure up to 1850°F for two hours. The explosion/impact tests are conducted at slightly less time and heat.

350-1 gives one hour of protection up to 1700°F, and a slightly less vigorous explosion/impact test.

Computer media storage classifications are for containers which do not allow the internal temperature to go above 150°F.

Table 8-1
FIRE RESISTANT CONTAINERS

UL Record Safe Classifications

Classification	Temperature	Time	Impact	Old Label
350-4	2000°F	4 hrs.	yes	A
350-2	1850°F	2 hrs.	yes	B
350-1	1700°F	1 hr.	yes	C
350-1 (Insulated Record Container)	1700°F	1 hr.	yes	A
350-1 (Insulated Filing Device)	1700°F	1 hr.	no	D

UL Computer Media Storage Classification

Classification	Temperature	Time	Impact
150-4	2000°F	4 hrs.	yes
150-2	1850°F	2 hrs.	yes
150-1	1700°F	1 hr.	yes

UL Insulated Vault Door Classification

Classification	Temperature	Time	Impact
350-6	2150°F	6 hrs.	no
350-4	2000°F	4 hrs.	no
350-2	1850°F	2 hrs.	no
350-1	1700°F	1 hr.	no

Insulated vault door classifications are much the same as for safes, except that they are not subject to an explosion/impact test.

UL testing for burglary-resistance in safes does not include the use of diamond core drills, thermic kance or other devices yet to be developed by the safecracker.

In some businesses, a combination consisting of a fire-resistant safe with a burglary-resistant safe welded inside may serve as a double protection for different assets (See Figure 8-10), but in no event must the purposes of these two kinds of safes be confused if there is one of each on the premises. Most record safes have combination locks, relocking devices and hardened steel lockplates to provide a measure of burglar resistance. It must be re-emphasized that record safes are designed to protect documents and other similar flammables against destruction by fire. They provide only slight deterrence to the attack of even unskilled burglars. Similarly, the resistance provided by burglar-resistance is powerless to protect contents in a fire of any significance.

Money Safes. Burglary-resistant safes are nothing more than very heavy metal boxes which offer degrees of protection against various forms of attack (See Figure 8-9). A safe with a UL rating of TL-15, for instance, weighs at least 750 pounds and

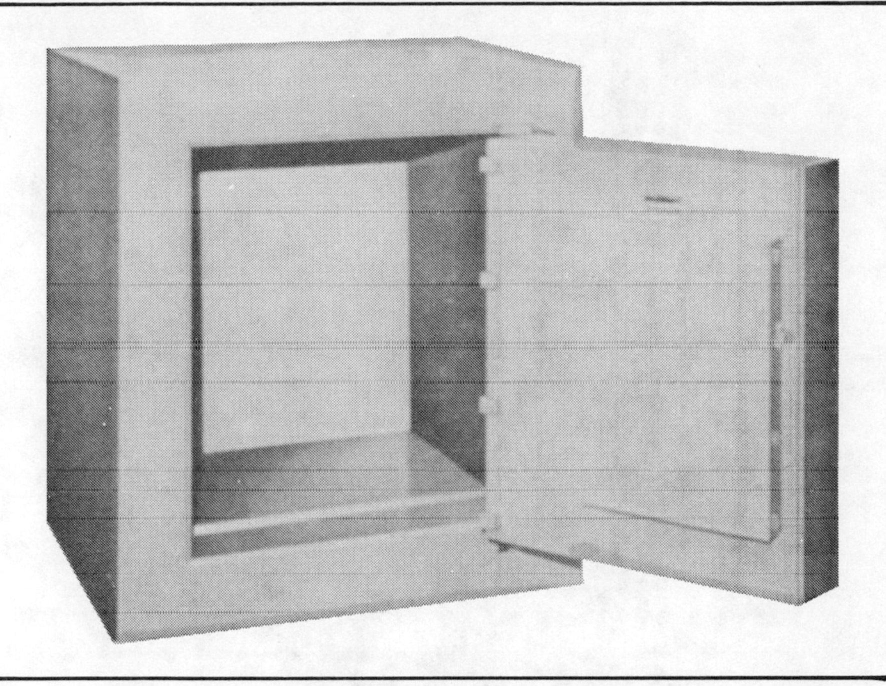

Figure 8-9. This is a UL classified TRTL-30X6 money safe. According to new UL requirements, this safe must resist attack on all six sides with tools and torches for 30 minutes. The walls and door are constructed by nesting two metal shells of different sizes and pouring the new composite, concrete-like material in between. (Courtesy of the Mosler Safe Company.)

its front face can resist attack by common hand and elective tools for at least fifteen minutes. Other safes will resist not only attack with tools but also attack with torches and explosives (See Table 8-2).

Since burglary-resistant safes have a limited holding capacity, it is always advisable to study the volume of the items to be secured. If the volume is sufficiently large, it might be advisable to consider the installation of a burglary-resistant vault which, although considerably more expensive, can have an enormous holding capacity.

Securing the Safe. Whatever safe is selected must be securely fastened to the structure where it is located. Police reports are filled with cases where unattached safes, some as heavy as a ton, have been stolen in their entirety—safe and contents—to be worked on in uninterrupted concentration. A study of safe burglars in California showed that the largest group (37.3 percent) removed saves from the premises to be opened elsewhere.[47]

Figure 8-10. Record safes are available with a variety of interiors. Note the money safe welded into place on the bottom surface. (Courtesy of the Mosler Safe Company.)

A convicted criminal recently told investigators how he and an accomplice had watched a supermarket to determine the cash flow and the manager's banking habits. They noted that he built up cash in a small, wheeled safe until Saturday morning, when he banked. Presumably, he felt secure in this practice, since he lived in an apartment above the store and perhaps felt that he was very much on top of the situation in every way. One Friday night, the thief and his friend rolled the safe into their station wagon. They pried it open at their leisure to get the $15,000 inside.

Pleased with their success, the thieves were even more pleased when they found that the manager replaced the stolen safe with one exactly like it and continued with the same banking routine. Two weeks later, our man went back alone and picked up another $12,000 in exactly the same way as before.

It is becoming a common practice to install the safe in a concrete floor where it offers great resistance to attack. In this kind of installation only the door and its combination are exposed. Since the door is the strongest part of a modern safe, the chances of successful robbery are considerably reduced.

Table 8-2
UL MONEY SAFE CLASSIFICATION

Classification	Description	Construction
TL-15	Tool resistant	Weight: At least 750 lbs. or anchored. Body: At least one inch thick steel or equal. Attack: Door and front face must resist attack with common hand and electric tools for 15 minutes.
TL-30	Tool resistant	Weight: At least 750 lbs. or anchored. Body: At least one inch thick steel or equal. Attack: Door and front face must resist attack with common hand and electric tools plus abrasive cutting wheels and power saws for 30 minutes.
*TRTL-30	Tool and torch resistant	Weight: At least 750 lbs. Attack: Door and front face must resist attack with tools listed above, and oxy-fuel gas cutting or welding torches for 30 minutes.
TRTL-30X6	Tool and torch resistant	Weight: At least 750 lbs. Attack: Door and entire body must resist attack with tools and torches listed above, plus electric impact hammers and oxy-fuel gas cutting or welding torches for 30 minutes.
TXTL-60	Tool, torch and explosive resistant	Weight: At least 1000 lbs. Attack: Door and entire safe body must resist attack with tools and torches listed above, plus eight ounces of nitroglycerine or its equal for 60 minutes.

*As of January 31, 1980, UL stopped issuing the TRTL-30 label, replacing it with the TRTL-30X6 label which requires equal protection on all six sides of the safe. Some manufacturers, however, continue to produce safes meeting TRTL-30 standards in order to supply lower priced containers, which provide moderate protection against tool and torch attack.

Vaults

Vaults are, essentially, enlarged safes. As such, they are subject to the same kinds of attack and must look at the same basic principles of protection as safes.

Since it would be prohibitively expensive to build a vault out of shaped and welded steel and special alloys, the construction, except for the door, is usually of high quality, reinforced concrete. There are many ways in which such a vault can be constructed, but however it is done, it will always be extremely heavy and, at best, a difficult architectural problem.

Typically, vaults are situated at or below ground level so they do not add to the stresses of the structure housing them. If a vault must be built on the upper stories of a building, it must be supported by independent members which do not provide support for other parts of the building. And, it must be strong enough to withstand the weight imposed upon it if the building should collapse from under it as the result of fire or explosion.

The doors of such vault are normally 6″ thick, and they may be as much as 24″ in the largest installations. Since these doors present a formidable obstacle to any criminal, an attack will usually be directed at the walls, ceiling, or floor, which must match the strength of the door. As a rule, these surfaces should be twice as thick as the door and never less than 12″.

If at all possible, a vault should be surrounded by narrow corridors which will permit inspection of the exterior, but which will be sufficiently confined to discourage the use of heavy drilling or cutting equipment by attackers. It is important that there be no power outlets anywhere in the vicinity of the vault.

Container Protection

Since no container can resist assault indefinitely, it must be supported by alarm systems and frequent inspections. Capacitance and vibration alarms are the types most generally used to protect safes and file cabinets. Ideally, any container should be inspected at least once within the period of its rated resistance. CCTV surveillance can, of course, provide constant inspection and, if the expense is warranted, is highly recommended.

By the same token, safes have a greater degree of security if they are well-lighted and located where they can be seen readily. Any safe located where it can be seen from a well-policed street will be much less likely to be attacked than one which sits in a darkened back office on the upper floors.

Continuing Evaluation

Security containers are the last line of defense, but in many situations, they should be the first choice in establishing a sound security system. The containers must be selected with care after an exhaustive evaluation of the needs of the facility under examination. They must also be reviewed regularly for their suitability to the job they are to perform.

Just as the safe manufacturers are continually improving the design, construction, and materials used in safes, so is the criminal world improving its technology and technique of successful attack. Because of the considerable capital outlay involved in providing the firm with adequate security containers, many businessmen are reluctant to entertain the notion that these containers may someday become outmoded—not because they wear out or cease to function, but because new tools and techniques have nullified their effectiveness.

In selecting security containers it is important that the equipment conform to the needs of the risk, that it be regularly re-evaluated and, if necessary, brought up to date, however unwelcome the additional outlay may be.

TRAFFIC CONTROL

Controlling traffic in and out and within a facility is essential to its security program. Perimeter barriers, locked doors, and screened windows prevent or deter the entry of unauthorized visitors. But, since some traffic is essential to every operation, no matter how highly classified it may be, provision must be made for the control of this movement.

Specific solutions will depend upon the nature of the business. Obviously, retail establishments, which encourage high volume traffic and which regularly handle a great deal of merchandise both in and out, have a problem of a different dimension from the industrial operation working on a highly classified government project. Both, however, must work from the same general principles toward providing the greatest possible security within the efficient and effective operation of the job at hand.

Controlling traffic includes the identification of employees and visitors and directing or limiting their movement, the control of all incoming and outgoing packages, and control of trucks and private cars.

Visitors

All visitors to any facility should be required to identify themselves. When allowed to enter after establishing themselves as being on an authorized call, they should be limited to predetermined, unrestricted areas.

If possible, sales, service, and trade personnel should receive clearance in advance upon making an appointment with the person responsible for their being there. Although this is not always possible, most businesses deal with such visitors on an appointment basis and a system of notifying the security personnel can be established in a majority of cases.

Businesses regularly called upon, unannounced, by salesmen or other tradespeople should set aside a waiting room which can be reached without passing through sensitive areas. In some cases, it may be advisable to issue them a pass which clearly designates them as visitors. If they will be escorted to and from their destination, a pass system is probably unnecessary.

Ideally, all traffic patterns involving visitors should be short, physically confined to keep them from straying, and capable of being observed at all points along the route. In spread-out industrial facilities, they should take the shortest, most direct route that will not pass through restricted, sensitive, or dangerous areas, and will pass from one reception area to another.

To achieve security objectives without alienating visitors and without in any way interfering with the operation of the business, any effective control system must be simple and understandable. It must incorporate certain specific elements in order to accomplish its aim. It must limit entry to those people who are authorized to be there, and it must be able to identify such people. It must have a procedure by which persons may be identified as being authorized to be in certain areas. It must prevent theft, pilferage, or damage to assets of the installation. And it must prevent injury to the visitor.

Employee Identification

Small industrial facilities and most offices find that personal identification of employees by guards or receptionists is adequate protection against intruders entering under the guise of employees. In plants of over 50 employees per shift, or in high turnover businesses, this type of identification is inadequate. The opportunity for error is simply too great.

The most practical and generally accepted system is the use of badges or identification cards. Generally speaking, this system should designate when, where, how, and to whom, passes should be displayed, what is to be done in case of the loss of the pass, procedures for taking badges from terminating employees, and a system for a cancellation and re-issue of all passes, either as a security review or when a significant number of badges have been reported lost or stolen.

To be effective, badges must be tamper-resistant, which means that they should be printed or embossed on a distinctive stock which is worked with a series of designs difficult to reproduce. They should contain a clear and recent photograph of the bearer, preferably in color. The photograph should be at least one inch square, and should be updated every two or three years or when there is any significant change in facial appearance, such as the growing or removal of a beard or moustache. It should, additionally, contain vital statistics, such as date of birth, height, weight, color of hair and eyes, sex and both thumbprints. It should be laminated and of sturdy construction. In cases where there are areas set off or restricted to general employee traffic, it might be color coded to indicate those areas to which the bearer has authorized access.

If a badge system is established, it will only be as effective as its enforcement. Facility guards are responsible to see that the system is adhered to, but they must have the cooperation of the majority of the employees and the full support of management. If the system is simply a *pro forma* exercise, it becomes a useless annoyance and could better be dispensed with.

Package Control

Every facility must establish a system for the control of packages entering or leaving the premises. However desirable it might seem, it is simply unrealistic to suppose that a blanket rule forbidding packages either in or out would be workable. Such a rule would be damaging to employee morale and, in many cases, would actually work against the efficient operation of the facility. Therefore, since the transporting of packages through the portals is a fact of life, they must be dealt with in order to prevent theft and misappropriation of company property.

If it is deemed necessary, the types of items that may be brought in or taken out may be limited. If such is the case, the fact must be publicized and clearly understood by everyone.

Packages brought in should be checked as to content. If possible, where they are not to be used during work, they should be checked with the guard to be picked up at the end of the day. In most cases, spot checking will suffice.

Whatever the policy concerning packages, whether they are to be checked or inspected, that policy must be widely publicized in advance. This is to avoid the appearance of discrimination against those whose packages are opened and examined or those that are denied entrance in conformity to the company policy.

Vehicle Control

Vehicular traffic within the boundaries of any facility must be carefully controlled, for safety as well as to control the transporting of pilfered goods from the premises.

Merchandise and materials can be readily concealed in or under employee cars, company and outside trucks, railroad cars, or any vehicle that has access to the facility. The more readily these vehicles may drive through the area or approach operating zones, the more acute the problem can become.

It is important, therefore, to make every effort to separate the parking area from all other areas of the facility. This area must be protected from intruders, but the parking lot itself must be separated from production, distribution, and storage areas. Employees and visitors going to and from their cars will pass through pedestrian gates where they can be identified by security personnel.

It would be well to make a most careful evaluation of this vital need. Even the high cost of industrial real estate should not prevent the allocation of space for a parking facility which is separated by a barrier from the rest of the premises. Time and again, firms have reported losses disappearing to almost nothing when they have isolated the parking lot. This is an essential feature of any industrial operation today.

INSPECTIONS

In spite of all defensive devices, the possibility of an intrusion always exists. The highest fence can be scaled and the stoutest lock can be compromised. Even highly sophisticated alarm systems can be contravened by a knowledgeable professional. The most efficient system of physical protection can, at some time or another, be foiled.

It is necessary, therefore, to continually support each element of the system with another—to remember the concept of defense in depth. The ultimate backup—surveillance—must never let down.

Guard Patrols

Visual inspections by irregular patrols through office spaces or through an industrial complex, or constant CCTV surveillance of these same areas, are vital to the success of the security program.

It is equally important to "sweep" the facility after closing time. "Hide-ins" are common in offices or retail establishments. These are thieves who conceal themselves in a closet or utility room and wait for the establishment to close and everybody to go home. After the "hide-in" takes what he is looking for, his only challenge is to break out. The chances of catching such a thief in a premise protected only by perimeter alarms is remote indeed. He must be picked up on the "sweep" when guards go through the entire facility from top to bottom or from east to west to see that everyone required to do so has left.

Specific duties of guards on patrol are discussed elsewhere, but, in general, it should be noted that patrols should be made at least once each hour, and more often if the area and the size of the guard force permit.

Particular attention must be paid to any signs of tampering with locks, gates, fences, doors, or windows. The presence of any rubbish or piles of materials should be noted for the possibility of concealment—particularly if they are near the perimeter barrier or in the vicinity of storage areas.

In the patrol of office buildings, it is wise to stop occasionally for a long enough period of time to listen for any sounds that might indicate the presence of an intruder.

It is equally important that patrols in any facility be alert to any condition that might prove to be hazardous. These might be anything from an oil slick on a typically trafficked area to a heater left on and unattended. Those conditions presenting an immediate danger must be corrected immediately; others must be reported for correction. All of them must be noted in the log and on the appropriate form.

ALARMS

In order to balance the cost factors in the consideration of any security system, it is necessary to evaluate the security needs and then to determine how that security can or, more importantly, should be provided. Since the employment of security personnel can be costly, methods must be sought to improve their efficient use and to extend the coverage they can reasonably provide.

Protection provided by physical barriers is usually the first area to be stretched to its optimum point before looking for other protective devices. Fences, locks, grilles, vaults, safes and similar means to prevent entry or unauthorized usage are employed to their fullest capacity. Since such methods can only delay intrusion rather than prevent it, security personnel are engaged to inspect the premises thoroughly and frequently enough to interrupt or prevent intrusion within the time span of the deterrent capability of the physical barriers. In order to further protect against entry, should both barrier and guard be circumvented, alarm systems are frequently employed.

Such systems permit more economical usage of security personnel, and they may also substitute for costly construction of barriers. They do not act as a substitute for barriers per se, but they can support barriers of lesser impregnability and expense, and they can warn of movement in areas where barriers are impractical, undesirable, or impossible.

In determining whether a facility actually needs an alarm system, a review of past experience of robbery, burglary or other crimes involving unauthorized entry should be part of the survey to be evaluated in the formulation of the ultimate security plan. Such experience, viewed in relation to national figures and the experience of neighboring occupancies and businesses of like operation, may well serve as a guide to determine the need for alarms.

Kinds of Alarm Protection

Alarm systems provide for three basic types of protection in the security system:

1. *Intrusion alarms* signal the entry of persons into a facility or an area while the system is in operation.
2. *Fire alarms* operate in a number of ways to warn of fire dangers in various stages of development of a fire, or respond protectively by announcing the flow of water in a sprinkler system, indicating either that the sprinklers have been activated by the heat of a fire or that they are malfunctioning. (Fire alarms will be discussed in detail in the following chapter.)
3. *Special use alarms* warn of a process reaching a dangerous temperature, either too high or too low, warn of the presence of toxic fumes, or warn

that a machine is running too fast. Although such alarms are not, strictly speaking, security devices, they may require the immediate reaction of security personnel for remedial action, and, thus, deserve mention at this point.

Alarms do not, in most cases, initiate any counter action. They serve only to alert the world at large or, more usually, specific reactive forces to the fact that a condition exists which the facility was alarmed.

Alarm systems are of many types, but they all have common elements. Each one has a sensor of some kind which is designed to respond to a certain condition such as the opening of a door, movement within a room, or the rapid rise of heat. Whenever such condition occurs, the sensor activates a circuit. These circuits can be turned to any use that is desired. They can be designed to sound a bell, dial a phone, or punch a tape. The capability of the circuit is limited only by the design of the system deemed most effective.

The questions that must be answered in setting up any alarm system are:

1. Who can respond to an alarm fastest and most effectively?
2. What are the costs of such response as opposed to a response of somewhat lesser efficiency?
3. What is the predicted loss factor as between these alternatives?

Alarm Systems

The response systems currently available are:

1. *The Central Station*—This is a facility set up to monitor alarms indicating fire, intrusion, and industrial processes. Such facilities are set up in a location as centrally as possible to a number of clients, all of whom are serviced simultaneously. Upon the sounding of an alarm, a team of security men is dispatched to the scene and the local police or fire department is notified. Depending on the nature of the alarm, on-duty plant or office protection is notified, as well. Such a service is as effective as its response time, its alertness to alarms, and the thoroughness of inspection of alarmed premises.
2. *Proprietary System*—This functions in the same way as a central station system except that it is owned by, operated by and located in the facility. Response to all alarms is by the facility's own security or fire personnel. Since this system is monitored locally, the response time to an alarm is considerably reduced.
3. *Local Alarm System*—In this case, the sensor activates a circuit which, in turn, activates a horn or siren or even a flashing light located in the im-

mediate vicinity of the alarmed area. Only guards within sound or hearing can respond to such alarms, so their use is restricted to situations where guards are so located that their response is assured.

Such systems are most useful for fire alarm systems, since they can alert personnel to evacuate the endangered area. In such cases, the system can also be connected to local fire departments to serve the dual purpose of alerting personnel and the company fire brigade to the danger as well as calling for assistance from public fire-fighting forces.

4. *Auxiliary System*—In this system, installation circuits are led into local police or fire departments by leased telephone lines. The dual responsibility for circuits and the high incidence of false alarms have made this system unpopular with public fire and police personnel. In a growing number of cities, such installations are no longer permitted as a matter of public policy.
5. *Local Alarm-by-Chance System*—This is a local alarm system in which a bell or siren is sounded with no predictable response. These systems are used in residences or small retail establishments which cannot afford a response system. The hope is that a neighbor or a passing patrol car will react to the alarm and call for police assistance, but such a call is purely a matter of chance.
6. *Dial Alarm System*—This system is set to dial a predetermined number or numbers when the alarm is activated. The number selected might be the police or the subscriber's home number, or both. When the phone is answered, a recording states that an intrusion is in progress at the location so alarmed. This system is relatively inexpensive to install and operate, but, since it is dependent on general phone circuits, it could fail if lines were busy or if the phone connections were cut.

Alarm Sensors

The selection of the sensor or triggering device is dependent on many factors. The object, space or perimeter to be protected is the first consideration. Beyond that, the incidence of outside noise, movement or interference must be considered before deciding on the type of sensor that will do the best job.

A brief examination of the kinds of devices available will serve as an introduction to a further study of this field:

1. *Electromechanical Devices* are the simplest alarm devices used. They are nothing more than switches which are turned on by some change in their attitude. For example, an electromechanical device in a door or a window, their most common application, would be held in the open or non-contact position by a plunger on a spring when the door or window was closed.

Opening either of these entrances would release the plunger, which, under the action of the spring, would move forward, engaging the contacts in the device and thus activating the alarm.

Such devices operate on th principle of breaking the circuit. Since these devices are simply switches in a circuit, they are normally used to cover several windows and doors in a room or along a corridor. Opening any of these entrances opens the circuit and activates the alarm. They are easy to circumvent in most installations by jumping the circuit, or they can be defeated from within by tying back the plungers with string or rubber bands.

2. *Pressure Devices* are also switches, activated by pressure applied to them. This same principle is in regular use in buildings with automatic door openers. In security applications, they are usually in the form of mats. These are sometimes concealed under carpeting or, when they logically fit in with existing decor, are placed in a strategic spot without concealment. Wires leading to them would, naturally, be hidden in some way.
3. *Taut Wire Detectors* are most often used in perimeter defense, where they are stretched along the top of the perimeter barrier in such a position that anyone climbing the fence or wall would almost certainly disturb them. The tension of the wire is scientifically calibrated, and any change in this tension (either more or less) will activate the alarm. These devices may be used in roof access protection or, occasionally, in the closed area inside the perimeter barrier.
4. *Photoelectric Devices* use a beam of light transmitted for as much as 500 feet to a receiver. As long as this beam is directed into the receiver, the circuit is inactive. As soon as this contact is broken, however briefly, the alarm is activated. These devices, too, are used as door openers.

   In security applications the beam is modulated so that the device cannot be circumvented by a flashlight or some other light source, as can be done in non-security applications. For greater security, ultraviolet or infrared light is used, although even these can be spotted by an experienced intruder unless an electronic flicker device is incorporated into the device. Obviously the device must be undetectable, since once the beam is located, it is an easy matter to step over or crawl under it.

   In some applications a single transmitter and receiver installation can be used, even when they are not in a line of sight, by a mirror system reflecting the transmitted beam around corners or to different levels. Such a is difficult to maintain, however, since the slightest movement of any of the mirrors will disturb the alignment and the system will not operate.
5. *Motion Detection Alarms* operate with the use of radio frequency transmission or with the transmission of ultrasonic waves (See Figure 8-11).

   The *radio frequency* motion detector transmits waves throughout the protected area from a transmitting to a receiving antenna. The receiving

Space Protection Guide

Call for Three-Way Coverage

Space protection differs from perimeter protection by providing volumetric, or three-dimensional coverage. Infrared, ultrasonic, and microwave detectors utilize sensors which can detect an intruder through the height, width and depth of an area.

Unlike perimeter protection devices, they can detect an intruder who may already be inside the building when the system is armed. Even the professional burglar finds space protection devices more difficult to recognize and harder to circumvent. Whether he attempts entry through the ceiling, a wall, the floor or a ventilating duct, as soon as he enters the space, his presence is known.

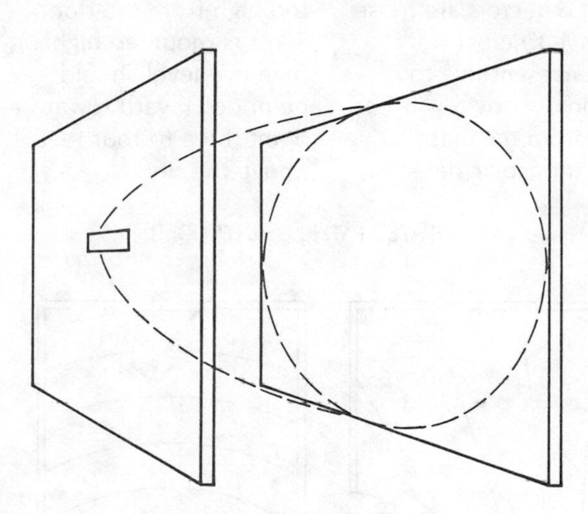

Before Installing, "Case the Joint"

Planning for proper space protection calls for an installer to think like a thief. If the actions of a burglar can be predicted, space protection devices can be positioned where (See figure next page.)

Figure 8-11. Space protection guide. Copyright, 1970, *Security Distributing and Marketing*, a Cahners Publication. Reprinted with permission.

(Figure 8-11 continued)

they have the best chance of catching him.

Depending on the floor plan of the premises, hallways or entrances are likely places to trap intruders. Visualize the movement of a burglar from likely points of entry toward the most valuable objects on the premises. Position a detector so that it is most sensitive to the expected motion to the intruder.

For example, passive infrared detectors are most sensitive to cross walk and least sensitive to motion directly toward or away from the unit.

An ultrasonic detector, on the other hand, responds best to movement directly toward it. Angle the sensor so that it points directly toward the expected motion of the intruder through the entryway or toward the valuables he is likely to steal.

After the horizontal angle of the sensor is determined, the vertical adjustment can be set. The sensor is aiming for a man-sized target, so the vertical angle must not be too high from the floor. Sensors mounted higher than eye level should point downward toward a point three to four feet above the floor.

ANTICIPATED PATHS OF INTRUDER

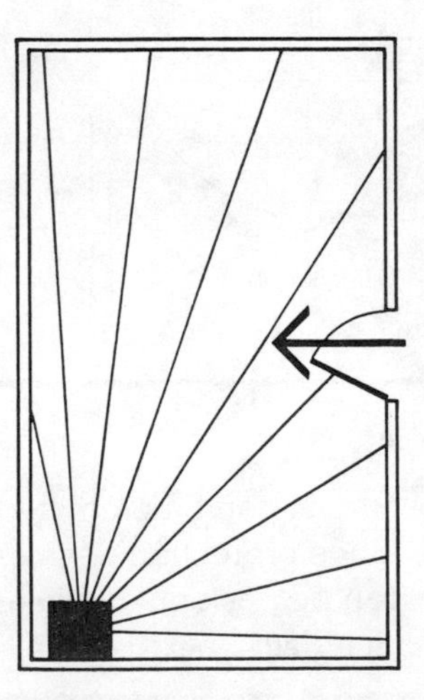

Passive infrared

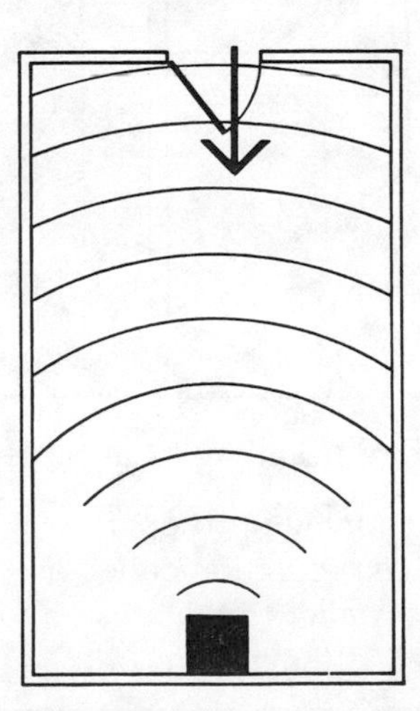

Ultrasonic

(Figure 8-11 continued)

Wide Demand for Space Protection

In the past, space protection was restricted to high-risk customers and premises subject to attack by professional burglars. That market is bigger than ever, but space protection has also spread to small businesses, institutions and homes.

Space protection offers three advantages to customers. It can be used in place of specific protection measures for certain openings. It detects stay-behind criminals and it detects forced entry through unprotected walls, floors and ceilings. Low cost systems designed for low to medium risk customers are proliferating, creating a much wider market for security dealers who sell space protection. (Also see figure, next page.)

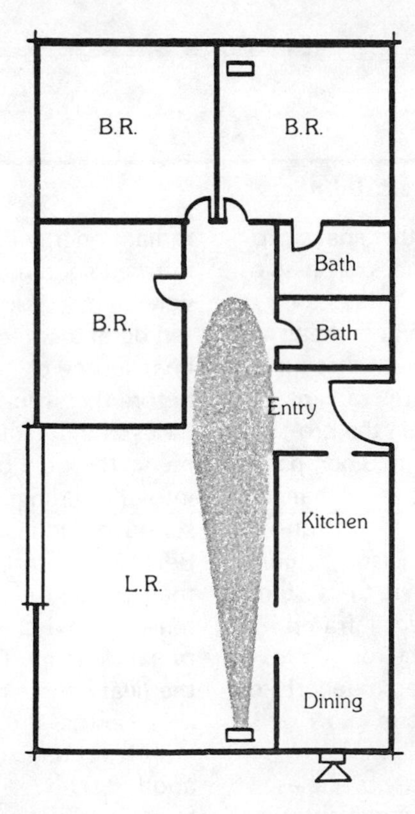

(Figure 8-11 continued)

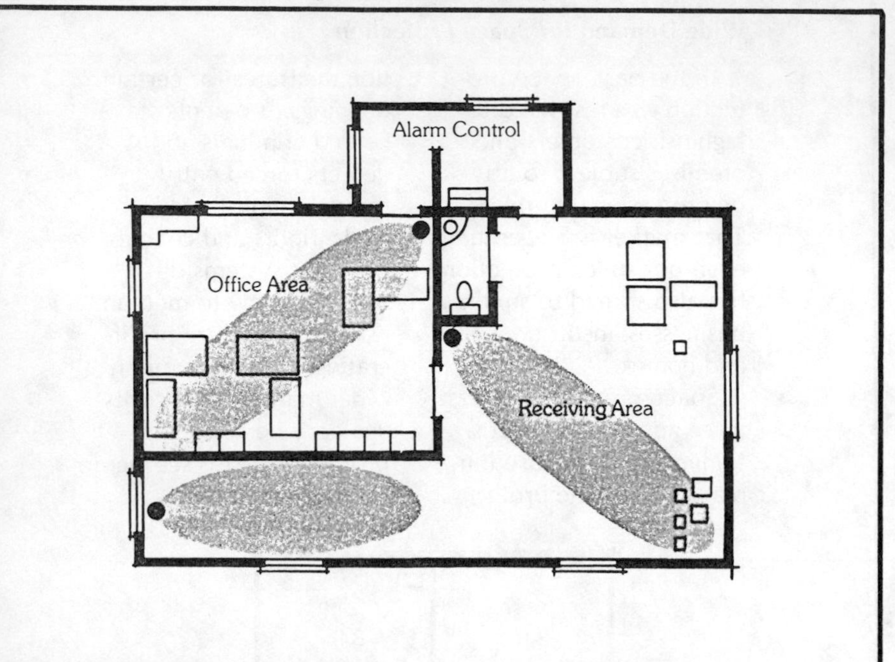

How PIR 'Sees' Intruders

Unlike other space protection devices, a passive infrared (PIR) detector responds to body heat, not motion. It takes a reading on the radiant energy within the protected area and compares that norm to any change in the infrared environment. Any person entering the area while the system is armed adds infrared energy to the room, causing a rapid departure from the norm and triggering an alarm.

Inanimate objects as well as humans transmit radiant energy in the infrared spectrum. That is why many passive infrared detectors require a heat source to move before they trip. In many cases, the detector segments the field of view into alternating zones, sensitive and insensitive. Before an alarm, more than one sensitive zone must sense a change in radiant energy. That is, the heat source must move across the field of view from one zone to another. The result is fewer inadvertent alarms.

(Figure 8-11 continued)

Special Spots for Infrared Units

Because a passive infrared detector transmits no energy, it may be the best choice when the contents of the room absorb ultrasonic or microwave energy.

Clothing on display in a department store can absorb transmitted energy and reduce the efficiency of motion detectors. Hard merchandise can reflect transmitted energy.

In art galleries and museums, where the owners are concerned about works of art being exposed to high frequency energy, passive infrared is an alternative.

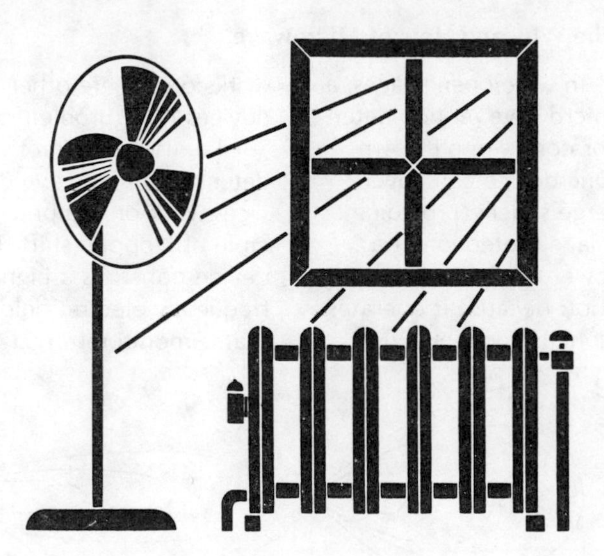

For Best Results . . .

The strong suit for a passive infrared detector is its immunity to disturbances that can cause motion detectors to trigger false alarms.

As heat sensors, PIR devices are sensitive to any rapid changes in infrared energy. While vibration on movement within the protected area cause no problems, environmental factors must be considered in every installation.

Sun: Passive infrared units

(Figure 8-11 continued)

installed in spring may operate without difficulty until winter. Then intense sun reflecting off the snow can shine infrared energy through a window and trigger an alarm.

Auto Headlights: If 'PIR is directed away from windows, headlights from a parking lot reflected inside will not be able to set off the alarm.

Heaters and Boilers: Careful positioning will direct PIR units away from sources of rapid heat rise.

Air Conditioning Ducts: A drop in temperature can affect operation of passive infrared devices too.

Small Animals: Position the infrared beam a few feet above the floor to prevent alarms caused by rodents or pets.

The Why and How of Microwave

In wide open spaces, a microwave motion detector comes into its own. One device can cover large spaces, providing space protection at a lower cost per square foot. Because it operates at high frequency, it works out where other devices have problems.

Like ultrasonic motion detectors, microwave devices work on the principle of Doppler shift. The device transmits a high frequency electric field that is monitored by a

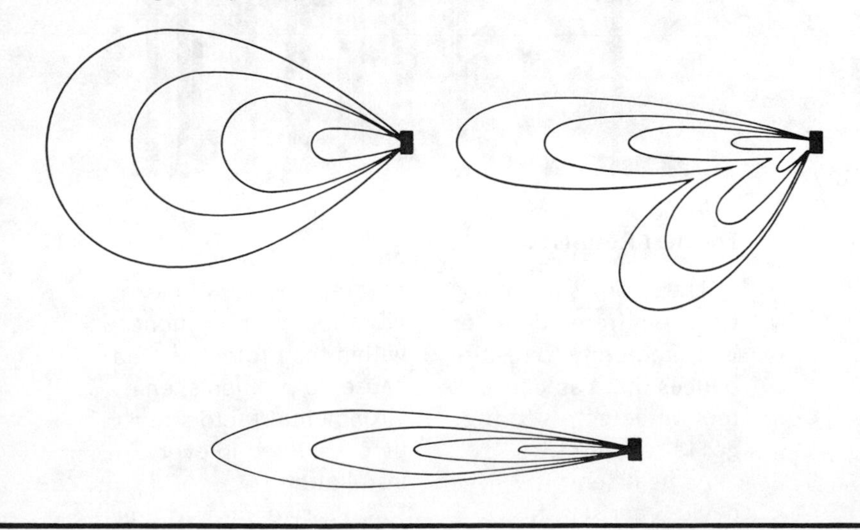

(Figure 8-11 continued)

receiving antenna. Objects within the protected area partially reflect microwave energy back to the detector. Then electronic circuitry compares the transmitted frequency to the received frequency and alarm when a change is detected.

Protection coverage patterns and range adjustment are extremely important because microwave energy passes through most materials used in building construction. A sensor set at the wrong pattern or range can detect motion beyond the area to be protected.

Unlike glass, wood, plastic and many other materials, metal reflects microwave energy. Metal surfaces or steel reinforced walls can distort the pattern of coverage, causing alarms.

Preferred Places for Microwave

It makes economic sense to take advantage of the large area coverage that microwave devices can provide. For example, where one unit can do the work of many ultrasonics, installation time and dealer costs are cut. One manufacturer recommended the use of microwave motion detectors in these locations:

- industrial assembly areas
- power plants
- schools
- hallways
- laboratories
- warehouses
- museums
- exhibit halls
- auditoriums
- theaters

Another supplier suggested using microwave devices where the equipment must be hidden. A microwave unit can do its job even if it is installed in a closet. The corrosive atmosphere of a factory also lets a microwave unit excel. It can be encased and sealed against corrosion and still provide reliable space protection.

(Figure 8-11 continued)

Install Ultrasonic Right, Keep Costs Low

Ultrasonic detectors remain the mainstay of space protection because of their relatively low cost. One detector can flood an area with high frequency sound inaudible to the human ear. Movement within field of ultrasonic energy changes the frequency of the reflected signal and triggers an alarm.

If ultrasonic detectors are chosen at random or installed improperly, efficiency is lost. A moving curtain or squealing brakes from nearby traffic can change the frequency of the signal enough to cause an unnecessary alarm and a service call. Although circuitry is often designed to reduce the effects of the environment, installation is crucial to trouble-free operation.

Placement Key to Motion Detectors

Motion detectors can be upset by background disturbances, that is, factors that stimulate an intruder. If ultrasonic or microwave devices are to perform most effectively, the installer must remove sources of disturbance, shield the detector from them or position detectors to minimize disturbance effects.

Sometimes a problem can be eliminated. A ringing bell can be moved from the room, or a fan can be covered with wire mesh. But proper placement is often the only feasible solution.

Drafts and Ventilating Ducts: Position ultrasonics to avoid air turbulance.
Bells, Hissing Radiators: Ultrasonic may respond to sound waves above the audible range.
Large Metal Objects: Metal reflects microwave energy. Coverage can be distorted.
Soft Goods: Range of ultrasonics can be reduced by absorption.
Small Animals: Ultrasonics and microwaves will trigger an alarm. Infrared beams can be aimed a few feet above the floor.
Radio Interference: Only severe cases can affect

(Figure 8-11 continued)

space protection equipment. Radar close by can affect microwave devices.

Use common sense as a guide to possible problems with motion detectors. In a supermarket, suspended signs are likely to sway in a draft. Be sure to adjust detectors to compensate for legitimate movement.

antenna is set or adjusted to a specific level of emission. Any disturbance of this level by absorption or alteration of the wave pattern will activate the alarm.

The false alarm rate with this device can be high, since the radio waves will penetrate the walls and respond to motion outside the designated area unless the walls are shielded. Some such devices on the market permit an adjustment whereby the emissions can be tuned in such a way as to cover only a single area without leaking into outside areas, but these require some skill to tune them properly.

The *ultrasonic* motion detector operates in much the same way as the radio frequency unit, except that it consists of a transceiver which both transmits and receives ultrasonic sound waves. One of these units can be used to cover an area, or they may be used in multiples where such use is indicated. They can be adjusted to cover a single, limited area, or broadened to provide area protection.

The alarm is activated when any motion disturbs the pattern of the sound waves. Some units come with special circuits that can distinguish between inconsequential movement, such as flying moths or moving drapes, and an intruder.

Ultrasonic waves to not penetrate walls and are, therefore, unaffected by outside movement. They are not affected by audible noise per se, but such noises can sometimes disturb the wave pattern of the protective ultrasonic transmission and create false alarms.

6. *Capacitance Alarm Systems,* also referred to as the proximity alarm, are used to protect metal containers of all kinds. This alarm's most common usage is to protect a high security storage area within a fenced enclosure. To set the system in operation, an ungrounded metal object, such as the safe, file, or fence mentioned above, is wired up to two oscillator circuits which are set in balance. An electromagnetic field is created around the protected object. Whenever this field is entered, the circuits are thrown out of balance and the alarm is initiated. The electromagnetic field may project several feet from the object, but it can be adjusted to operate only a few inches from it where traffic in the vicinity of the object is such that false alarms would be triggered if the field extended too far.
7. *Sonic Alarm Systems,* which are known variously as noise detection alarms, sound alarms or audio alarms, operate on the principle that an intruder will make enough noise in a protected area to be picked up by microphones which will, in turn, activate an alarm. This system has a wide variety of uses, limited only by the problems of ambient noise levels in a given area. The system consists simply of a microphone set in the protected area and connected to an alarm signal and receiver. When a noise activates the alarm, a monitoring guard turns on his receiver and listens in on the prowler.

   Such a system must be carefully adjusted to avoid alarming at every noise. Usually adjustment is turned to alarm at sounds above the general level common to the protected area. The system is not useful in areas where background noise levels are so high that they will drown out the anticipated sounds of surreptitious entry. This device, too, may come with a sound discriminator which evaluates sounds to eliminate false alarms.
8. *Vibration Detectors* provide a high level of protection against attack in specific areas or upon specific objects. In this system, a specialized type of contact microphone is attached to an object, such as works of art, safes, files or to a surface such as a wall or a ceiling. Any attack upon or move-

ment of these objects or surfaces causes some vibration. This vibration is picked up by the microphone, which, in turn, activates an alarm. The unit may be adjusted as to sensitivity, which will be set according to its application and its environment. Here again, a discriminator unit is available to screen out harmless vibrations. This unit is very useful in specific application, since its false alarm rate is very low.

Certain additional alarm devices are currently in use as perimeter protection. These are:

9. *The Electronic Fence,* which consists of from three to nine wires stretched along a cleared space inside the perimeter barrier. These wires are spaced well apart and serve as antennae which, when acting to conduct an electric circuit, create an electromagnetic field as in the capacitance system. Any approach to this fence disturbs the balance of the field and initiates an alarm. It is important that this fence be protected from false alarms, which could be triggered by animals, tall grass or any other casual and innocent disturbance.
10. *The Radio Frequency System* is a system in limited use. This system operates somewhat like a photoelectric system, except that radio waves are used in a narrow beam projected from a transmitter to an antenna. Any disturbance of energy thus transmitted activates the alarm.

Cost Considerations

The costs involved in setting up even a fairly simple alarm system can be substantial, and the great part of this outlay is nonrecoverable should the system prove to be inadequate or unwarranted. Its installation should be predicated on exposure, concomitant need, and the manner of its integration into the existing, or planned, security program.

This last is important to consider, because the effectiveness of any alarm procedure lies in the response it commands. As elementary as it may sound, it is worth repeating that an alarm takes no action; it only notifies us that action should be taken. There must, therefore, be some entity near at hand that can take that action. Too often, otherwise effective alarm systems are set up without adequate supportive or responding personnel. This is at best wasteful and at worst dangerous.

In many instances, alarm installations are made in order to reduce the size of the guard force. This is sometimes possible. At least, if the current guard force cannot be reduced in number, those additional guards needed to cover areas now alarmed will no longer be required.

Good business practice demands that the expense of alarm installations be undertaken only after a carefully considered cost and effectiveness analysis of all the elements. If the existing force of security personnel can cover the security requirements of the facility, no alarm system is needed. If they can cover the ground but not in a way that will satisfy security standards, then more guards will be needed, or the existing force must be augmented by an alarm system that will effectively extend their coverage and their protective ability. The costs and effectiveness must be studied together with an eye toward the efficient achievement of stated objectives.

SURVEILLANCE

Surveillance of a facility is normally conducted by patrolling security personnel who watch for any signs of criminal activity. If they spot any, they are in a position to take such action as necessary.

CCTV

Patrols cannot, however, watch everying at once. In order to extend this area of surveillance, closed-circuit television cameras (CCTV) are used to keep corridors, entrances and designated security areas under observation.

Tape recorders are frequently used in conjunction with these cameras for file and reference purposes. Such tape recordings can be invaluable in identifying persons and for reviewing the circumstances of any event.

CCTV cameras available today can be highly effective in almost total darkness. Such installations are expensive, but since one man can sit at a control panel and watch the output of many cameras covering large parts of the facility, the savings in security personnel, as well as the added security, may make such an investment practical.

Other Cameras

For surveillance after the fact, especially in those cases where a series of thefts has taken place, motion picture or sequence cameras may be used.

Motion picture cameras using high-speed 16mm film and fast lenses can be set up to take pictures in normal light. These cameras can be activated by an alarm or by a switch. The coverage of events is limited by the amount of film in the camera. This is not a totally satisfactory device, since cameras of this kind are never completely silent and need at least normal room light levels to record legibly.

Sequence cameras record still pictures at regular intervals, or they can, by switch control, take a prearranged number of pictures in rapid succession. The time interval between pictures can be adjusted. Pictures can be taken in almost total darkness with infrared sensitive film and in infrared emission.

With either of these methods, cameras should be concealed or at least out of reach and secured.

Both of these methods are considerably cheaper than a CCTV installation, but neither is nearly so effective.

A Word of Caution

There is much equipment available to the security manager today. Some of it is useful, some not, but none of it is better than the use to which it is put or the system into which it is integrated. No equipment stands on its own. It can be used only if it is employed properly, fully and effectively. In the end, it must return, by some kind of estimation, more than the investment that installed it.

REVIEW QUESTIONS

1. What factors need to be considered when purchasing and installing locking devices for security purposes?
2. Describe the basic principles of an effective key control plan.
3. What are the elements that are necessary for an effective visitor access control system?
4. Explain the characteristics of each of the common types of alarm systems.
5. Give an example of a situation where a motion detection alarm might be deployed effectively. Under what circumstances would an ultrasonic system be chosen over a radio frequency system?

Chapter 9

FIRE PROTECTION AND EMERGENCY PLANNING

No facility protection program is complete without clear, well-defined policies and programs confronting the possible threat of fire, explosion, bomb threat, or any other natural or man-made disaster. While planning for such contingencies is a top management responsibility, in most situations the task of carrying out the emergency response falls specifically upon security.

Fire and emergency planning are designed, first, to anticipate what might happen to endanger people or physical property, and to take the necessary preventive measures; and, second, to make provision—through appropriate hardware and/or personnel response—for prompt and effective action when an emergency does occur.

While the emphasis in this chapter (as in most actual practice) is on physical safeguards, it is important to emphasize the human aspect of fire and emergency protection. Disastrous losses often occur not from the failure or absence of physical safeguards, but from human error—the failure to close a fire door, the failure to maintain existing protection systems in good working condition, the failure to inspect or to report hazards, and, at the management level, the failure to ensure, through continous employee education and training, that the organization remains prepared at any time for any emergency.

FIRE PREVENTION AND PROTECTION

While a variety of special perils might be of particular concern in a given situation, the threat of fire is universal. Because it is also one of the most damaging and demoralizing hazards, fire prevention and control must be a cornerstone of any comprehensive loss prevention program.

It should be noted here that any defense against fire must be viewed in two parts. Fire *prevention*, which is usually the major preoccupation, embodies the control of the sources of heat and the elimination or the isolation of the more obviously dangerous fuels. This commendable effort to prevent fire must not, however, be undertaken at the expense of an equal effort for fire *protection.*

Fire protection includes not only the equipment to control or extinguish fire, but also those devices which will protect the building, its contents, and particularly its occupants, in the event of fire. Fire doors, fire walls, smoke-proof towers, fire safes, non-flammable rugs and furnishings, fire detector systems—all are fire protection matters, and are essential to any fire safety program.

Vulnerability to Fire

There are no fireproof buildings, however frequently the term may be misapplied. There are *fire-resistant* buildings. But since even these are filled with tons of combustible materials such as furnishings, panelling, stored flammable materials and so on, they can become an oven that does not itself burn but can generate heat of sufficient intensity to destroy everything inside it. Eventually such heat can even soften the structural steel to such an extent that part or all of the building may collapse. By this time, however, the collapse of the building endangers only outside elements, since everything inside, with the possible exception of certain fire-resistant containers and their contents, will have been destroyed.

The particular danger of this situation is that, while wooden or wood frame construction can be recognized for the fire hazard they represent, many otherwise knowledgeable people are oblivious to the potential dangers from fire in steel and concrete construction. And the danger can be one which is largely unrecognized because it may grow slowly.

The degree of fire exposure in any fire-resistant building is dependent upon its fire-loading—the amount of combustible materials that occupy its interior spaces. In the case of multiple occupancies, such as large office buildings, no one office manager can control the fire-loading. Hence the risk, since the safety of anyone's premises is dependent upon the fire load throughout the entire building. In such an environment, new furniture, decorative pieces, drapes, carpeting, unprotected insulated cables, or even volatile fluids for cleaning or lubricating are piled in every day. And the classic triangle of fire grows larger with each such addition.

The Nature of Fire

The classic triangle so frequently referred to in describing the nature of fire consists of heat, fuel, and oxygen. If all three exist, there will be fire; remove or reduce any one and the fire will be reduced or extinguished.

Certainly fuel and oxygen are always present. It would be difficult to imagine any facility that had no combustible items exposed, and air must be present. Only sufficient heat is missing, and that is readily supplied by a careless cigarette or faulty wiring, two of the most common causative factors.

There are, in fact, an almost infinite number of heat sources which can complete the deadly triangle and start a fire raging in virtually any facility. Every fire prevention program must start by controlling the amount and nature of the fire load or fuel, and by instituting programs to prevent the occurrence of any heat build-up, whether from careless smoking or sparks from a welding torch.

By-Products of Fire

Contrary to popular opinion, flame or visible fire is rarely the killer in the approximately 12,000 deaths from fire that occur annually in this country. These are usually caused by smoke or heat, or from gas, explosion or panic. Several such by-products accompany every fire; all must be considered when defenses are being planned.

Smoke will blind and asphyxiate—and in an astonishingly short time. Tests have been conducted in which smoke in a corridor reduced the visibility to zero in two minutes from the time of ignition. A stairway *two feet* from a subject in the test was totally obscured.

Gas, which is largely carbon dioxide and carbon monoxide, collects under pressure in pockets in the upper floors of the building. As the heat rises and the pressure increases, explosions can occur.

Heat expands and creates pressure. It ignites more materials, explodes gas.

Expanded air created by the heat creates fantastic pressure, which will shatter doors and windows and travel at crushing force and speed down every corridor and through every duct in the building.

All of these elements move upward and therefore permit some control over the direction of the fire—if the construction of the building has been planned with proper fire measures in mind.

Classes of Fire

All fires are classified in one of four groups. It is important that these groups and their designation be widely known, since the use of various kinds of extinguishers is dependent upon the type of fire to be fought.

Class A. Fires in ordinary combustible materials such as waste paper, rags, drapes, furniture. These fires are most effectively extinguished by water or water fog. It is important to cool the entire mass of burning materials to below the ignition point to prevent rekindling.

Class B. Fires fueled by substances such as gasoline, grease, oil, or volatile fluids. Such fluids are used in many ways and may be present in virtually any facility. Here a smothering effect such as carbon dioxide (CO_2) is used. A stream of water on such fires would simply serve to spread the substances with disastrous results. Water fog is excellent, since it cools without spreading the fuel.

Class C. Fires in live electrical equipment such as transformers, generators, or electric motors. The extinguishing agent must be non-conductive to avoid danger to the fire fighter.

Class D. Fires involving certain combustible metals, such as magnesium, sodium, potassium. Dry powder is usually the most, and in some cases the only, effective extinguishing agent. Because these fires can occur only where such combustible metals are in use, they are fortunately rare.

Extinguishers

The security department must evaluate the fire risk for each facility and determine the types of fires to which it might be exposed. Every operation is probably potentially subject to Class A and Class C fires, and most would be additionally threatened by Class B fires to some degree.

Having made such a determination, security must then select the types of fire extinguishers most likely to be needed. The choice of extinguisher is not difficult, but it can only be made after the nature of the risks are determined. Extinguisher manufacturers can supply all pertinent data on the equipment they supply, but the types in general use should be known.

Soda and Acid. These water-base extinguishers are effective on small Class A fires. They are, however, heavy and somewhat clumsy, and the stream is small.

Dry Chemical. These were originally designed for Class B and Class C fires. The new models now in general use are also effective on Class A fires, since the chemicals are flame-interrupting and in some cases act as a coolant.

Dry Powder. Used on Class D. fires. Smothers and coats.

Foam Extinguishers. Effective for small Class A and Class B fires where blanketing is desirable.

CO_2. Generally used on Class B or Class C fires, they can be useful on Class A fires, though the CO_2 has no lasting cooling effect.

Carbon Tetrachloride. These are still in fairly wide use, though they are no longer recommended in the National Fire Protection Association Extinguisher Standard. They are usually rated for use on Class B and Class C fires. The liquid carbon tetrachloride vaporizes when exposed to heat and smothers the fire by oxygen exclusion. These extinguishers should be used only in the open. Their use in closed spaces could be very dangerous—quite possibly fatal—to the operator.

Water Fog. Fog is one of the most effective extinguishing devices known for dealing with Class A and Class B fires. It can be created by a special nozzle on the hose or by an adjustment of an all-purpose nozzle similar to those found on a garden hose.

It might be useful here to look briefly at the advantages fog has over a solid stream of water.

1. It cools the fuel more quickly.
2. It uses less water for the same effect, so water damage is reduced.
3. Because fog reduces more heat more rapidly, atmospheric temperatures are quickly reduced. Persons trapped beyond a fire can be brought out through fog.
4. The rapid cooling draws in fresh air.
5. Fog reduces smoke by precipitating out particulate matter as well as by actually driving the smoke away from the fog.

After extinguishers have been installed, a regular program of inspection and maintenance must be established. A good policy is for security personnel to visually check all devices once a month (See Fig. 9-1), and to have the extinguisher service company inspect them twice a year. In this process, the serviceman should re-tag and, if necessary, recharge the extinguishers and replace defective equipment.

Fire Alarm Signaling Systems

Fire alarm systems, like intrusion alarms, can be viewed as consisting of the signaling device and the sensor. The sensor warns of fire or its danger by activating a circuit, whereas the signaling device notifies those concerned of the danger.

The signaling system is activated by a sensor, a water-flow switch, or a manual alarm. The signal so activated is transmitted to an alarm receiver at a monitoring panel, which activates appropriate alarms and, at the same time, notifies fire protection personnel of the location of the alarm. In many areas the alarm, when sounded, will be in such a series of siren sounds or bell strokes that the location of the fire is identified.

FIRE EXTINGUISHER CHECKLIST

No.	Location	Description and size	Fully charged	Operable	Sealed	Comments
1	Sheet metal shop	CO_2 # 15	yes	yes	yes	Hanging bracket should be replaced

Date 5-14-80 Signed J McCarthy

Figure 9-1. Fire extinguisher safety check lists.

The manual fire alarm stations are of two different types:

1. *Local Alarms* are clearly identified at the pull box as such. This alarm is designed to alert personnel in the building, or in the area where such alarm sounds, that a fire (or in some cases a drill) is in progress. Depending on the prescribed procedures, personnel so alerted will stand by for further instructions or immediately evacuate the building. It is important to remember, however, that such alarms sound locally only. They do not notify anyone beyond the sound of the alarm of the fire threat. Unless someone should happen to notify the fire department by phone, there will be no response to the alarm. It will be necessary, as part of procedure, to designate the person responsible for notifying the fire department.
2. *Remote Signal Stations* transmit a signal directly to the monitoring panel. This performs manually essentially what the automatic signaling devices do automatically. Typically, a local signal is sounded at the same time the signal is being transmitted to the receiver.

Fire Alarm Sensors

After the combustion point is reached, a fire goes through four stages in its development, and each stage is more dangerous than its predecessor. The four types of sensors are designed to be activated at one of each of the four stages of a fire:

1. The germinal or incipient stage—invisible products of combustion are given off. No smoke or flame is visible. There is no detectable heat at this point.
2. Smoldering stage—smoke is now visible. No flame or appreciable heat.
3. Flame stage—an actual visible fire is born. Flame and smoke are visible. Low heat still at this point for a brief moment.
4. Heat stage—at this stage, heat is intense and growing, and the air is expanding rapidly and dangerously.

Since each sensor performs a different function at these different stages of a fire's development, each one has a place in an effective fire protection system.

1. *The Ionization Detector* operates when exposed to those invisible products of combustion which are emitted by a fire in its earliest stage of development. Since a fire may be in progress for as much as two or three hours before it gives any otherwise detectable evidence of its presence, this sensor is extremely useful in giving early warning of an incipient fire. Invisible products of combustion—largely hydrocarbons—interfere with a current passing between two plates in this device and activate an alarm.
2. *The Photoelectric Smoke Detector* operates on the basic principle of every photoelectric device, in which a beam of light is transmitted to a cell which is in balance as long as the light source is constant. When that source is in any way interrupted, the balance is disturbed and the unit alarms. These detectors will alarm on a concentration of smoke of from two to four percent.
3. *The Infrared Flame Detector,* as its name suggests, reacts to the infrared emissions from flame.
4. *Thermal Detectors* sense the temperature in the protected area. Some are set to alarm when the temperature reaches a fixed level; others respond to a rapid climb in the temperature. This latter type, known as "rate-of-rise detectors, cannot be used in industrial areas where the temperature fluctuates strongly.

   Some detectors combine the features of both the fixed temperature and rate-of-rise units and are generally considered to be more useful than either unit alone.

Automatic Sprinkler Systems

A second level of alarm is activated by the flow of water in an automatic sprinkler system. Such an alarm is important in what areas covered by sprinkler systems are rarely otherwise alarmed. When the heat level in such areas rises sufficiently, the sprinkler system is activated by the melting of a metal seal and water flows through the system upon the release of sprinkler head valves. This activates an alarm which indicates the flow of water.

Education in Fire Prevention and Safety

Educating employees about fire prevention, fire protection, and evacuation procedures should be a continuous program. Ignorance and carelessness are the causes of most fires and of much loss of life. An ongoing fire safety program will inform all employees and help to keep them aware of the ever-present, very real danger of fire.

Such a program would ideally include evacuation drills. Since such exercises require shutting down operations for a period of time and lead to the loss of expensive man-hours of productive effort, management is frequently cool toward them.

However, indoctrination sessions for new employees and regular review sessions for all personnel are essential. Such sessions should be brief and involve only a small group. They should include the following subjects as well as any others that may have particular application to the specific facility:

1. Walk to primary and secondary fire exits, and demonstrate how such exits are opened. Emphasize the importance of closing exit stairwells. If possible, employees should walk down these stairs.
2. Explain how to report a fire. Emphasize the need to report first before trying to put out the fire.
3. Distribute a simple plan of action in the event of fire.
4. Explain the alarm system.
5. Explain the need to react quickly and emphasize the need to remain calm and avoid panic.
6. Explain that elevators are *never* to be used as an emergency exit.
7. Point out the danger of opening doors. Explain that doors must be felt before being opened. Opening a hot door is usually fatal.
8. Demonstrate available fire fighting equipment, or show manufacturers' films or fire department films on similar equipment.
9. Describe what should be done if escape is cut off by smoke or fire:
   a. Move as far from fire as possible.
   b. Move into building perimeter area with a solid door.
   c. Remove readily flammable material out of that area, if possible.
   d. Since expanded air exerts enormous pressures, barricade the door with heavy, non-combustible materials.
   e. Open top and bottom of windows. Fire elements will be exhausted through the top while cool air will enter through the bottom.
   f. Stay near the floor.
   g. Hang something from windows to attract firemen.

Employees in Fire Fighting

Since the danger of fire with its concomitant risk to life and property affects every employee, many experts feel that the responsibility in case of fire is a shared one. Few disagree that everyone must be educated in the principles of fire prevention and fire protection, including indoctrination in evacuation procedures and how to report a fire. But beyond this, there is little agreement on what employees should be asked to do.

Some business offices set up a system of floor wardens whose job it is to pass the word for evacuation, and who then sweep their area of responsibility to see that it is clear of personnel, that papers are deposited in fireproof containers and that high value, portable assets are removed from the premises.

Other take the view that their employees were not hired to act as emergency supervisors and do not expect them to act as such. Many firms of this latter persuasion ask that certain minimal functions be performed by those persons who are on hand but do not assign roles to specific people. Examples of this might be a policy of returning tapes to a fire-proof container in the computer area or securing fire-resistant safes in the accounting or cashier's office before evacuation. This responsibility would fall on the personnel in these areas at the time of the alarm and would or should take little time to accomplish. When the signal for evacuation is given, no time should be lost in vacating the building.

Many professionals feel that office employees should never be asked to do more than see to their own safety by beating an orderly retreat along predetermined escape routes. Only in the most extreme emergency—and then only if they are otherwise trapped—should employees engage in fighting a fire of any magnitude. They can be expected, as a normal reaction, to make an effort to put out a wastebasket fire or a small blaze in a broom closet, but even in these cases the alarm must be given as first priority. Any fire that threatens to involve a major part of an office of other parts of the building should be left to professionals.

Obviously, all such situations are matters of on-the-spot judgment. Policies concerning every situation are difficult, if not impossible, to predetermine.

Industrial Fire Brigades. The situation is quite different in industrial fire operations. In such facilities, the formation of a fire brigade composed of a few selected and trained employees is fairly general practice. There is general agreement that the nature of their employment in industrial areas makes these employees more competent to handle fire-fighting assignments, which are, in many cases, not that far removed from their regular work.

The exact size of each fire protection organization will vary according to the size and nature of each plant. Very large facilities, or those whose fire risk is high because of the nature of the operation, may have a full-time fire department. In smaller or less hazardous facilities, regular employees are organized into a fire brigade, which is broken down usually by departments or areas into fire companies.

These companies are assigned to a given area for purposes of fire protection, fire prevention, and fire fighting. They are also available as a part of the fire brigade in any other area of the plant if a fire occurs that requires more manpower than the assigned company can handle.

The size of the brigade will depend upon the size of the plant and the nature of the risk involved. It will also be affected by the general availability, size, competence, and response capability of the public fire-fighting facilities in the neighboring areas. Whatever the size, however, it must consist of men sufficiently well-trained and familiar with the plant operation and layout to fight fires effectively in any part of the facility, if the need arises.

The plant engineer and the maintenance crew should certainly be included in the brigade. Their knowledge in servicing valves, pumps, and other machinery is invaluable in an emergency situation.

The Brigade Chief. The brigade must have a clear chain of command, headed by a chief who is qualified by experience, training, or background in an allied field, or preferably all three. He will command the brigade in all fire operations.

The brigade chief will establish training programs for brigade members, he will maintain contact with local fire departments, and he will have clear authority in all matters where fire risk is concerned, including the storage of particularly flammable materials, fire fighting equipment, fire escapes, and fire doors. He will be involved in the planning of new construction or remodeling of existing facilities. He will appoint an assistant chief to act in his absence, and captains for each of the fire companies in the brigade. He will, additionally, establish procedures for the inspection and maintenance of all fire fighting equipment on a regular basis.

Finally, his assignment as brigade chief should be his primary duty. Members of the brigade, unless it is a proprietary fire department exclusively, will be employees who are assigned to fire protection as a collateral duty; the chief should have such duty as his primary responsibility.

Evacuation

Evacuating Industrial Facilities. Evacuation plans for an industrial facility are relatively simple, since most buildings within the perimeter are one-, two-, or in rare cases three-story buildings. Because they are occupied by personnel of a single company or under that company's jurisdiction, a single plan involving all personnel can be drawn up. And because, in most cases, aesthetic considerations are not a prime concern in the design of the industrial building, fire escapes can be constructed in any way for the greatest safety.

Although many of these buildings have elevators, most of which serve the dual purpose of hauling freight and personnel, these elevators are not necessarily the prime means of moving to and from the upper level, as they are in a high-rise office building.

Generally, these buildings can be cleared in minutes. This is not to say that an evacuation plan is not needed. It is—and it must be widely distributed and clearly understood. Most industrial buildings are more open, the exits are more visible—more a part of the unconscious orientation of the employees and, therefore, more a part of the natural traffic flow than in many other types of construction.

Evacuating High-Rise Buildings. Evacuation from high-rise urban buildings is quite a different story. In this situation, employees come to work and leave regularly by elevator. Yet if there should be a fire, they are told they must *not* use the only means of entrance or exit they really know.

In most cases they have never been on the fire stairs. They may have only a vague idea where those stairs are located, but in a time of emergency—a time of anxiety bordering on panic, when instinctual behavior would be most natural—they are asked to vacate the premises in a way which to them is very unnatural indeed.

Even in wide-open industrial facilities where orientation is quick and easy, there will always be some people who will pass a clearly marked exit to get to the employee door they are used to using. This is much more likely to happen in buildings with windowless corridors and fire exits, however clearly lighted and marked, well off the normal traffic pattern used by the employees!

To overcome this problem—which must be overcome until such time as elevators are made safe to use in the case of fire—it is advisable, as a drill, to walk every employee from his desk to the nearest fire exit and to the nearest secondary exit which he would use in case his first escape route was cut off by fire or smoke.

This could be done over a period of time with small groups at each drill. It is important that the drill actually start at the desk or office of the employee, so that the route as well as location of the exit is made clear.

Planning and Training. Evacuation plans must be based on a well-considered system and on thorough and continuing education. They should also be based on indoctrinating employees in the principles of fire safety, stressing that they are to make their own way to the proper exit and leave as quickly and calmly as possible.

Adults do not respond to being lined up like children at a fire drill and marched down the fire stairs. Whereas they might be inclined to follow a leader under many circumstances, when it comes to a concept as simple as vacating the premises, a leader has no purpose or place. They will rebel or even panic if they feel restrained or regimented in their movements toward the exit.

In setting up plans for evacuation, it might be well to review and evaluate the circumstances of a given facility and then ask a few questions:

1. Are routes to exits well lighted, fairly direct, and free of obstacles?
2. Are elevators posted to warn against their use in case of fire? Do these signs point out the direction of fire exits?
3. Are handicapped persons provided for?
4. Do corridors have emergency lighting in the event of power failure?

5. Who makes the decision to evacuate? How will personnel be notified?
6. Who will operate the communication system? What provisions have been made in case the primary communication system breaks down? Who is assigned to provide and receive information on the state of the emergency and the progress of the evacuation? By what means?

BOMBS AND BOMB THREATS

Any business, industry or institution can become the victim of a bombing or bomb threat. Most telephone bomb threats—approximately 98 percent—turn out to be hoaxes. The target of the threat, however, has no way of knowing whether a real bomb has been planted. Contingency planning is necessary for an organization to be able to protect its personnel and property from the hazards of an explosion. In the absence of a specific response plan, the bomb threat will often cause panic. This may be the precise result the caller seeks.

Controlling access to the facility, having adequate perimeter barriers and lighting, checking all parcels and packages, locking areas such as storerooms, equipment rooms, and utility closets, and taking note of any suspicious persons or anyone not authorized to be in an area, are all measures that can thwart the bomb planter as well as the thief or other intruder.

Contingency plans should specify who will be responsible for handling the crisis and delegating authority in the event a bomb threat is received. The officer in charge of responding to a bomb threat must be someone who will be available 24 hours a day. A control center or command post, with provision for communication with all parts of the facility and with law enforcement agencies, should also be designated. All personnel who will be involved in the bomb threat response must receive training in their assignments and duties. Plans should be in writing.

Telephone Operator's Response

The telephone operator's role is critical in handling the bomb threat call. The operator should receive training in the proper response so as to elicit from the caller as much information as possible. The two most important items of information to be learned are the expected time of the explosion and the location of the bomb. The operator should remain calm and attempt to keep the caller talking as long as possible in hopes of gaining information or clues that will aid investigators. The caller's accent, tone of voice, and any background noises should be noted. Many organizations provide telephone operators with a Bomb Threat Report Form for recording all information (See Figure 9-2).

General Services Administration

BOMB THREAT INFORMATION

Date:

	Received	Ended
Time Call:		

EXACT WORDS OF CALLER:

(Continue on reverse)

QUESTIONS TO ASK:

1. WHEN IS BOMB GOING TO EXPLODE?
2. WHERE IS BOMB RIGHT NOW?
3. WHAT KIND OF BOMB IS IT?
4. WHAT DOES IT LOOK LIKE?
5. WHY DID YOU PLACE BOMB?

DESCRIPTION OF CALLER'S VOICE:

☐ Male ☐ Female

☐ Young ☐ Middle-Aged ☐ Old

TONE OF VOICE

ACCENT

BACKGROUND NOISE

IS VOICE FAMILIAR?

☐ Yes. ☐ No.

IF "YES", WHO DID IT SOUND LIKE?

ADDITIONAL COMMENTS:

Name of Person Receiving Call

Organization & Location

Home Address

Office Phone

Home Phone

Figure 9-2. Telephone bomb threat information form.

After receiving a bomb threat call, the operator must inform the designated authority within the organization (Chief of Security, for example). Law enforcement authorities, and others in the organization, will be notified in turn, according to the written contingency plans.

Search Teams

A decision must be made whether to conduct a search of the premises, and how extensive the search should be. If possible, the search should be conducted by employee teams rather than police or fire department officers. Employees are familiar with their work area and can recognize any out-of-place object. An explosive device may be virtually any size or shape. Any foreign object, therefore, is suspect.

Basic techniques for a two-man search team include the following:

1. Move slowly and listen for the ticking of a clockwork device. (It is a good idea to pause and listen *before* beginning to search an area to become familiar with the ordinary background noise which is always present.)
2. Divide the room to be searched into two halves. Search each half separately, in three layers: floor to waist level, waist to eye level, and eye to ceiling level.
3. Starting back to back and working toward each other, search around the walls at each of the three height levels; then move toward the center of the room.

If a suspicious object is found, *it must not be touched.* Its location and description should be reported immediately to designated authorities. A clear zone with a radius of at least 300 feet should be established around the device (including the floors above and below). Removal and disarming of explosive devices should be left to professionals.

Those assigned to search a particular area should report to the control center after completing their search.

Evacuation

Evacuating a facility for any reason, particularly in response to a bomb threat, is a drastic reaction to the potential danger. There clearly are situations when such an extreme course is indicated, but such a decision can never be undertaken lightly. It is essential that a thorough and exhaustive dialogue relative to this complex problem be undertaken at the earliest opportunity so that plans and policies may be formulated prior to any actual pressure from such an emergency. Many experts in the field argue that a total evacuation is rarely, if ever, indicated. The argument concerns the risk of exposing a greater number of people to the blast when the location of the bomb is unknown. Whenever personnel are moved about in large groups, the exposure to injury is increased. Moreover, the movement of large numbers under the threat of bombing can create panic—a very dangerous situation.

A bomber who has shown his familiarity with the facility by placing a bomb on the premises presumably can be assumed to be familiar with normal and perhaps even abnormal or emergency traffic patterns. He has probably placed his bomb in such a way as to create the greatest possible injury and havoc. In such a situation, total evacuation might serve only to expose the greatest number of employees to injury and death.

It has also been found that hoaxers or mischief-makers may be encouraged by a mass evacuation to repeat such calls and subject the facility to a string of bomb threats.

The decision whether to evacuate will be made by management of the threatened facility, often in conjunction with law enforcement officials. It may be decided to evacuate the entire facility, or only the areas in the vicinity of the suspected bomb, or (unless a device has actually been discovered) not to evacuate at all. Detailed plans are necessary to insure safe and orderly evacuation. Personnel should leave through designated exits and assemble in a predetermined safe area. Elevators should not be used during evacuation. Doors and windows should be left open for increased venting of the explosive force.

When authorities have determined that the bomb threat emergency has ended, all personnel who have been notified of the threat should be informed that normal operations are to resume.

LABOR DISPUTES

If a facility is faced with a deteriorating labor-management relationship in face of new union contracts, it may be necessary to review those security measures designed to protect personnel and property during unusual periods.

The security director should review every element of these plans with his department to assure himself that his staff is totally familiar with the job to be performed, and that every member is adequately trained to perform as necessary. Be certain that all security personnel are aware that their role is to protect personnel and property only, and that they are not to participate in any way in the elements of the dispute which is strictly between management and the participating union.

In clarifying policy and getting approval of specific plans of action from responsible management, foresight is indicated. The earlier the protection system is established, the better. Many elements will require time and considerable concentration from those executives who will later be occupied with handling the labor problem immediately at hand, and may not be available for consultation.

A checklist of some possible measures is listed below as a guide to the development of a more thorough plan which will apply more specifically to a given facility:

1. Secure all doors and gates not being used during the strike and see that they remain secured.
2. Remove all combustibles from the area near the perimeter—both inside and outside.
3. Remove any trash and stones from the perimeter that could be used for missiles.
4. Change all locks and padlocks on peripheral doors of all buildings to which keys have been issued to striking employees.
5. Recover keys from employees who will go out on strike.
6. Nullify all existing identification cards for the duration and issue special cards to workers who are not striking.
7. Check all standpipe hoses, fire extinguishers, and other firefighting equipment after striking workers have walked out.
8. Test sprinkler systems and all alarms—both fire and intrusion—after striking workers have walked out.
9. Consider construction of barriers for physical protection of windows, landscaping, lighting fixtures.
10. Move property most likely to be damaged well back from the perimeter.
11. Be certain all security personnel are familiar with the property line and stay within it at all times when on duty.
12. Guards are not to be armed nor are guards to be used to photograph, tape, or report on the conduct of strikers. The only reports will relate to injury to personnel or property.
13. Notify employees who will continue to work to keep windows of their automobiles closed and their car doors locked when moving through the picket line.
14. Consider the establishment of a shuttle bus for non-striking employees.
15. Establish, in advance, which vendors or service persons will continue to service the facility, and make arrangements to provide substitute services for those unwilling to cross the picket line.
16. Keep lines of communications open.
17. Since functional organizational lines may be radically changed during the walkout, find out who is where, and who is responsible for what.

EMERGENCY PLANNING

The first thing that must be said about emergency planning is that there must be a *plan,* a detailed set of policies and procedures that take into account any reasonably foreseeable emergency or disaster that would affect the safety of people and the assets of an organization. The plan should be in writing, and it should spell out, in as detailed a manner as possible, the steps to be taken in a given emergency and *by whom*.

Note that there are two key elements in any contingency plan: first, what is to be done, and second, who is to do it.

Having a detailed plan of action ensures that the right people, equipment, and facilities will be available in a crisis, and that everyone—management and employees—will know what to expect, what to do, and who is in charge. In a serious crisis response must often be swift and sure. Confusion and panic are all too common. If the twin objectives of preventing or minimizing injury or accident to people, and preventing or minimizing the loss of assets are to be achieved in an emergency, key people in the organization must be able to act quickly and responsibly.

Advance planning has other advantages. It makes it possible to enlist the thinking of those who will be involved in the crisis situation. Since all facets of the organization will, typically, be affected, all should have some input in the planning. The committee approach, in fact, is most commonly used in developing emergency plans. This cooperative effort enables the organization to anticipate problems in different areas of a facility and to prepare for them.

Types of Emergencies

While it is impossible to predict all emergencies that might occur, it is possible to make reasonable estimates of vulnerabilities for a given facility in a particular geographical location. During the past decade the most common emergencies have been fire, bomb threat, and labor dispute, with the result that most organizations have specific contingency plans for these hazards, which have already been discussed. Other threats may be relatively predictable for a given site, such as the possibility of explosion, a spill, or a chemical leak in a chemical plant. In developing emergency and disaster plans, priority should naturally be given to those areas of most immediate concern.

The types of emergencies for which plans should be drawn up might include:

1. Airplane crash
2. Bomb and bomb threat
3. Building collapse
4. Civil disturbance
5. Earthquake
6. Fire
7. Flood
8. Riot
9. Sabotage

10. Strikes and pickets
11. Utility failure
12. Windstorm

While ''game plans'' should exist for each emergency, insofar as possible standard procedures should be adopted, since there are many common elements in the emergency response to any crisis. Wherever possible, the same chain of command should exist, the same communications, the same command post and control center, and the same first aid center. Evacuation routes and exits should be identical for a bomb threat or a fire. An important rule in all such planning is *keep it simple*. Employing common elements in all plans helps to eliminate confusion and uncertainty during the frantic first moments of a disaster.

Emergency Personnel and Equipment Records

We have already said that the emergency plans should be in writing, preferably in an Emergency Planning Manual, copies of which should be immediately available to designated members of management, the Emergency Team, security, finance, and other key employees with specific responsibilities in the emergency response. Plans should be reviewed periodically and kept up-to-date.

Permanent records, kept in a secure location (such as the security office), should include:

1. Names and phone numbers of management personnel to be notified.
2. Names of emergency forces (assignments, location, phone numbers).
3. Names of backup emergency forces.
4. List of emergency equipment and supplies, including type, location, quantity, backup and outside support.
5. Building plans.
6. Mutual-aid agreements.
7. Outside organizations (police, fire, hospital and ambulance), locations and phone numbers.
8. Emergency Planning Manual.

Where a permanent location for the emergency command post exits, copies of all emergency personnel and equipment records, building plans, and other relevant documents should be stored at that location. Where this is a site other than the security offices, an additional backup file should be maintained by security.

Elements of the Emergency Plan

Developing an emergency plan will, in effect, create the organization necessary to carry it out. It should provide for:

1. Designating authority to declare an emergency and to order shutdown and evacuation, total or partial.
2. Establishing an emergency chain of command.
3. Establishing reporting responsibilities and channels.
4. Designating an emergency headquarters, or command post.
5. Establishing and training Emergency Teams for each shift.
6. Establishing specific asset protection and life-saving steps.
7. Designating equipment, facilities, and locations to be used in an emergency.
8. Communicating necessary elements of the emergency response plan to all affected personnel.
9. Communication with outside agencies.
10. Public relations and release of information

While the authority to declare an emergency and to order shutdown and/or evacuation normally rests with someone in higher management, this sometimes is exercised by a plant manager in an industrial facility, for example, or by the designated emergency plan chairman or administrator. Whoever is so designated in the plan must be someone who is available or on call at all times.

Someone must also be in charge of executing the plan in an emergency, with backup to ensure coverage around the clock. He will be responsible for initiating action according to the plan in an emergency, ordering and directing shutdown and evacuation when so authorized, making emergency announcements, and coordinating emergency responses. Often this emergency plan administrator or chief is the plant or facility manager, and he will take advantage of existing organizational chains of command, with orders and directions filtering down through department heads to supervisors, foremen, and working personnel.

As indicated, specific responsibilities in the plan will depend on the nature of the facility and the threat. These should take advantage of special skills and knowledge (of electricians, engineers, maintenance crew). The important factor is to have these responsibilities spelled out in advance in the emergency plan.

Continuity in planning is essential. Provision should be made for alternates for each individual given responsibility in the emergency plan. Also, since an emergency can occur at any hour of the day, the plan should designate emergency responsibilities for each shift or for open and closed hours, depending on the situation.

Security Responsibilities

While it is not possible in this generalized review to specify the duties of all those involved in the emergency response, the commonly accepted responsibilities of the security department in an emergency are essentially an expansion of normal security functions, including the preservation of order, protection of life and property, vehicle and pedestrial control. These duties, however, should not simply be "understood," but, as with all aspects of emergency planning, should be designated in the plan.

1. Establish communications with police.
2. Activate emergency steps to protect people, property and valuable information or other assets.
3. Mobilize an emergency guard force.
4. Control movement of personnel and others.
5. Control entrances and exits.
6. Control classified and dangerous areas.
7. Control evacuation as ordered.

In some circumstances, depending on the size of the Emergency Team, security may also be involved in fire fighting and rescue, first aid and other assistance as needed. The security director should always be a member of any emergency planning committee as well as the Emergency Team. Emergency and disaster planning is a total management responsibility, but by its very nature security will always play an active and prominent role.

REVIEW QUESTIONS

1. What are the four classes of fire, the fuels needed to ignite each and the extinguishing agents that can be used on each class?
2. In what ways is an IONIZATION detector different from a SMOKE, INFRA-RED or THERMAL detector?
3. What are the key elements of any emergency plan?
4. What should be the role of security in the emergency response?
5. When management is developing a plan for emergency evacuation, what things need to be considered?

Chapter 10

SAFETY PLANNING AND SUPERVISION: THE ROLE OF OSHA

Safety consciousness in business and industry did not begin with the passage of OSHA—the Occupational Safety and Health Act—in 1970, but it is largely a product of the twentieth century. Prior to the Industrial Revolution, the worker was an independent craftsman. If he suffered economic loss because of an accident or because of illness rising out of his prolonged exposure to a particular work environment, the problem was his, not the employer's. This attitude generally prevailed during the rapid expansion of the factory system in America throughout the 19th century. Only toward the latter part of this century did it begin to become obvious that, "While the factories were far superior in terms of production to the preceding small handicraft shops, they were often inferior in terms of human values, health, and safety."[48]

The atmosphere of reform which gained impetus after the turn of the century resulted in, among other new laws, the first effective Workmen's Compensation Act in Wisconsin in 1911.[49] Compulsory laws on workmen's compensation followed in many states after the Supreme Court unheld their constitutionality in 1916. Even the most hard-headed employers found that their costs dictated compliance with the spirit of the law.

As a result of this growing concern for industrial safety, there followed a long downward curve in work-connected accidents and injuries that lasted through the period between the two World Wars and continued into the 1950's. By 1958 this trend had leveled off, and by 1968, for the first time in over 50 years, the curve began to rise again.

Fourteen thousand occupational fatalities and over two million disabling work-connected injuries each year seemed to be considerably more than the number that might one day be arrived at as the irreducible minimum. The result, through the 1960s, was increasing federal concern with establishing standards of occupational safety and health. Prior to that decade, only a few federal laws, such as the Walsh-

Healey Public Contracts Act, had been enacted, with most legislation in this area being left to the states. During the 1960's, a number of laws were passed—the McNamara-O'Hara Service Contracts Act, the Federal Construction Safety Act, the Federal Coal Mine Health and Safety Act, among others—all dealing with safety and health standards in specific fields and under specific circumstances. Public Law 91-596, known as OSHA, which was signed into law on December 29, 1970, was the first legislation which attempted to apply standards to virtually every employer and employee in the country.

REQUIREMENTS OF OSHA

Extent of Coverage

General rules governing safety and health standards are laid down in OSHA to apply to industry as a whole, supplemented by a vast literature relating to employee safety in virtually every industry or process. These specific applications concern themselves with everything from the operation of machines and equipment to the handling of compressed air; from dry grinding and buffing operations to non-ionizing radiation protection.

In order to assure the broadest possible application of the Act, Congress worded it to state that every employer "affecting commerce" would be subject to its provisions. This means that if any devices, tools, materials or equipment used on the job were manufactured in another state, the employer using such elements is "affecting commerce."

Thus almost all employers in every industry are covered. About 60 million workers in five million establishments are protected by the Act. And it follows that the 60 million workers so protected must comply with all rules, regulations and orders issued pursuant to the Act that are in any way applicable to their actions and conduct.

The great mass of material specifying the standards, the various reports to be made and filed, and the procedures that must be followed in posting notices, notifying employees, and filing for variances, may seem overwhelming—indeed, even impossible—to comply with. Yet these requirements must be met. And a closer examination suggests that the routine is not as difficult as it appears. What it does require is planning to set up a reasonable system for compliance with the letter and the spirit of this landmark legislation.

It should be noted that federal OSHA standards are not compulsory where the state's own safety and health standards are equal to or in excess of those required by federal OSHA. In general, where this situation applies, federal inspectors do not come in. At the time of this writing, some 32 states have adopted the federal OSHA standards. Other states have preferred to carry out their own programs (thus sacrificing the substantial government grants available under the federal law).

Administration/Research and Development

Compliance with the wide-ranging OSHA legislation is administered by 10 regional and over 100 area offices and field stations, under the policy guidance and administration of the Department of Labor and the Department of Health and Human Resources.

In addition to enforcing the law, the Department of Health and Human Resources is responsible, through the National Institute for Occupational Safety and Health (NIOSH), for an ongoing program to develop criteria for establishing safety and health standards, to research health matters, to undertake special studies, make determinations and report on toxicity, and to undertake and promulgate professional training and educational programs. NIOSH has already proposed many standards, particularly in the use of hazardous materials, reflecting an increasing emphasis on the health aspect of OSHA. In the pursuit of this research and development program, NIOSHA has wide authority to require special reports and to interrogate employers and employees alike.

OSHA requires that the Department of Labor establish standards with respect to toxic agents and other harmful physical agents. These standards must be of such a stringency that, as far as is humanly possible, no employee will suffer any kind of impairment or disability no matter what his exposure (provided, of course, that all other safety precautions are complied with).

Description of Standards

Generally speaking, OSHA requires that an employer provide a safe and healthful place to work for his employees. This is spelled out in great detail in the Act to avoid leaving the thrust of the legislation in any doubt.

Though much of the language in the Act is technical in nature and largely couched in legalese, the thrust of the legislation is absolutely clear and unambiguous in what is known as the "General Duty" clause which states that each employer "shall furnish to each of his employees . . . a place of employment . . . free from recognized hazards that are causing or likely to cause death or serious physical harm to his employees" and that, further, he "shall comply with all occupational safety and health standards promulgated under this Act." Much of the rest of the Act deals with procedures and standards of safety and is, in places, difficult to follow.

It speaks of free and accessible means of egress, of aisles and working areas free of debris, of floors free from hazards. It gives specific requirements for machines and equipment, materials, and power sources. It specifies fire protection by fixed or portable systems, clean lunch rooms, environmental health controls, and adequate sanitation facilities. Whereas, in past years, employers might contend in all sincerity that their facilities met community standards for safety and cleanliness, with the enactment of OSHA these standards have been formalized to describe minimum levels of acceptability. They might also contend that some specific demands of the Act were unclear, but there is no mistaking what the Act is getting at." The Congress declares to be its purpose and policy . . . to assure so far as possible every working man and woman in the nation safe and healthful working conditions and to preserve our human resources."

Perhaps the strongest resistance to OSHA in its first years has been the complaint that some of the basic standards went too far or were unnecessary. Recent emphasis by the OSHA administration has been upon the elimination of "Mickey Mouse" standards that have no direct bearing on improving safety in the workplace.

Records and Reports

Record keeping and reporting is an essential element of OSHA, both for its enforcement and for developing information regarding the causes and the prevention of occupational accidents and illnesses. This latter function, although it may appear burdensome to some employers, is vital to OSHA's role in helping to accumulate data which will substantially reduce the annual toll of injuries and illnesses on the job, which has stubbornly refused to abate since 1958.

In the past, certain information concerning illness and injury was submitted for insurance purposes under the provisions of workmen's compensation acts. Unfortunately, this material was neither complete nor compiled in a uniform manner on a national scale. As a result, its accuracy was suspect. It was neither useful, nor used to develop an approach to preventing injury and illness among workers.

Under OSHA, the centralized collection of this information and its careful analysis is a major effort. The data will be used to identify the problem areas which up to now have been imperfectly catalogued. It will identify the types of problems, direct the countermeasures to be taken, and maintain an annual accounting of results.

The Act spells out the types of records that must be kept by all employers. These records report on all work-related deaths, injuries and illnesses. The records must be kept up to date, and they must be available to Labor Department inspectors (called compliance officers) on request. They will be maintained on the forms provided, but they may be recorded on some private equivalent form providing the data is as readable and comprehensible as the form itself.

OSHA Form 200. This form replaces previous OSHA Forms 100 and 102. The left-hand portion of the form provides for a continuous log of all recordable injuries and illnesses. The log must be maintained in each work establishment, and each injury or illness must be entered in it as early as possible, but in no case later than six working days. The right-hand portion of the form is an annual summary of injuries and illnesses. This must be complete and ready for examination no more than one month after the close of the calendar year. This portion of Form 200 must be posted in a conspicuous place (where notices to employees are normally posted) no later than

February 1st, and must remain posted for 30 consecutive calendar days. This posting provides employees and managers at the local level with a picture of their own achievements or problems in safety and health.

OSHA Form 101. This is a supplementary record for more detailed information on individual accidents. This record, too, must be completed and available for inspection no more than six working days after receiving information of a recordable case.

Forms 200 and 101 are for records only. They are not submitted to OSHA. They must be accurately recorded and kept up-to-date at all times, however. They must be available at the work establishment for inspection, and they must be kept at the work establishment for five years following the end of the year which they relate to.

OSHA Form 103. Unlike the other two forms described, Form 103 does not apply to all employers. In the early stages, only a sample of about 250,000 employers are being asked to submit this annual report each year. These are firms falling within a statistical sample determined by the Bureau of Labor Statistics (BLS). The information received is used by the BLS for the furtherance of its statistical studies. Any employer receiving Form 103 from the BLS must complete it and return it to the Bureau.

Other Requirements

In addition to complying with specified standards and keeping records, every employer covered by the Act must:

1. Keep a careful record of all employee exposure to potentially toxic materials or harmful physical agents that are required to be monitored. Employees must be able to observe both the monitoring and the records. In the event that any employee has been exposed to these materials or agents in concentrations beyond those established by the standards, he will be notified immediately.
2. Post a notice informing employees of their protection and obligations under the Act. Copies of this poster are available on request from OSHA regional and area offices. He must also have available to all employees a description of the standards relative to the particular establishment. These standards should be available in the production manager's office to any employee who wishes to inspect them.
3. Post any citation issued for a violation. This citation must be prominently displayed at the place where the violation occurred.
4. Notify employees of the application for a variance from the standards. This must be posted on the bulletin board and a copy delivered to authorized employee representatives.

False statements by an employer or by employees may result in a $10,000 fine or six months imprisonment, or both. In addition, any failure to post notices as required will result in civil penalties for the employer.

Variances

In view of the fact that standards are still being established in industries and processes where none have existed up to now, employers may, under certain circumstances, ask for variances from the standards eventually promulgated. These are:

1. *Permanent Variance*—The employer must prove that all the conditions, operations and processes which he is using will be as safe and healthful as those under the standard.
2. *Temporary Variance*—Interim variances from a standard for as long as two years may be allowed if the employer can prove that, although he is doing everything possible to protect his employees from the dangers covered in the standard, he is unable to effect immediate compliance, either because qualified personnel or materials are not available, or because he cannot complete the necessary physical changes in time, but is making every effort toward speedy compliance.

   Notice to the employees that a temporary variance is being requested must be posted so that they may reply if so inclined.
3. *Experimental Variance*—In cases where an employer wishes to experiment to validate new techniques for safeguarding employees, a variance may be granted upon approval of the experiment contemplated.
4. *National Defense Variance*—Certain variances may be granted in order to facilitate or avoid impeding the national defense effort.

   Predictions are that variances will not be widely granted—and then only after the most thorough examination of the existing conditions and an evaluation of the proposed operation.

Inspection

In order for OSHA to be effective, there must be some inspection for violations of standards of health and safety, and a means whereby these violations will be mandatorily corrected. This is particularly important where legislation is so wide-ranging and, as in this case, where a new concept of standards has been established.

Violations stemming from ignorance, misunderstanding, oversight, or willful non-compliance can be expected, especially when it is considered that these regulations apply to millions of workplaces that have never before been obliged to meet any—or, at best, the barest minimum—safety standards.

The provisions of the law regarding inspection for the purpose of identifying violations and enforcing compliance, although modified in one respect by a recent Supreme Court decision (see below), are stringent. They allow few opportunities for compromise. The law is designed to eliminate, insofar as possible, all possible dangers to health or safety that exist under present standards, and to collect data to revise, update and tighten those standards in future years.

Priorities for Inspection. Certain inspections can be anticipated according to priorities for their scheduling. These priorities are:

1. Catastrophes and incidents resulting in any fatalities or five or more hospital cases.
2. Employee complaints.
3. High injury and health hazard industries.
4. Routine random inspections.

The reason for the first priority is obvious. In the second category, whenever an employee or group of employees file a formal, written complaint specifying circumstances that would indicate a violation, there must be an investigation. If none is made, or if no citation is issued after an investigation, the government is required by law to respond and to explain its position fully in writing. It can be anticipated that this may well become the principal motivation for many inspections.

At the third level of priority are those companies with an accident or illness rate considerably higher than its industry norm. In addition to these companies, most, or perhaps all, of those that deal with lead, carbon monoxide, asbestos, or similar health-hazardous materials can expect inspections. Statistics may begin to identify other materials as hazardous from time to time. Companies dealing with such materials will then come under Labor's examination.

Nature of the Inspection. In order to accomplish the goals of OSHA, Department of Labor compliance officers are authorized to examine machinery, processes, materials, equipment, and the overall condition of the work establishment. They may question any employer, employee, contractor, or sub-contractor, either in public or in private at the sole discretion of the inspector. Interference with the inspector in the performance of any examination he feels necessary is subject to stiff penalties.

It is significant to note that these inspections were designed to view the operation of the facility in its day-to-day condition. For this reason, during most of OSHA's first decade, it has been the practice to make most inspections *unannounced*. The obvious purpose of surprise inspections is to prevent employers from correcting violations or performing temporary cosmetic routines for the benefit of a pre-announced visit by a compliance officer. It remains to be seen how this aspect of inspections will be affected by the Supremee Court's 1978 ruling that the Occupational Safety and Health Administration could not inspect businesses without first obtaining warrants. (See discussion below.)

The inspection itself is characterized by extreme openness at all times. Before it begins, the inspector and the employer will meet to establish the compliance officer's identity and to discuss the purpose and scope of the inspection. At this time the officer will show copies of any complaints his office has received. (Any employee complaint must be submitted to the employer at the time of the inspection. However, the complainant's name may be blocked out at his request.)

Both employer and employee representatives may accompany the inspector on his rounds. If employees have filed a complaint, they may employ the services of technicians or experts (at their own expense) to represent them during the inspection. Employees may designate how many of their group will walk through the facility, subject to the inspector's agreement that the designated employees are necessary to a thorough and open examination. (He will usually keep the party down to a minimum.) Employee time off from regular duties to accompany the inspection party is strictly an intramural problem. The government has made no provisions requiring that the employer pay, or not pay, for the time so spent.

The employer should describe the safety practices of the company to the inspector, and he should specify the voluntary efforts that have been undertaken to conform to the standards. Whatever else he does, the employer must note everything that occurs during the inspection in order to analyze the validity of the compliance officer's views and to determine whether or not to dispute citations and penalties.

Employer Cooperation. Until recently, it was the position and practice of the OSHA administration that no employer could refuse entrance to an OSHA inspector unless there was good—and probably extraordinary—reason for the refusal. If denied entrance or interfered with in any way, the inspector could apply for an inspection warrant to conduct his examination under the authority of that instrument, but warrants were seldom required.

The right of OSHA inspectors (and comparable officers in state programs) to search private commercial premises without a warrant was challenged in 1975 in what became a landmark case. The president of a Pocatello, Idaho, heating, plumbing and electrical installation firm, Barlow's Inc., refused to allow an OSHA inspector onto his premises without a warrant on two occasions. He then sued, claiming that the Fourth Amendment precluded warrantless searches. His contention was upheld by the Idaho 9th District Court. That decision was appealed.

On May 23, 1978 in *Marshall vs. Barlow's* 98 S.Ct. Rptr. 1816 (1978) the U.S. Supreme Court ruled that the Constitutional protection against unreasonable searches protects commercial buildings as well as private homes. The Court held that officials of the OSHA administration cannot inspect businesses without first obtaining warrants, except for a few specialized industries "with a long history of close governmental supervision" such as firearms and liquor industries. The warrant requirement can be lifted when the businessman consents to the inspection.

The impact of this ruling in the long run is difficult to predict with accuracy. In some cases it will mean delay; in others, an added complication in carrying out OSHA's mandate to uphold and enforce safety standards. However, it is not necessary for OSHA inspectors to show "probable cause" to obtain a warrant, as it is in criminal cases. The inspectors are merely required to show that the proposed inspection is part of a general administrative plan to enforce the safety laws. Since this is, in fact, the general purpose of the inspection procedure, it seems probable that in most cases, the issuing of warrants will be essentially a formality. The inspector will simply obtain a warrant prior to arrival at the workplace and proceed with the inspection as in the past. This would particularly be true of the first three categories of inspection priority—accidents resulting in any fatality or multiple hospital cases, employee complaints, and high hazard industries.

Violations

These inspections can be relatively routine or traumatic for the employer, depending on how he has prepared his facility to comply with OSHA. He may have decided that the odds for an inspection were very small—after all, the fiscal 1974 budget called for 80,000 investigations of complaints in the nation's 5,000,000 covered establishments, or only 1.6 percent. He may have been unaware of some of the problems in his plant, or he may have failed to fully understand the provisions of the Act. In any case, such an employer could find himself in great difficulty if he is inspected. Violations can carry penalties of up to $10,000 or six months in jail or both.

Violations of specified safety standards must be promptly cited in writing by the inspecting officer. He must, at the same time, fix a reasonable period for abatement or correction of the condition creating the violation. Employers and employees may appeal the length of time allowed for the abatement by application to the area director.

Violations are classified as follows:

1. *Serious Violation*—In any situation where there is a reasonable possibility that serious physical harm or death could result from the conditions or practices in operation, a serious violation exists. Such a citation provides for a mandatory fine of up to $1,000 for each violation. An employer can be excused from the penalty (though not the obligation to correct the condition) if he is able to establish that he could not have known of the condition after a conscientious effort on his part.

2. *Non-Serious Violation*—Cases where violation of a standard would not cause serious injury but could have an effect on the health or safety of employees are adjudged non-serious violations. These will be evaluated in terms of the number of employees exposed to the hazard, or the seriousness or nature of the injury or illness that might result. The number of incidents of violation would also be taken into account.
3. *Willful Violation*—If an employer consciously and intentionally engages in an operation or practice which he knows to be a violation, he may be cited for a willful violation. It is important to note that the commission of the violation need not be established as malicious or hostile. It need only be proved that it was voluntary and intentional, as opposed to accidental or even negligent. It is possible that a single offense might not be judged "willful." Repeated offenses in these circumstances almost certainly would be.

   Willful violations may be penalized by a fine of not more than $10,000 for each violation. If death to an employee results from such a violation, there could be a six-month jail sentence in addition to the fine.
4. If several instances of a violation of the same standard exist, they will be regarded as one violation. The number of such instances will, however, have an effect on the penalty.
5. Generally speaking, violations will be cited only when they are actually observed by the inspector. This will not apply in cases where it can be established that a violation existed and contributed to an accident.
6. There is no provision in the Act for the issuance of citations or the levying of penalties on employees who fail to comply with standards relevant to their own conduct. Even though compliance to rules, standards and orders is required of each employee, wherever there is non-compliance it is the employer who is held responsible and is subject to citation. If, for example, an employee refuses to wear protective glasses, the employer may be cited. It is essential, therefore, that employers make compliance with all safety and health regulations and standards under OSHA a condition of employment.
7. Circumstances frequently exist where more than one employer and his employees are engaged in the same or a related job or area. Responsibility for violations in such a situation is frequently very difficult to assign. However, no employer is relieved at any time of the responsibility for the safety of his own employees.

   If an employer creates or permits a violation, and that violation affects either his own or other employees in the same area, he will be cited. On the other hand, if an employer knows, or should know, of an existing violation and still permits his employees to work, he can be cited, even though he had no direct responsibility for creating the violating condition.

This means that employer "A" could be cited for creating or permitting a violation, while employer "B" is similarly cited, not for any responsibility for the violation but because, in effect, he overlooked the danger it might create.

Of course, if two or more employers jointly create or permit a violation, both will be cited.

8. *De Minimis Violation*—In situations where a violation would have no probable effect on health or safety, a notice of *de minimis* violation will be issued without penalty. This notice will be mailed to the employer for his information, but he is neither required to post it nor to report corrective action taken, since any such action in this case would be at his own discretion.

Except for the *de minimis* notice (which is not classified as a citation), all citations must be prominently posted at or as near as possible to the place where the violation occurred. They must remain posted for a period of three days or until the condition is corrected, whichever is longer. The employer may post such comments as he wishes to make alongside the posted citation.

Citations or penalties may be appealed by applying to the Secretary of Labor within 15 working days of receipt of the notice. The cited employer must also, within this period, notify the area director that he wishes to appeal. If he does not appeal within this time period, the citation and penalty become final and cannot be reviewed by any court or agency.

Employees who feel the assigned abatement period is excessive or unreasonable may file with the Secretary within a 15-day period. The OSHA Review Commission will conduct a hearing into any such complaints.

Corrective Action

After a citation has been issued and become final, and the abatement period has been established, the employer must take corrective action. He has no alternatives. At this point there is no avenue of review or appeal. If he fails to abate within the prescribed period of time, he may be assessed a penalty of not more than $1,000 for each day that the violation continues beyond the time allowed.

If an inspector should discover a condition of such gravity and immediacy that it could cause death or serious injury before it could be handled through routine procedures, he must immediately notify affected employees and the employer. If corrective action is not immediately taken, the Secretary of Labor may petition the U.S. District Court for a restraining order. This order may specify and require immediate remedial action. It may also prohibit employees or others, except those involved in correcting the condition, from entering the threatened area.

Classes of Injuries and Illness

Any illness, injury, or death must be entered on the appropriate records and, in some cases, reported. Since this record will be inspected, it is important that procedures be established for handling this recording function efficiently. In all cases the employer should establish for himself (a) that there was an injury, (b) whether it was work-related, and (c) whether it should be reported.

The classes of injuries and illnesses can be broken down into 6 categories.

Fatalities. Any death resulting from on-the-job or job-related accidents or illnesses must be reported within 48 hours to the Area Director. It may be reported by phone or telegram if that would expedite the matter. The date of death must be entered in the log and in the annual summary.

Multiple Injury. All accidents resulting in the hospitalization of five or more employees must be reported within 48 hours to the Area Director.

Lost Work Days. Any injury or work-connected illness which results in lost work days must be recorded. This computation is not confined to those work days when the employee is bedridden or recuperating at home. It also includes days when the employee is assigned to a temporary job because of his condition, days when, although assigned to his regular job, he cannot perform all its functions, and days on which he is unable to put in full time on his regular job.

It is the employer's responsibility to determined the status of the employee with regard to his ability to perform in whole or in part. In many cases, he may wish to be advised by a physician. Each case should be investigated to insure the accuracy of its recording in the correct category.

It is sometimes difficult to determine when days are no longer lost as a result of accident or illness. As a guide, some criteria may be helpful:

- If a worker is transferred to another job, even though he still could not perform the original, his lost days are terminated.
- If a worker continues in his old job with the work re-defined, eliminating the need for those parts he can no longer perform, lost work days are no longer recorded.
- If the worker leaves the company entirely, he is no longer employed, hence no longer officially losing work days.

There are many grey areas in this recording procedure, and in many cases the determination of the best way to file the data will not be easy. When in doubt, consult the OSHA Area Director.

Medical Treatment. Any case requiring medical treatment, but not otherwise recorded, must be filed. Medical treatment means care provided or administered by a physician. It does not include cases "which do not ordinarily require medical care" such as minor scratches, cuts, burns, blisters, and so forth, even though a physician may have provided treatment.

Employees are not required to report injuries or illnesses within any given period of time. However, the employer must record the case (if he determines it is recordable) within six working days after he learns of it.

Non-Fatal Cases Without Lost Work Days. Any illness or injury must be reported if it results in termination of employment, loss of consciousness, any restrictions on work or motion, or transfer to another job.

Diagnosed Occupational Illness. Any illness diagnosed as "occupational" must be recorded.

Occupational illness is an abnormal condition, other than one resulting from injury, caused by any kind of exposure to those factors attendant to an occupation. These are normally chronic illnesses caused by contact, inhalation, absorption or ingestion. Categories of such illnesses are:

1. Occupational skin diseases or disorders.
2. Dust diseases of the lungs.
3. Respiratory conditions due to toxic agents.
4. Poisoning or systemic effects of toxic materials.
5. Disorders due to physical agents other than toxic materials. These could include heat stroke, frostbite, effects of ionizing radiation such as X-ray, effects of non-ionizing radiation such as welding flash or microwaves.
6. Disorders due to repeated trauma. These might be noise-induced hearing loss or conditions caused by repeated motion, vibration or pressure.
7. All other occupational diseases, including anthrax, brucellosis (undulant fever), infectious hepatitis, food poisoning.

Because it is difficult to analyze the accident and illness experience among the varied activities undertaken by many (especially larger) companies when records are kept company wide, and because it is difficult to analyze these rates by size of establishment when the records are handled by central offices, employers are directed to keep these records *at the lowest level of operations* at the work site or establishment where the workers are actually employed.

This has a practical necessity, since these records must be available at the work site for examination by the compliance officer.

PLAN FOR COMPLIANCE

The importance of OSHA to every business cannot be overstated. This is an area in which all management is concerned from one viewpoint or another. From design to accounting, from research and development, to merchandising, no part of the company remains untouched by OSHA.

Since it impinges on so much of company policy, the responsible element in an OSHA compliance program should be a key member of management. He must have upper-echelon authority, since he will frequently be called upon to cut through interdepartmental disputes. The safety professionals are essential to the program, but they may no longer administer it. That supervision will come from the highest level of management.

In general, any plan for OSHA compliance should direct the program supervisor and his staff to:

1. Become familiar with the applicable law and what it requires.
2. Create a very specific plan for required record-keeping as specified.
3. Get a *Federal Register** and learn how to use it.
4. Locate violations within the facility.
5. Find administrative violations.
6. Develop an extensive corrective action plan and program.
7. Develop a system using established management and supervisory methods that ensures continued compliance.

SAFETY AND LOSS CONTROL

With the advent of OSHA, the focus of attention on safety in the workplace created many new attitudes about the place of loss control within the organization. Many companies that had, at best, paid lip service to the concepts of safety that are commonplace today came to see that safety, like security, was good business and that a well managed loss control program would produce gratifying savings in a potentially costly area of company operations. But it must be noted that a well-managed safety program goes well beyond simply complying with applicable OSHA standards.

For all of its progressiveness in extending mandated safety conditions into tens of thousands of companies hitherto untouched by any safety regulation, OSHA has, in some respects, taken safety administration a step backwards, in that its focus is almost exclusively on unsafe or unsanitary conditions. For some years, the safety professionals agreed that unsafe *acts* were the major cause of accidents or incidents.

**Federal Registers* are available at any of the OSHA field offices. One copy is free. Additional copies are available at: Superintendent of Documents, Government Printing Office, Washington, D.C. 20402.

H.W. Heinrich, an outstanding pioneer in safety studies, held that unsafe *acts* caused 85 percent of all accidents, unsafe *conditions* caused the remaining 15 percent. Therefore, if these acts could be modified, the accidents would be sharply reduced. Today, safety supervisors agree that unsafe acts are the principal villain and that the system's approach to safety is the only way to really control losses. It is necessary, however, for management to get the system together and implement a strong, active program for it to be effective. Safety problems are caused, they don't just happen, and each one of these problems can be identified and controlled.

SETTING UP THE PROGRAM

In setting up a loss control program, it is important to look at the concept of loss in the broadest sense. While the effort is thought of purely as a "safety" program, the conclusion is that it concerns itself only with accidents, or even more narrowly, with accidents resulting in injury to a person or persons. To limit the program in this way would be to lose much of its value. Frank E. Bird, Jr., in his informative book, *Management Guide to Loss Control*[50], refers to any undesired or unwanted event that degrades (or could degrade) the efficiency of the business operation as an "incident". Thus, the field of view is broadened immeasurably. Incidents could be anything from production problems to bad inventory control; from serious injury to a breakdown in quality control. An accident, on the other hand, is an undesired event resulting in physical harm to a person or damage to property. Thus, an accident is an incident, but an incident is not always an accident. The distinction is important, especially in view of a 1969 study of industrial accidents undertaken by the Insurance Company of North America. In this study, 1,753,498 accidents reported by 297 cooperating companies were analyzed. These companies represented 21 different industrial groups, employing 1,750,000 employees who worked over three billion man-hours during the period under the study. This massive study indicated a ratio of six hundred no-damage or near-miss incidents to thirty damage-to-property, to ten minor injuries, to one serious or disabling injury. As Bird points out, property damage accidents cost billions of dollars a year, and yet they are frequently reported erroneously as "near-miss" accidents. Even though property damage resulted, the absence of any personal injury in the case removed it from the category of accident in the view of the reporting agency. This is a throwback to earlier attitudes which related the term "accident" to injury only.

Accidents, by our definition, refer to property damage as well, and in aggregate, can amount to substantial costs for the company that fails to keep these accidents under control. In fact, an effective loss control program can be an organization's best money-maker when it can be seen that the actual cost of accidents may be anywhere from six to 50 times as much as the money recovered from insurance. Uninsured costs in building damage, production damage, wages to the injured for lost time, clerical costs, cost of training new workers, supervisors, extra time, all mount up. By controlling such incidents, the profit picture is immeasurably improved. In a company operating at a four percent profit margin, the sales department would have to generate sales of $1,250,000 just to compensate for an annual loss of $50,000 in incidents.

ASSESSMENT

In order to set up for acceptable performance or to take action to bring a facility up to an acceptable level of loss control effectiveness, an assessment of the situation is necessary. This can be done in two parts and, preferably, by one well-qualified man. The team approach may be used, but it is usually not as effective in the long run as the one-man approach.

The first order of business is an attitude survey. This consists of a private, one-on-one interview with all line supervisors from the facility manager to various first-line supervisors or foremen. Questions directed at each individual should elicit his attitudes about current safety conditions, the need for a safety program, his attitudes about safety management, generally, and his feelings about some of the techniques (or absence of techniques) in the loss control effort. Questions might be along the lines of:

1. Who do you think is responsible for safety?
2. Does a sloppy loss control program affect your job? How?
3. How would you improve the safety record of this facility?
4. How can top management improve the safety record of this facility?
5. Have you done everything in the past six months to improve the safety of the facility? Your crew?
6. How much authority do you have to correct unsafe conditions?
7. What supervisory safety training have you had?

After the questionnaire has been drawn up and the interviews have been conducted with supervisors, a tabulation of responses should be very revealing. They should indicate the management levels where deficiencies exist, and they should point up existing problems both in plant safety and in the program in general.

After the attitude survey, the assessment should undertake a review of all accident/incident records of the past three years. This should be classified by type such as burns, bruises, broken bones. From this, it will be easy to see the types of accidents that have been occurring and perhaps pinpoint an area or process or condition that is particularly unsafe. When this list has been drawn up, it will be necessary to assign cause and responsibility to each accident. There will usually be several causes in each case. In developing the accident-cause correlation chart, assign causes as: management responsibility, supervisor responsibility, employee error, mechanical design, mechanical failure.

The next step in the assessment is to determine those accidents caused by lack of personal protective equipment. These accidents are eye injuries and foot injuries. It must be determined whether the problems arise from lack of protective equipment or an unwillingness to use it, and corrective action must be taken.

Next determine accidents by job title.

Finally, list accidents under each supervisor.

These last two categories should immediately suggest corrective action, whether it be changing job specifications or educating supervisors.

FINDING THE CAUSE OF ACCIDENTS

The cause of accidents should be determined before they occur. Since accidents are caused, the conditions that cause can be known and controlled. It is, therefore, of the greatest importance that management deal vigorously with what *can* cause an accident. Unsafe acts and unsafe conditions will ultimately cause accidents if they are allowed to continue.

Unsafe acts will be discovered and corrected only when the immediate supervisor is alert to the problem. He must set up a system for closely observing all workers, especially those in hazardous jobs, in operation. To do this, he must have a job safety analysis at his disposal. This analysis breaks each job down into component parts, and each part is studied for the hazards it may present.

Unsafe conditions are uncovered by constant inspections. Such conditions do not disappear entirely because they were taken care of once. Unsafe conditions are continuously created by the operation of the facility. Normal wear and tear, careless housekeeping, initial bad design, or simply the deterioration that results from inadequate maintenance caused by a cost-cutting management all create unsafe conditions which have a high potential loss factor. Early discovery of unsafe conditions is essential to good loss control, and the procedure is simply inspections, inspections, inspections.

THE OPERATION

Whatever the overall integration of safety and the security operation, the safety function can operate in any one of three modes:

1. In a staff capacity, where its experts offer advice, make recommendations to upper echelon management and develop policy for management approval. Line supervisors bear full responsibility for safety in their areas.
2. The safety department is both staff and line in that it performs all staff functions as above and it will also help out on especially hazardous jobs. It will hold some safety meetings and training sessions.
3. The department holds all safety meetings, training sessions, accident in-

vestigations and actively perform in all areas of safety. A good case can be made for operating in any one of the three methods, though it would appear that the combination of staff and line method is generally the most effective.

Management Leadership

Management's attitude toward safety filters down through the entire company. Top management's concern will be reflected in that of the supervisors; in turn, the supervisors' attention to safety will affect the individual employee's attitude.

Mangement is responsible not only for basic policy of providing a work environment free of hazards—which should be embodied in an executive policy statement—but also for active leadership. This can be expressed by holding subordinates responsible for accident prevention, and in such visible ways as plant tours, letters to employees, safety meetings, posters, prompt accident investigations, and personal example. (In a "hard hat" area, the president of the company should also put on a hard hat.)

General safety rules must be established and published, as in the employee handbook or manual. Safety rules should be continually reviewed and updated.

Assignment of Responsibility

Responsibility for the safety program should be clear and personal. In the small company it may rest on the owner, himself, if he is the acting supervisor. Otherwise, it will generally be an added responsibility of the supervisors in companies with fewer than 100 employees.

In larger companies, safety should be an assigned responsibility of a ranking member of management. He may delegate the authority to oversee the program to a safety director (who may be called the safety professional, safety engineer or safety supervisor, depending on his qualifications and the nature of the operation). In many companies safety is a responsibility of the security director, who will often have a safety specialist working under him. (In virtually all circumstances, there is a close relationship between safety and security.

Identification and Control of Hazards

OSHA standards (or equivalent state standards) provide the baseline for the company safety program. A bewildering catalog of standards has already developed (Cal/OSHA, in California, has more than 6,000 standards), and new ones are constantly being added. Checklists (available from OSHA, the National Safety Council and other sources) can provide the starting point for detailed inspections designed to identify hazards. The confusion that might accompany a consideration of all the standards begins to sort itself out when inspections zero in only on those that apply to specific operations and conditions.

A safety program should include *periodic* inspections scheduled at regular intervals. Figure 10-1 is an example of a monthly checklist for inspection. In addition, looking for safety hazards and violations should be part of the day-to-day activity of both safety professionals and security personnel. Some hazards which might be present in any business facility are shown in Figure 10-2.

Training

All employees must be initially and periodically trained both in general safety principles and in safe work practices in their specific jobs. Safety rules, such as the wearing of protective clothing (gloves, headgear, respirators, shoes, eye protection) should be clearly explained and promptly enforced. The importance the company attaches to safety should particularly be emphasized in new employee training, but it is also important to pay attention to regular employees—including the ''old timers'' who did not grow up with safety awareness as part of their conditioning.

In addition to the above, there are specific training requirements in the OSHA standards (such as those involving the operation of certain types of equipment). Employer and employees should be aware of those standards that apply in the specific workplace.

Record Keeping

OSHA injury/illness record keeping requirements have already been reviewed, but the basic steps deserve repetition:

1. Obtain a report on every injury requiring medical care.
2. Record each injury (on OSHA Form 200).
3. Prepare a supplementary record of occupational injuries and illnesses, either on OSHA Form 101 or workmen's compensation reports giving the same information.
4. Prepare an annual summary (also part of Form 200), and post it for 30 days beginning no later than February 1.
5. Maintain these records for at least five years.

With or without OSHA requirements, the value of accurate and cumulative records of accidents or occupational illness has become increasingly obvious in recent

Monthly Safety Check

Dept. ________ Date ________
Supervisor ________________
Indicate discrepancy by ☒

General Area	
Floor Condition	
Special Purpose Flooring	
Aisle, Clearance/Markings	
Floor Openings, Require Safeguards	
Railings, Stairs Temp./Perm.	
Dock Board (Bridge Plates)	
Piping (Water-Steam-Air)	
Wall Damage	
Ventilation	
Other	
Illumination—Wiring	
Unnecessary/Improper Use	
Lights on During Shutdown	
Frayed/Defective Wiring	
Overloading Circuits	
Machinery Not Grounded	
Hazardous Location	
Other	
Housekeeping	
Floors	
Machines	
Break Area/Latrines	
Waste Disposal	
Vending Machines/Food Protection	
Rodent, Insect, Vermin Control	
Vehicles	
Unauthorized Use	
Operating Defective Vehicle	
Reckless/Speeding Operation	
Failure to Obey Traffic Rules	
Other	
Tools	
Power Tool Wiring	
Condition of Hand Tools	
Safe Storage	
Other	

First Aid	
First Aid Kits	
Stretchers, Fire Blankets, Oxygen	
Fire Protection	
Fire Hoses Hung Properly	
Extinguisher Charged/Proper Location	
Access to Fire Equipment	
Exit Lights/Doors/Signs	
Other	
Security	
Doors/Windows Etc. Secured When Required	
Alarm Operation	
Dept. Shut Down Security	
Equip. Secured	
Unauthorized Personnel	
Other	
Machinery	
Unattended Machines Operating	
Emergency Stops not Operational	
Platforms/Ladders/Catwalks	
Instructions to Operate/Stop Posted	
Maint. Being Performed on Machines in Operation	
Guards in Place	
Pinch Points	
Material Storage	
Hazardous & Flammable Material Not Stored Properly	
Improper Stacking/Loading/Securing	
Improper Lighting, Warning Signs, Ventilation	
Other	

Figure 10-1. Sample monthly safety checklist.

1. **Floors, aisles, stairs and walkways**
   - Oil spills or other slippery substances which might result in an injury–producing fall.
   - Litter, obscuring hazards such as electrical floor plugs, projecting material, or material which might contribute to the fueling of a fire.
   - Electrical wire, cable, pipes, or other objects, crossing aisles which are not clearly marked or properly covered.
   - Stairways which are too steep, have no non-skid floor covering, Inadequate or non-existent railings, or those which are in a poor state of repair.
   - Overhead walkways which have inadequate railings, are not covered with non-skid material, or which are in a poor state of repair.
   - Walks and aisles which are exposed to the elements and have not been cleared of snow or ice, which are slippery when wet or which are in a poor state of repair.

2. **Doors and emergency exits**
   - Doors that are ill fitting, stick, and which might cause a slow down during emergency evacuation.
   - Panic–type hardware which is inoperative or in a poor state of repair.
   - Doors which have been designated for emergency exit but which are locked and not equipped with panic–type hardware.
   - Doors which have been designated for emergency exit but which are blocked by equipment or by debris.
   - Missing or burned out emergency exit lights.
   - Non-existent or poorly marked routes leading to emergency exit doors.

3. **Flammable and other dangerous materials**
   - Flammable gases and liquids which are uncontrolled in areas which they might constitute a serious threat.
   - Radioactive material not properly stored or handled.
   - Paint or painting areas which are not properly secured or which are in areas that are poorly ventilated.
   - Gasoline pumping areas located dangerously close to operations which are spark producing or in which open flame is being used.

4. **Protective equipment or clothing**
   - Workmen in areas where toxic fumes are present who are not equipped with or who are not using respiratory protective apparatus.
   - Workmen involved in welding, drilling, sawing and other eye endangering occupations who have not been provided or who are not wearing protective eye covering.
   - Workmen in areas requiring the wearing of protective clothing, due to exposure to radiation or toxic chemicals, who are not using such protection.
   - Workmen engaged in the movement of heavy equipment or materials who are not wearing protective footwear.

Figure 10–2. Common safety hazards. (From Finneran, Eugene, *Security Supervision: A Handbook for Supervisors and Managers,* Woburn: Butterworth (Publishers) Inc., 1981.)

- Workmen who require prescription eyeglasses who are not provided or are not wearing safety lenses.

5. Vehicle operation and parking

- Forklifts which are not equipped with audible and visual warning devices when backing.
- Trucks which are not provided with a guide when backing into a dock or which are not properly chocked while parked.
- Speed violations by cars, trucks, lifts and other vehicles being operated within the protected area.
- Vehicles which are operated with broken, insufficient or non-existent lights during the hours of darkness.
- Vehicles which constitute a hazard due to poor maintenance procedures on brakes and other safety-related equipment.
- Vehicles which are parked in fire lanes, blocking fire lanes or blocking emergency exits.

6. Machinery maintenance and operation

- Frayed electrical wiring which might result in a short circuit or malfunction of the equipment.
- Workers who operate presses, work near or on belts, conveyors and other moving equipment who are wearing loose fitting clothing which might be caught and drag them into the equipment.
- Presses and other dangerous machinery which are not equipped with the required hand guards or with automatic shut off devices or dead man controls.

7. Welding and other flame– or spark–producing equipment

- Welding torches and spark–producing equipment being used near flammable liquid or gas storage areas or being used in the vicinity where such products are dispensed or are part of the productive process.
- The use of flame– or spark–producing equipment near wood shavings, oily machinery or where they might damage electrical wiring.

8. Miscellaneous hazards

- Medical and first aid supplies not properly stored, marked or maintained.
- Color coding of hazardous areas or materials not being accomplished or which is not uniform.
- Broken or unsafe equipment and machinery not being properly tagged with a warning of its condition.
- Electrical boxes and wiring not properly inspected or maintained, permitting them to become a hazard.
- Emergency evacuation routes and staging areas not properly marked or identified.

Figure 10–2 continued

years. A high incidence of a particular type of injury, for example, is a clear warning of a hazardous situation that requires corrective action. Failure to act on such warnings is costly to the employer, as well as the employee.

Emergency Care

Under OSHA, all businesses are required, in the absence of an infirmary or hospital in the immediate vicinity, to have a person or persons trained in first-aid available, along with first-aid supplies. Where employees are exposed to corrosive materials, procedures for drenching or flushing of the eyes and body should be provided in the work area.

Procedures should be established for handling injury accidents without confusion or delay. The extent of these preparations will, of course, depend on the nature of the business and the type of hazards.

Employee Awareness and Participation

Developing safety and health awareness is one of the primary goals of OSHA. Active steps by management, such as those suggested above, are essential to involve all employees in the need to create a safe work environment.

Safety awareness has an added benefit for both the employer and employees in that it tends to carry over into a concern for off-the-job safety. Accidents away from the work environment account for more than half of all injuries, and the ratio of deaths is three-to-one higher in off-the-job accidents. Carrying over safety practices from the job to activities away from it is an aspect of safety training that is receiving increasing emphasis from today's safety professionals.

A Hazardous Materials Program

In addition to the seven steps in safety planning outlined above, particular types of businesses dealing with hazardous substances should have a hazardous materials program. As a minimum, it is necessary to:

1. Identify what hazardous materials you have and where.
2. Know how to respond to an accident involving hazardous materials.
3. Know how to deal with spills.
4. Set up appropriate safeguards.
5. Train employees in dealing with hazardous materials.

Materials Safety Data Sheets are designed specifically to help identify the nature of potential hazards. These Data Sheets, obtainable from vendors of hazardous materials or equipment, include such information as chemical composition, health hazard rating, protective gear needed, reactivity data, fire data (such as flash point), disposal procedures, and Threshold Limit Value (or TLV, which is the amount of exposure an individual can have to a specific chemical).

CONCLUSION

The impact of the passage of OSHA has been tremendous. In many ways it represents a social revolution that cuts across every element of society, affecting all workers whether they are covered immediately or not. Yet the basic message of OSHA is not really very unusual at all. In its long and careful way, it says that every worker has the right to work in a safe and sanitary environment, that he has a right to be protected from injuries and illness caused by correctible hazardous conditions, and that he is obligated to cooperate in his own protection.

The threat of accident or illness has never improved any company's operation. Most employers have lived with a certain level of risk because they did not think anything could or should be done about it. That is no longer possible.

If the traditional question—"Do you want it quick or do you want it safe?—does not disappear entirely, it will have a different dimension—and require a different answer. What yesterday might have been thought of as an eminently acceptable risk, may today be a very dubious one.

From now on employers, grudgingly or willingly, will be testing the theory that a business will improve substantially when accidents and the threat of them are reduced or eliminated. Technical improvements frequently result from the obligatory overhaul of a production line or process. Improved morale leads to increased production and work efficiency. Certainly there should be a saving in insurance costs, and a substantial reduction of costly and damaging lawsuits.

Today, every employer is in the safety business up to his neck. If he doesn't swim, he will surely sink like a stone.

REVIEW QUESTIONS

1. Under what circumstances are states not required to follow OSHA standards?
2. Describe the content and purpose of each of the OSHA forms?
3. In order, what are the priorities for OSHA inspections?
4. How are violations classified by OSHA?
5. According to OSHA, what are the classes of illnesses or injuries that must be recorded?

Chapter 11

INTERNAL THEFT CONTROLS

It is sad, but true, that virtually every company will suffer losses from internal theft—and these losses can be enormous. "A well-informed security superintendent of a nationwide chain of retail stores recently estimated that it takes between forty and fifty shoplifting incidents to equal the annual loss caused by one dishonest individual inside the organization."[35] The American Management Association, in a 1977 estimate, put employee theft in the "best guess" range of from $5 to $10 billion annually. Obviously, a problem of such magnitude must be vigorously dealt with.

THE DISHONEST EMPLOYEE

Unfortunately, there is no sure way by which potentially dishonest employees can be recognized. Proper screening procedures can eliminate applicants with an unsavory past or those who seem unstable and, therefore, possibly untrustworthy. There are even tests that purport to measure an applicant's honesty index. But, tests and employee screening can only indicate potential difficulties. They can screen out the most obvious risks, but they can never truly vouch for the performance of any prospective employee under the circumstances of new employment or under changes that may come about in his life apart from his job.

WHY EMPLOYEES STEAL

Since there is no fail-safe technique for recognizing the potentially dishonest employee on sight, it is important to try to gain some insight into the reasons employees may steal. If some rule of thumb can be developed which will help to identify the patterns of the potential thief, it would provide some warning for an alert manager.

There is no simple answer to the question of why heretofore honest men and women suddenly start to steal from their employer. The mental and emotional process that leads to this is complex, and motivation may come from any number of sources.

Some employees steal because of resentment of real or imagined injustice, which they blame on management indifference or malevolence. Some feel that they must maintain status and steal to augment their income after financial problems. Some may steal simply to tide themselves over in a genuine emergency. They rationalize the theft by assuring themselves that they will return the money after the current problem is solved. Some simply want to indulge themselves, and many, strangely enough, steal to help others.

THE OPPORTUNITY TO STEAL

A high percentage of employee thefts begin with the opportunities that are regularly presented to them. If systems are lax or supervision is indifferent, the temptation to steal that which is improperly secured or unaccountable may be too much to resist by any but the most resolute employees.

Many experts agree that the fear of instant discovery is the most important force in deterring internal theft. When that likelihood is eliminated, theft is bound to follow. Threats of dismissal or prosecution of any employee found stealing are never as effective as the conviction that management supervision is such that discovery will almost certainly follow any theft.

DANGER SIGNS

The stated, and even the more deeply seated, root causes of such thefts are many and varied, but there are certain signs that can indicate that a hazard exists.

The conspicuous consumer is perhaps the most obvious and the easiest risk to identify. An employee who habitually or suddenly is seen in flashy cars, flashy clothes and with flashy companions is a man to watch. His habits are visibly extravagant and he gives the impression of having found a money tree growing in his backyard. Even though he may not be stealing to support his expensive tastes, he is likely to run into financial difficulties at his pace. Then he may feel obliged to look beyond his salary check to support his life style.

Any employee who shows a pattern of financial irresponsibility is a potential risk. Many people are simply incapable of handling their own affairs. They may do their job with great skill and efficiency, but they are in constant difficulty in their private lives. These people are not necessarily compulsive spenders, nor do they necessarily have expensive tastes. (They probably live quite modestly, since they have never been able to manage their affairs effectively enough to live otherwise.) They are simply people unable to come to grips with their own economic realities.

Garnishments or inquiries by creditors may identify such an employee for you. If there seems a reason to make one, a credit check might reveal the tangled state of affairs.

The employee caught in a genuine financial squeeze is also a possible problem. If he has been hit with financial demands from an illness in the family or possibly a heavy tax lien, he may find the pressures too great to bear. If such a situation comes to the attention of management, counseling is in order. Many companies maintain funds which are designated to make low interest loans in such cases. Alternatively, some arrangement might be worked out through a credit union. In any event, an employee in such extremities needs help fast. He should get it both as a humane response to his needs and as a means of protecting company assets.

In addition to these general categories, there are specific danger signals which should be noted:

- Gambling on or off premises.
- Excessive drinking or signs of alcoholism.
- Obvious extravagance.
- Persistent borrowing.
- Requests for advances.
- Bouncing personal checks or checks post-dated.

What Employees Steal

The employee thief will take anything that he may consider useful to him or that has a resale value. He can get at the company funds in many ways—directly or indirectly—through collusion with vendors, collusion with outside thieves or hijackers, fake invoices, receipting for goods never received, falsifying inventories, payroll padding, false certification of overtime, padded expense accounts, cash register manipulation, overcharging, undercharging, or simply by gaining access to a cashbox.

This is only a sample of the kinds of attack that are made on company assets using the systems set up for the operation of the business. It is in these areas that the greatest losses can occur, since they are frequently based on a systematic looting of the goods and services in which the company deals, and the attendant operational cash flow.

Significant losses do occur, however, in other, sometimes unexpected, areas. Furnishings frequently disappear. In some firms with indifferent traffic control procedures, this kind of theft can be a very real problem. Desks, chairs, paintings, rugs—all can be carried away by the enterprising employee thief.

Office supplies can be another problem if they are not properly supervised. Beyond the anticipated attrition in pencils, paper clips, note pads and rubber bands, these materials are often stolen in case lots. Many firms which buy their supplies at discount are, in fact, receiving stolen property. The market in stolen office supplies is a brisk one and becoming more so as the prices for this merchandise soar.

The office equipment market is another active one, and the inside thief is quick to respond to its needs. Typewriters always bring a good price, and with miniaturization, calculators and mini-computer units make tempting and easy targets.

Personal property is also vulnerable. Office thieves do not make fine distinctions between company property and that of their fellow workers. The company has a very real stake in this kind of theft, since personal tragedy and decline in morale follow in its wake.

Although security personnel cannot assume responsibility for losses of this nature, since they are not in a position to know about the property involved or to control its handling (and should so inform all employees), they should make every effort to apprise all employees of the threat. They should further note, from time to time, the degree of carelessness the staff displays in handling of personal property and send out reminders of the potential dangers of loss.

Methods of Theft

Since it is estimated that somewhere between seven and 10 percent of business failures annually are the result of some form of employee dishonesty, there is a very real need to examine the shapes it frequently takes. There is no way to describe every kind of theft, but some examples here may serve to give some idea of the dimensions of the problem:

1. Payroll and personnel employees collaborating to falsify records by the use of nonexistent employees or by retaining terminated employees on the payroll.
2. Padding overtime reports, part of which extra unearned pay is kicked back to the authorizing supervisor.
3. Pocketing unclaimed wages.
4. Splitting increased payroll which has been raised on checks signed in blank for use in the authorized signer's absence.
5. Maintenance personnel and contract servicemen in collusion to steal and sell office equipment.
6. Receiving clerks and truck drivers in collusion on falsification of merchandise count. Extra unaccounted merchandise is fenced.
7. Purchasing agents in collusion with vendors to falsify purchase and payment documents. The purchasing agent issues authorization for payment on goods never shipped after forging receipt of shipment.

8. Purchasing agent in collusion with vendor to pay inflated price.
9. Mailroom and supply personnel packing and mailing merchandise to themselves for resale.
10. Accounts payable personnel paying fictitious bills to an account set up for their own use.
11. Taking incoming cash without crediting the customer's account.
12. Paying creditors twice and pocketing the second check.
13. Appropriating checks made out to cash.
14. Raising the amount of checks after voucher approval or raising the amount of vouchers after their approval.
15. Pocketing small amounts from incoming payments and applying later payments on other accounts to cover shortages.
16. Removal of equipment or merchandise with trash.
17. Invoicing goods below regular price and getting a kickback from the purchaser.
18. Underringing on a cash register.
19. Issuing (and cashing) checks on returned merchandise not actually returned.
20. Forging checks, destroying them when returned with a statement from the bank, and changing cash books accordingly.

These are a few of the techniques that have been used to defraud an employer. Each one is workable. Unless a concerted effort is made to control them, they or others like them can destroy a company.

Management Failure in Loss Prevention

Many security specialists have speculated that the enormous losses in internal theft come largely from firms which have refused to believe that their employees would steal from them. As a result, their loss prevention systems are weak and ineffective.

Managers of such companies are not naive. They are aware that the criminally inclined employee exists—and in great numbers—but he exists in other companies. Such men are truly shocked when they discover how extensive employee theft can be. This doesn't mean that a high percentage of the employees are involved (though they may be). Two or three employees, given the right opportunities, can create havoc in any business in a very short time—and the great majority of them go undetected.

National crime figures do not necessarily concern each individual manager. He can shrug them off as an enormous problem that someone else must deal with, one that has no effect on his operation—and, hopefully, he is right. But he simply cannot afford to close his eyes to the potential damage that he faces. He must be convinced that one of his employees *might* steal from him. He must be persuaded that there is, at least, a *possibility* that he is exposed to embezzlement and that maybe sooner or later someone will be unable to resist the temptation to take what is so alluringly available.

It may be surprising to find that there are shrewd businessmen who are still willing to ignore the threat of internal theft and who are, therefore, unwilling to take defensive steps against it. But, the fact is they do exist in large numbers. Many security men report that the biggest problem is in convincing management that the problem is there and that steps must be taken for the protection of employer and employee alike.

There is some evidence, and more speculation, that these managers are reluctant to, in effect, declare themselves as suspicious of the employees—to make overt action implying distrust of men and women, many of whom are perhaps old and trusted members of the corporate family. Such an attitude does the manager honor but it is somewhat distorted. It fails to take into account the dynamics of all our lives. The man who today would stand firm against the most persuasive temptation, tomorrow finds himself in different circumstances—perhaps with different resultant attitudes. Whereas, today the thought of embezzlement is repugnant to him, tomorrow he might treat the notion with a different view.

The Contagion of Theft

Theft of any kind is a contagious disorder. Petty, relatively innocent pilferage by a few spreads through the facility. As more people participate, others will follow, until even the most rigid break down and join in. Pilferage becomes acceptable—even respectable. It has a general social acceptance which is reinforced by almost total peer participation. Few men make independent ethical judgments under such circumstances. In this microcosm, the act of petty pilferage is no longer viewed as unacceptable conduct. It has become not a permissible sin, but a right.

The docks of New York City were an example of this progression. Forgetting for the moment the depredations of organized crime and the climate of dishonesty that characterized that operation for so many years, even longshoremen not involved in organized theft had worked out a system all their own. For every so many cases of whisky unloaded, for example, one case went to the men. Little or no attempt was made to conceal this. It was a tradition, a right. When efforts were made to curtail the practice, labor difficulties arose. It soon became evident that certain pilferage would have to be accepted as an unwritten part of the union contract under the existing circumstances.

This is not a unique situation. The progression from limited pilferage to its acceptance as normal conduct to the status of an unwritten right has been repeated time and again. The problem is, it doesn't stop there. Ultimately, pilferage becomes serious theft, and then the real trouble starts.

Even before pilferage expands into larger operations, it presents a difficult problem to any business. Even where the amount of goods taken by any one individual is small, the aggregate can represent a significant expense. With the costs of materials, manufacture, administration and distribution rising as they are, there is simply no room for added, avoidable expenses in today's competitive markets. The business that can operate the most efficiently, and offer quality goods at lower prices because of the efficiency of its operation, will have a huge advantage in the marketplace. When so many companies are fighting for their very economic life, there is simply no room for waste—and pilferage is just that.

Moral Obligation to Control Theft

When we consider that internal theft accounts for five times the loss caused by automobile thieves, burglars and armed robbers combined, we must be impressed with the scope of the problem facing today's businessman. Fortunately, there are steps that can be taken to control internal theft. Losses can be cut to relatively insignificant amounts by setting up a program of education and control which is vigorously administered and supervised.

It is also important to observe that management has a moral obligation to its employees to protect their integrity by taking every possible step to avoid presenting open opportunities for pilferage and theft that would tempt even the staunchest of men to take advantage of the trust placed in them.

This is not to suggest that each company should assume a paternal role toward its employees and undertake their responsibilities for them. It is to suggest strongly that the company should keep its house sufficiently in order to avoid inciting employees to acts that could result in great personal tragedy as well as in damage to the company.

PERSONNEL POLICIES FOR INTERNAL SECURITY

A plan for internal security must have the whole-hearted support and involvement of management from the highest echelon. Any initial reluctance to institute such a plan must be overcome with tact and conviction. Once that has been accomplished, the nature of the commitment must be fully understood. No security program, or any other, for that matter, can be effective without management's full and continuing support.

This aspect of the security program, like every other one, must be integrated into the whole which, as frequently noted, must be a total system involving all aspects of the facility's security needs. Neither internal security nor access control nor any other part of the security program can be dealt with as an independent unit. Each must be a part of a cohesive whole in the development of a systems approach to the objective.

Personnel Screening

The objective of a program encompassing internal security is to prevent theft by employees. If all the employees were of such character that they could not bring themselves to steal, the security personnel would have little to do. If, on the other hand, thieves predominate in the mix of employees, the system will be sorely tried, if indeed it can be effective at all. Basic to its effectiveness is the cooperation of that majority of honest personnel who perform as assigned and, in so performing, refuse to initiate or collaborate in conspiracies to steal. Without this dominant group, the system is in trouble from the start.

The best place to start, then, is in the personnel office, where bad risks can be screened out on the basis of reasonable security procedures.

In some industries, especially those with high technical requirements, this can be a problem, since qualified personnel may be difficult to find. There can be resistance from the employment office to the disqualification of an otherwise qualified applicant on marginal grounds of security. Here, a job of education must be undertaken to convince those objectors that a man who may later embezzle from the company is a poor risk from many viewpoints, no matter how highly qualified he may be in the specific skills for which he has been considered.

Rejection of bad risks must be on the basis of standards that have been carefully established in cooperation with the personnel director. Once established, these standards must be met in every particular, just as proficiency standards must be met. Obviously the standards must be reviewed from time to time to avoid dealing with applicants unjustly and to avoid placing the company at a competitive disadvantage in the labor market by demanding more than is available. Even here, however, a bottom line must be drawn. After a certain point, compromises and concessions can no longer be made without inviting damage to the company.

Such a careful, selective program may add some expense to the employment procedure, but it can pay for itself in reduced losses, better people and lower turnover. And the savings in crimes that never happened, though unknowable, could be thought of as enormous.

Employment History and Reference Checking

The key to reducing internal theft lies in the quality of employees employed by the facility. The problem will not be stamped out in the hiring process, no matter how carefully and expertly the selection of hires from applicants may be. The systems of theft prevention and the program of employee motivation are ongoing efforts which must recognize that the elements of availability, susceptibility and opportunity are dynamic and in a constant state of flux. However, the point at which the problem must be taken in hand is at the beginning—in the very process of selecting the personnel who will work in the facility. And in this process, there is an immense amount of vital information about the prospective employee that can be developed by a knowledgeable screener simply by knowing what to look for in the employment application. The answers are not as obvious as they once were, and the ability to perceive and evaluate what is on the application is more important than ever now that the forms are more circumstantial than they have been in the past.

Privacy legislation of the 1970's, together with fair employment laws, has drastically changed the employment application forms in businesses across the country. What was once considered reasonable and appropriate to ask job applicants is now forbidden by law. As generally interpreted, federal law prohibits asking:

1. Marital status
2. Number of children in the family
3. The identity of persons living with the applicant
4. Whether the applicant owns or rents his home
5. Maiden name (in case of female applicants)
6. Church affiliation
7. Whether applicant's wages have ever been garnished.
8. Whether the applicant has ever been arrested.
   (In most, but not all, states, the applicant may be asked, but only if it is job-related). Convictions may always be questioned.

In some aspects, these regulations have had a stream-lining effect in eliminating irrelevant questions and confining these questions exclusively to those matters which are related to the job applied for. The subtler kinds of discrimination on the basis of age, sex, national origin, have been largely eliminated from the employment process. In making these changes to protect the applicant, state and federal law has created new dilemmas for employers and their security staffs.

Department of Justice Order 601-75 prohibits any criminal justice agency funding through non-criminal justice agencies or even to confirm or deny the existence of such information for employment purposes. The Fair Credit Reporting Act requires that a job applicant must give his written consent to any credit bureau inquiry. Many more such regulations are under consideration. In 1976, there were over 150 privacy bills in various stages in 49 states.[36]

Understandably, there is some confusion with regard to the rules governing employment screening. In spite of such confusion, the pre-employment inquiry is one of the most useful security tools available against employee dishonesty and continued profit drains that an employer can use. Therefore, he should consult legal counsel to determine those laws that relate to him in his locality, establish firm and precise policies regarding employment applications and hiring practices, and get on about the business of getting and keeping his house in order.

Generally speaking, look for and be wary of applicants who:

1. Show signs of instability in personal relations.
2. Lack job stability. The grasshopper does not make for a good job candidate.

3. Show a declining salary history, or are taking a cut in pay from the last job than the job under consideration pays.
4. Show unexplained gaps in employment history.
5. Are clearly overqualified.
6. Are unable to recall or are hazy about names of supervisors in the recent past, or who forget the address where they lived in the recent past.

If the job applied for is one involving the handling of funds, it would be advisable to get the applicant's consent to make financial inquiry through a credit bureau. Be wary if such an inquiry turns up a history of irresponsibility in financial affairs, such as living beyond one's means.

It is important that the application form ask for a chronological listing of all previous employers, among other things, in order to provide a list of firms to be contacted for information on the applicant, as well as to show continuity of career. Any gaps could indicate a jail term which had been overlooked in filling out the application. When checking with previous employers, dates on which employment started and terminated should be verified.

References submitted by the applicant must be contacted, but they are apt to be biased. After all, they were names submitted by the man to be investigated; they are not likely to be negative or hostile to his interests. It is important to contact someone—preferably an immediate supervisor—at each previous employer, and such contact should be made by phone or in person.

The usual and easiest system of contact is by letter, but this leaves much to be desired. The relative impersonality of a letter, especially one in which a form or evaluation is to be filled out, leads to impersonal and, essentially, uncommunicative answers. Since many companies, as a matter of policy, stated or implied, are reluctant to give a man a bad reference except in the most extreme circumstances, a written reply to a letter will, in some cases, even be misleading.

On the other hand, phone or personal contacts may become considerably more discursive and provide shadings in the tone of voice that can be important. Even when no further information is forthcoming, this method may indicate that a more exhaustive investigation is required.

Backgrounding

It may be desirable to get a more complete history of a prospective employee—especially in cases where sensitive financial or supervisory positions are under consideration. Professional services providing this kind of investigation involve extra expense, but in many cases it can be well worthwhile. This backgrounding, which involves a discreet investigation into the past and present activities of the applicant, can be most informative.

Since it is estimated that as much as 93 percent of all persons ***known*** to have stolen from their employers are not prosecuted, a thorough investigation is certainly justified into the background of anyone considered for a job in which he will be responsible for significant amounts of cash or goods, or will be in a management position responsible for shipping, receiving, purchasing, or paying.

It is also significant to note that there is agreement among personnel and security experts that 20 percent of a given work force is responsible for 80 percent of the personnel problems of all kinds. If backgrounding can turn up this kind of record, it is well worth the expense.

Some firms state on their application form that applicants will be bonded or fingerprinted, or must give permission to undergo a polygraph test. Any criminal with a record will probably bow out at this point, since he knows his past is likely to disqualify him.

Backgrounding is also employed to investigate employees being considered for promotion to positions of considerable sensitivity and responsibility. Such a man may have been the very model of rectitude at the time of his employment but may since have fallen upon financial reverses that threaten his life style. If a background investigation uncovers such information, a company is in a position to offer assistance if it so desires, thus relieving him of the strain and need, both of which can lead to embezzlement. Such action can boost company morale, as well as reduce the potential for theft out of desperation.

Polygraph and PSE

Another approach to full background investigation is the use of the polygraph or of the psychological stress evaluator (PSE). In states where their use is legal, these machines can be useful tools in determining the past and current record of applicants for employment or promotion to positions of considerable sensitivity.

The use of this technique is controversial, as is the discussion among practitioners of the relative merits of each instrument. However, many security people look on them as invaluable tools, and generally agree that their use in the hands of a competent, trained professional can be most constructive.

These instruments differ. The polygraph measures the response of the heartbeat or pulse, the skin's galvanic action, and respiration. Sensors picking up these responses are attached painlessly to the subject by wires that run directly between him and the machine. He is questioned by the use of control questions, irrelevant and relevant inquiries. The readout of his pulse, respiration and skin action is recorded on a paper tape in the machine, and the responses to various questions are analyzed in terms of his charted reactions, as compared with his reactions to simple test questions which establish his normal response to lying.

The PSE works in a similar way operationally, but in this case a recording is made of the subject's replies to the questions put to him. These recordings analyze and reproduce the sound wave characteristics of his voice. By running these tapes through the instrument at different speeds, different aspects of the sound wave activity can be recorded on a graph for evaluation. Certain patterns produced by the autonomic (involuntary) nervous system indicate the truth or falsity of a given response.

Adherents of both systems claim a high degree of accuracy in such tests when they are conducted under the supervision of properly trained personnel.

It is important to determine the limitations on the use of such instruments in the various states. Some states forbid their use as a *requirement* for employment, but permit them to be used on a voluntary basis. Other states—Massachusetts, for example—forbid their use in any circumstance. There is a growing controversy over the use of such machines, and many states that exercised indifferent or no supervision over their employment are enacting legislation in which certain minimum training and qualification requirements are established.

Generally speaking, organized labor has lobbied diligently for legislation banning the use of so-called lie detectors in all industrial or commercial applications, and their efforts have borne some fruit in many states. The American Polygraph Association, while opposed to the use of the PSE essentially on the grounds that it has not yet proven itself, has endorsed much legislation setting stricter standards for polygraph operators but has, of course, fought labor's stand on the matter.

Many firms do, however, make use of polygraph or PSE examinations in hiring some of their personnel, and they make wide use of the approach in investigations of various kinds. Those firms that do use these machines have, in general, found that they have served a useful purpose.

Since such examinations are expensive, costing from $50 to $150, depending on the length of the test, not every firm will find them practical. Even those firms that do not find the expense acceptable usually limit their application to persons being hired or promoted to areas of particular sensitivity and responsibility. Each company must individually determine the need for such examinations, and their practicality, after they have decided whether the risk is such that the expense is warranted, and whether they can be satisfied by an evaluation based on other, cheaper methods.

Whatever the decision, the security manager would be well advised to consult a reputable firm handling polygraph examinations, as well as a PSE operation, to learn about what such examinations involve.

Continuity of Program

When a company has made a systematic and conscientious effort to screen out dishonest, troublesome, incompetent and unstable employees, it has taken a first and significant step toward reducing internal theft. It is important, then, that the program continue in effect on a permanent basis.

Care must be taken to avoid relaxing standards or becoming more superficial in checking applicants. There is a tendency to lose sight of the full dimensions of the problem if the security program makes substantial inroads on the loss factor. The past is too soon forgotten, and carelessness follows close behind. Active supervision is always necessary to maintain the integrity of this important aspect of every security program.

Hiring Ex-Convicts and Parolees

It should be strongly noted here that rigid exclusionary standards should never be applied to ex-convicts or parolees who openly acknowledge their past records. Such a policy would be, at best, unjust and, at worst, irresponsible. These men have done their time. Their records are available in situations where employment is being sought. To turn them away simply on the basis of their past mistakes would be to force them into a criminal pattern of life to survive.

Some such men might well be unacceptable in certain companies—but a rigid policy denying all of them employment would be to deny the company many potentially good men who deserve a chance for rehabilitation. Experience has shown that these men, knowingly hired in the right positions and properly supervised, are not only acceptable but, frequently, highly responsible and trustworthy. They should be given an opportunity to re-establish themselves in society.

Morale

In any organization which exists by the cooperative efforts of all its members, it is important that each one of those members feels that he is an important part of the operation, both as a contributor and as an individual.

This dual role is important to recognize. Each employee must feel that he is a significant, integrated part of the whole and identified with it, while still maintaining and protecting his basic importance as an individual apart from the structure. If he is denied reinforcement in either of these views, his sense of personal worth suffers. Anger, anxiety, frustration or feelings of inadequacy may follow. Any of these attitudes are damaging both to him and to the organization as a whole, and they must be headed off.

The best way to do that is to recognize that each employee is, in fact, an important member of the organization, or else he wouldn't be there. A function must be performed as a part of the larger function, and he is the one performing it. He is unique in that he is unduplicatable, and he is important in that uniqueness in that he is fellow man. His reinforcement comes from management, supervisors and peers. He is a vital, irreplaceable organ of the greater body. Of such recognition and respect is high morale created and maintained, and everyone benefits.

Every supervisor must be indoctrinated in the importance of the morale factor in every business. Just as no business can operate without employees, none can operate efficiently with a work force whose morale has been damaged and who respond with an attitude of listlessness and disinterest. Threats and coercion will not restore the optimum level of morale any more than directives and public notices can. This is a condition that comes from man's very being. It must be dealt with in those terms.

Additional factors are also important in this regard. Physical surroundings and appropriate rewards are part of the message that tells a man what the organization thinks of him.

In an atmosphere of concern, fairness and mutual respect, few men would be led to steal. It would be like stealing from themselves. It is important that such an atmosphere be developed and maintained for the good of all.

A program aimed at improving or maintaining employee morale might contain some or all of the following elements:

1. Clear statements of company policy which is consistently and fairly administered.
2. Regular review of wages and wage policy, updated to assure equitable wage levels.
3. House organ or newsletter and bulletin boards kept current.
4. Open, two-way avenues of communication between management and all employees.
5. Clear procedures, formal and informal, for airing grievances and personal problems with supervisors.
6. Vigorous training programs to improve job skills and pave the way for advancement.
7. Physical surroundings—decor, cleanliness, sound and temperate control, and general housekeeping at a high level.

SURVEY PROGRAM FOR INTERNAL SECURITY

As in other surveys, the first requirement before setting up protective systems for internal security is to survey every area in the company to determine the extent and nature of the risks. If such a survey is conducted energetically and exhaustively, and its recommendations for action are acted upon intelligently, significant losses from internal theft will be a matter of history.

Who Conducts the Survey

Such a survey should, typically, be undertaken by security personnel. In instances where losses have been particularly severe and troublesome, it might be wise to call in an independent security firm to work with staff security people. A separate survey of accounting procedures, including the flow of all documents authorizing and receipting shipments, inventorying, and other operations, might be assigned to a specialist in operational audits, probably one who works for the same firm that handles the company's annual audit. It must be remembered, however, that this survey is to examine the operation of accounting functions, and should not be considered as in any way related to normal auditing procedures.

Whoever conducts the survey must be fully familiar with all aspects of the operation in as much detail as possible. The pressures and tensions of a business should be examined as part of such a survey, as should the structure of both the formal and informal organization. The latter can be most easily identified as that unofficial chain of command or flow of work, or that acknowledged but unofficial series of power centers that operate sometimes parallel or sometimes even in opposition to the official organization of the company. It is easy to spot potential points of collusion when examining the organization chart. It is frequently less easy to spot such points in the informal organization, unless that aspect of the business is well known to the surveyor and thoroughly understood by him.

What is Covered

The survey should be conducted in a methodical manner, moving from department to department and covering every area of operation in each department before moving on. The basic question underlying this inspection is the loss potential inherent in each aspect of the routine. The greater the exposure, the more exacting the security procedures must be.

In the early stages, the survey team must try to think the way a thief would think. It must probe the routines currently in effect to find the weak links—to find those spots where a thief could strike with the greatest assurance of success and the least risk of exposure. Each member of the team must ask himself, "If I were so inclined, how would I go about stealing at this point or in that job?" This questioning and probing must also encompass every *possible* loss to theft and must not limit itself to considerations of likely areas of loss. Anything less than a total consideration of the possibilities will leave the job, to that degree, undone.

Since security, to be effective, requires a total systems approach, it must include every aspect of the business in its surveys. The surveys for the control of internal theft is no exception. Even though its recommendations will be incorporated into the overall security plan, and even though each department will be surveyed on its own and as a part of the total picture, each aspect of the business in any way affecting its internal security must be inspected for security flaws or weaknesses that could lead to employee theft.

As a practical matter, the company need not be reinspected for each aspect of the overall security plan, but it is important to consider different categories of danger individually before synthesizing them into a single protective procedure. The survey would include an examination of physical security, for example, or of the state of fire safety, or the condition and coverage of the alarm system, but at this stage each of these elements should be approached as separate problems. There will be time enough to coordinate the needs of each when all their requirements are consolidated into a single cohesive system.

In examining the needs of internal security, it will be necessary to consider hiring and promotion practices, the handling of receipts and disbursements, credits and rebates. Petty cash, payroll, bank deposits, handling of cash and negotiables, inventories, warehousing, data processing, purchasing, supplies, receiving, shipping, merchandise controls, inspection procedures, employee badging or identification policy, time card and key control, confidential data controls, and any other area or function where employee dishonesty could exist must be evaluated.

This evaluation will review the entire product of the survey, taking special note of those areas observed or surmised to be vulnerable to criminal manipulation. Such weaknesses must be eliminated or controls developed to protect against any possible breach of trust.

Need for Management Support

It is at this point, especially, that the strong support of top management may be needed most. In order to implement needed security controls, certain operational procedures may necessarily have to be changed. This will require cooperation at every level, and cooperation is sometimes hard to get in situations where a department manager feels his authority has been diminished in areas within his sphere of responsibility.

The problem is compounded when those changes determined to be necessary cut across departmental lines, and even serve, to some degree, to alter intradepartmental relationships. Effecting systems under such circumstances will require the greatest tact, salesmanship and executive ability. Failing that, it may be necessary to fall back on the ultimate authority vested in the security operation by top management. Any hesitation or equivocation on the part of either management or security at this point could damage the program before it has been initiated.

This does not, of course, mean that management must give security *carte blanche*. Reasonable and legitimate disagreements will inevitably arise. It does mean that proposed security programs based on broadly stated policy must be given the highest possible priority. In those cases where conflict of procedures exists, some compromise may be necessary, but the integrity of the security program as a whole must be preserved intact.

Communicating the Program

The next step is to communicate necessary details of the program to all employees. Many aspects of the system may be confidential or on a need-to-know basis, but since part of it will involve procedures engaged in by most or all of company personnel, they will need to know those details in order to comply. This can be handled in a series of meetings explaining the need for security, the damaging effects of internal theft to jobs, benefits, profit sharing and the future of the company. Such meetings can, additionally, serve to notify all employees that management is taking action against criminal acts of all kinds, at every level, and dishonesty will not be tolerated.

Such a forceful statement of position in this matter can be very beneficial. Most employees are honest men and women who disapprove of those who are criminally inclined. They are apprehensive and uncomfortable in a criminal environment, especially if it is widespread. The longer such conduct is condoned by the company, the more they lose respect for it, and a vicious cycle begins. As they lose respect, they lose a sense of purpose. Their work suffers, their morale declines, and at best their effectiveness is seriously diminished. At worst they reluctantly join the thieves. A clear, uncompromising policy of theft prevention is usually welcomed with visible relief.

Continuing Supervision

Once a system is installed, it must be supervised if it is to become, and remain, effective. Left to their own devices, employees will soon find short cuts, and security controls will be abandoned in the process. Old employees must be reminded regularly of what is expected of them, and new employees must be adequately indoctrinated in the system they will be expected to follow.

This must be a continuing program of education if expected results are to be achieved. With a national average turnover of 50 percent of the white-collar work force, it can be expected that the office force, which handles key paper work, will be replaced at a fairly consistent rate. This means that the company will have a regular influx of new people who must be trained in the procedures to be followed and the reasons for them.

Program Changes

In some situations, reasonable controls will create duplication of effort, cross-checking and additional paper work. Since each time such additional effort is required there is an added expense, procedural innovations requiring it must be avoided wherever possible. But most control systems aim for increased efficiency. Often this is the key to their effectiveness.

Many operational procedures, for a variety of reasons, fall into ponderous routines involving too many people and excessive paper shuffling. This may serve to increase the possibility of fraud, forgery or falsification of documents. When the same operational result can be achieved by streamlining the system, incorporating adequate security control, it should be done immediately.

Virtually every system can be improved, and every system should be evaluated constantly with an eye for such improvement, but these changes should never be undertaken arbitrarily. Procedures must be changed only after such changes have been considered in the light of their operational and security impact, and such considerations should further be undertaken in the light of their effect on the total system.

No changes should be permitted by unilateral employee action. Management should make random spot checks to determine if the system is being followed exactly. Internal auditors and/or security personnel should make regular checks on the control systems.

Violations

Violations should be dealt with immediately. Any management indifference to security procedures is a signal that they are not important, and, where work-saving methods can be found to circumvent such procedures, they will be. As soon as any procedural untidiness appears and is allowed to continue, the deterioration of the system begins.

It is well to note, too, that, while efforts to circumvent the system are frequently the result of the ignorance or laziness of the offender, a significant number of such instances are the result of an employee probing for ways to subvert the controls in order to divert company assets to his own use.

PROCEDURAL CONTROLS

Auditing Assets

Periodic audits by outside auditors are essential to any well-run security program. Such an examination will discover theft only after the fact, but it will presumably discover any regular scheme of embezzlement in time to prevent serious damage. If these audits, which are normally conducted once a year, are augmented by one or more surprise audits, even the most reckless criminal would hesitate to try to set up even a short-term scheme of theft.

These audits will normally cover an examination of inventory schedules, prices, footings and extensions. They should also verify current company assets by physical inventory sampling, accounts receivable, accounts payable (including payroll), deposits, plant and outstanding liabilities. In all these cases a spot check beyond the books themselves can help to establish the existence of legitimate assets and liabilities, not empty entries created by a clever embezzler.

Cash

Any business handling relatively few cash payments in and out is fortunate, indeed. Such a business is able to avoid much of the difficulty created by this security sensitive area, since cash handling is certainly the operation most vulnerable and the most sought after by the larcenous among the staff.

Cash by Mail. If cash is received by mail—a practice which is almost unheard of in most businesses—its receipt and handling must be undertaken by a responsible, bonded supervisor or supervisors. This administrator should be responsible for no other cash handling or bookkeeping functions. This official should personally see to it that all cash received is recorded by listing the amount, the payer, and such other pertinent information as procedures have indicated. This list should be made, in duplicate, on sequentially numbered forms, with both copies signed by the manager in charge of opening the mail as well as by the cashier who receipts for the money. The cashier will then keep one copy of the record for the file—the other will be forwarded to accounting. Both will verify the numbering on the cash receiving list.

There is clearly a danger here at the very outset. If cash is diverted before it is entered on any receipt, there is no record of its existence. Until it is channeled into company ledgers in some way and begins its life as a company asset, there is no guarantee that it won't serve some more private interest. This requires supervision of the supervisor. In the case of a firm doing a large catalogue business which receives large amounts of cash in spite of pleas for checks or money orders, it has sometimes been felt that the operation should be conducted in a special room reserved for the purpose.

Daily Receipts. All cash book entries must be checked against cash on hand at the end of each day. Spot checks on an irregular basis should also be conducted.

Cash receipts should be deposited in the bank and each day's receipts balanced with the daily deposit. Petty cash, as needed, should be drawn by check.

All bank deposits should be accompanied by three deposit slips, one of which is receipted by the bank and returned to the cashier by the person making the deposit. The second is mailed to the office accounting department and the third is the bank's.

Each day deposit slips should be balanced with the day's receipts.

Bank Statements. Bank statements must be received and reconciled by someone who is not authorized to deposit or withdraw funds, or to make a final accounting of receipts or disbursements. When bank statements are reconciled, cancelled checks should be checked against vouchers for alterations and for proper endorsement by the payee. Any irregularities in the endorsements should be promptly investigated. If the statement, itself, seems in any way out of order by way of erasure or possible alteration, the bank should be asked to submit a new statement to the reconciling official's special personal attention.

Petty Cash. A petty cash fund, set aside for that purpose only, should be established. The amount to be carried in such a fund will be based upon past experience. These funds must never be commingled with other funds of any kind and should be drawn from the bank by check only. They should never be drawn from cash receipts. No disbursements of any kind should be made from petty cash without an authorized voucher signed by the employee receiving the cash and countersigned by an authorized employee. No voucher should be cashed that shows signs of erasure or alteration. All such vouchers should be drawn up in ink or typed. In cases of typographical error, new vouchers should be prepared rather than correcting the error. If there is any reason for using a voucher on which an erasure or correction has been made, the authorizing official should initial the change or place of erasure.

Receipts substantiating the voucher should accompany it and should, if possible, be stapled or otherwise attached to it.

The petty cash fund should be brought up to the specified amount from time to time as prearranged, by check to the amount of its depletion. The vouchers upon which disbursements were made should always be verified by an employee other than the one in charge of the fund. All vouchers submitted and paid should be cancelled in order to avoid reuse.

Petty cash should be balanced occasionally by management, at which time vouchers should be examined for irregularities.

Separation of Responsibility

The principle of separation of responsibility and authority in matters concerning the company's finances is of prime importance in security management. This situation must always be sought out in the survey of every department. It is not always easy to locate. Sometimes even the employee who has such power is unaware of his dual role. But the security specialist must be sensitive to its existence and provide for an immediate change in such operational procedures whenever they appear.

An employee who is in the position of both ordering and receiving merchandise, or a cashier who authorizes and disburses expenditures, are examples of this double-ended function in operation. All situations of this nature are potentially damaging and should be eliminated. Such procedures are manifestly unfair to company and employee alike. To the company because of the loss that might incur; to the employee because of the temptation and ready opportunity they present. Good business practice demands that such invitations to embezzlement be studiously avoided.

It is equally important that cash handling be separated from the record-keeping function. When the cashier becomes his own auditor and bookkeeper, he has a free rein with that part of the company funds. The chances are he won't steal, but he could—and might. He might also make mistakes without some kind of double check on his arithmetic.

In some smaller companies this division of function is not always practical. In such concerns it is common for the bookkeeper to act also as cashier. If this is the case, a system of countersignatures, approvals and management audits should be set up to help divide the responsibility of handling company funds as well as accounting for them.

Promotion and Rotation

Most embezzlement is the product of a scheme operating over an extended period of time. Many embezzlers prefer to divert small sums on a systematic basis, feeling that the individual thefts will not be noticed and, therefore, the total loss is unlikely to come to management's attention.

These schemes are ultimately frustrated (if at all), either by some accident that uncovers the system or by the greed of the embezzler, who is so carried away by the success of his efforts that he steps up the ante. But while the theft is working and he is in control, it is usually difficult to detect. Frequently, he is in a position to alter or manipulate records in such a way that the theft escapes the attention of both internal and outside auditor. This can be countered by upward or lateral movement of employees.

Promotion from within, wherever possible, is always good business practice, and lateral transfers can be effective in countering possible boredom or the danger of reducing a function to rote, and thus diminishing its effectiveness.

Such movement also frustrates the embezzler. When he loses control of the books governing some aspect of the operation, he loses the opportunity to cover his theft. Discovery would inevitably follow a careful audit of books he could no longer manipulate. If regular transfers were a matter of company policy, no rational embezzler would set up a long-term plan of embezzlement unless he found a scheme that was audit-proof, and such an eventuality is highly unlikely.

To be effective as a security measure, such transfers need not involve all personnel, since every change in operating personnel brings with it changes in operation. In some cases, even subtle changes may be enough to alter the situation sufficiently to reduce the totality of control an embezzler has over his books. If such is the case, his swindle is over. He may avoid discovery of his previous looting, but he cannot continue in his grab without danger of being unmasked.

In the same sense, embezzlers dislike vacations. They are aware of the danger if someone else should handle their accounts, if only for the two or three weeks of vacation. So they make every effort to pass up the holiday.

Any manager who has a reluctant vacationer on his hands should recognize that he has a potential problem. Vacations are designed to refresh the outlook of everyone. No matter how tired they may be when they return to work, every vacationer has been refreshed emotionally and intellectually. His effectiveness in his job has probably improved and he is, generally speaking, a better man for it. The company benefits from his vacation as much as he does. No employee should be permitted to pass up his duly authorized vacation—especially one whose position gives him access to control over company assets.

Access to Records

Many papers, documents and records are confidential, or at least are available to only a limited number of people who need such papers in order to function. All other persons are deemed to be off limits. They have no need for the information. Such papers should be secured under lock and key—and, depending on their value or reconstructability, in a fire-resistant container.

Forms

Company forms are often extremely valuable to the inside as well as the outside thief. They should be secured and accounted for at all times. If it is at all possible or feasible to do so, they should be sequentially numbered and recorded regularly so that any loss can be detected at a glance.

Blank checks, order forms, payment authorizations, vouchers, receipt forms and all others which authorize or verify transactions, are prime targets for thieves and should, therefore, be accounted for.

Since there are many effective operational systems in use for the ordering, shipping or receipting of goods, as well as the means by which all manner of payments from petty cash to regular debt discharge are authorized, no one security system to protect against illegal manipulation within such systems would apply universally. It can be said, however, that since every business has some means to authorize transactions of goods or money, the means by which such authorizations are made must be considered in the security program. Security of such means must be considered as an important element in any company's defense against theft.

Generally speaking, all forms should be pre-numbered and, where possible, used in numerical order. Any voided or damaged forms should be filed and recorded and forms reported lost must be accounted for and explained. All such numbered forms of every kind should be inventoried and accounted for periodically.

In cases where purchase orders are issued in blocks to various people who have the need for such authority, such issuance must be recorded and dispositon of their use should be audited regularly. In such cases, it is customary for one copy of the numbered purchase order to be sent to the vendor, who will use that number in all further dealings on that particular order; another copy will be sent to accounting for purposes of payment authorization and accrual if necessary; and one copy will be retained by the issuing authority. Each block issued should be used sequentially, although, since some areas may have more purchasing activity than others, purchasing order copies as they are forwarded to accounting may not be in overall sequence.

Purchasing

Centralized Responsibility. Where purchasing is centralized in one department, controls will always be more effective. Localizing responsibility as well as authority reduces the opportunity for fraud, accordingly. This is not always possible or practical, but in areas where purchasing is permitted by departments needing certain materials and supplies, there can be confusion occasioned by somewhat different purchasing procedures. Cases have been reported where different departments pay different prices for the same goods and services and thus bid up the price the company is paying. Centralization of purchasing would overcome this problem.

Purchasing should not, however, be involved in any aspect of accounts payable or the receipt of merchandise other than informationally.

Competitive Bids. Competitive bids should be sought wherever possible. This, however, raises an interesting point that must be dealt with as a matter of company policy. Seeking competitive bids is always good practice, both to get a view of the market and to provide alternatives in the ordering of goods and materials, but it does not follow that the lowest bidder is always the vendor to do business with.

Such a bidder may lack adequate experience in providing the services bid for, or he may have a reputation of supplying goods of questionable quality, even though they may meet the technical standard prescribed in the order. A firm may also underbid the competition in a desperate effort to get the business, but then find it cannot deliver materials at that price, no matter what it has agreed to in its contract.

In order to function wisely and to be able to exercise good judgment in its area of expertise, purchasing must be permitted some flexibilty in its selection of vendors. This means that it will not always be the low bidder who wins the contract.

Now, since competitive bidding provides some security control in reducing favoritism, collusion, and kickbacks between the purchasing agent and the vendor, these controls would appear to be weakened or compromised in situations where the purchasing department is permitted to select the vendor on considerations other than cost. This can be true to some degree, but this is a situation in which business or operational needs may be in some conflict with tight security standards—and in which security should revise its position to accommodate the larger demands or efficiency and ultimate economy. After all, cheap is not necessarily economical.

Controls in this case could be applied by requiring that in all cases where the lowest bid was not accepted, a brief explanation in outline form be attached to the file along with all bids submitted. Periodic audits of such files could establish if any pattern of fraud seems likely. Investigation of the analysis or assumptions made by purchasing in assigning contracts might be indicated in some situations to check the validity of its stated reasoning in the matter.

Other Controls. Copies of orders containing the amount of merchandise purchased should not be sent to receiving clerks. These clerks should simply state the quantity actually received with no preconception of the amount accepted. Payment should be authorized only for that amount actually received.

Vendor invoices and receipts supporting such vouchers should be cancelled to avoid the possibility of their resubmission in collusion with the vendor.

Purchasing should be audited periodically, and documents should be examined for any irregularities.

Payroll

It is important that the payroll be prepared by persons who will not be involved in its distribution. This is consistent with the effort to separate the various elements of a particular function into its component parts, and then distribute the responsibility for those parts to two or more persons or departments.

Every effort should be made to distribute the payroll in the form of checks rather than cash, and such checks should be of a color different from those used in any other aspect of the business. They should also be drawn on an account set aside exclusively for payroll purposes. It is important that this account be maintained in an orderly fashion. Avoid using current cash receipts for payroll purposes.

Personnel Records. The payroll should be prepared from personnel records which, in turn, have come from personnel as each employee is hired. Such a record should contain basic data such as name, address, attached W-2 form, title, salary and any other information that the payroll department may need. The record will be countersigned by a responsible executive verifying the accuracy of the information forwarded.

This same procedure should be followed when an employee terminates his employment with the company. All such notifications should be consolidated into a master payroll list, which should be checked frequently to make sure that payroll's list corresponds to the employment records of personnel.

Unclaimed Checks. Unclaimed paychecks should be returned to the treasurer or controller after a reasonable period of time for redeposit in the payroll account. Certainly, such cases should be investigated to determine why the checks were returned or why they were issued in the first place. All checks so returned should be marked to prevent any re-use and filed for reference. Since payrolls usually reflect overtime and other payments in addition to regular salary disbursements, such payments should be supported by time sheets authorized by supervisors or department heads. Time sheets of this nature should be verified periodically to prevent overtime padding and kickback.

Time cards, themselves, should be marked to prevent re-use.

Payroll Audits. The payroll should be audited periodically for any irregularities, especially if there has been an abnormal increase in personnel or net labor cost.

To further guard against the fraudulent introduction of names into the payroll, distribution of paychecks should periodically be undertaken by the internal auditor, the treasurer or other responsible official. In large firms this can be done on a percentage basis, thus providing at least a spot check of the validity of the rolls.

Accounts Payable

As in the case of purchasing, accounts payable should be centralized to handle all disbursements upon adequate verification of receipt and proper authorization for payment.

These disbursements should always be by checks that are consecutively numbered and used in that order. Checks that are damaged, incorrectly drawn, or for any reason unusable must be marked as cancelled and filed for audit. All checks issued for payment should be accompanied by appropriate supporting data, including payment authorizations, before they are signed by the signing authority. It is advisable to draw the checks on a check-writing machine which uses permanent ink and is as identifiable to an expert as handwriting or a particular typewriter. Checks, themselves should be of safety paper which will show almost any attempted alteration.

Here, as in order departments, periodic audits must be conducted to examine the records for any sign of nonexistent vendors, irregularities in receipts or payment authorizations, forgeries, frauds or unbusinesslike procedures that could lead to embezzlement.

General Merchandise

Merchandise is always subject to embezzlement, particularly when it is in a transfer stage, as when it is being shipped or received. The dangers of loss at these stages are increased in operations where controls over inventory are lax or improperly supervised.

Separation of Functions. To control these sensitive aspects of any operation involving the handling of merchandise, it is desirable to separate the three functions. Receiving, warehousing and shipping should be the responsibility of three different areas. Movement of merchandise from one mode to another should be accompanied by appropriate documents which clearly establish the responsibility for specific amounts of merchandise passing from one sphere of authority to another.

Receipting for a shipment places responsibility for a correct count and the security of the shipment on the receiving dock. This responsibility remains his until he transfers it and receives a proper receipt from the warehouse man. The warehouse man must verify and store the shipment, which is his responsibility until it is called for (by the sales department, for example), or directed to be shipped by authorized voucher. The warehouse man assembles the goods, passes them along as ordered, and receives a receipt for those goods delivered.

In this process, responsibility is fixed from point to point. Various departments or functions take on and are relieved of responsibility by voucher and receipt. In this way a perpetual inventory is maintained as well as a record of responsibility for the merchandise.

All vouchers and requisitions must be numbered to avoid the destruction of records or the introduction or unauthorized transfers into the system. Additionally, stock numbers of merchandise should accompany all of its movement to describe the goods and thus aid in maintaining perpetual inventory records.

In small firms where this separation of duties is impractical, and receiving, shipping and warehousing are combined in one person, the perpetual inventory is essential for security, but it must be maintained by someone other than the person actually handling the merchandise. The shipper-receiver-warehouser should not have access to these inventory records at any time.

Inventories. Inventories will always be an important aspect of merchandise control, no matter what operations are in effect. Such inventories must be conducted by someone other than the person in charge of that particular stock. In the case of department stores, for purposes of inventory, personnel should be moved to a department other than their regular assigned one.

In firms where a perpetual inventory record is kept, physical counts on a selective basis can be undertaken monthly or even weekly. In this procedure a limited number of certain items randomly selected can be counted and the count compared with current inventory record cards. Any discrepancy can be traced back to the point of loss to determine its cause.

Physical Security. It is important to remember that personnel charged with the responsibility of goods, materials and merchandise must be provided the means to properly discharge that responsibility. Warehouses and other storage space must be equipped with adequate physical protection to secure the goods stored within. Authorizations to enter such storage areas must be strictly limited, and the responsible employee must have means to further restrict access in situations where he may feel that the security of the goods under his control is endangered.

Receiving clerks must have adequate facilities for storage or supervision of goods until they can be passed on for storage or for other use. Shipping clerks must also have the ability to secure goods in dock areas until they are received and loaded by truckers. Without the proper means of securing merchandise during every phase of its handling, assigned personnel cannot be held responsible for merchandise intended for their control, and the entire system will break down. Unreasonable demands, such as requiring a shipping clerk to handle the movement of merchandise in such a way that he is required to leave unprotected goods on the dock while he fills out the rest of the order, lead to the very reasonable refusal of personnel to assume responsibility for such merchandise. And when responsibility cannot be fixed, embezzlement is sure to result.

The Mailroom

The mailroom can be a rich field for a company thief to mine. Not only can it be used to mail out company property to an ally or to a set-up address, but it deals in stamps—and stamps are money. Any office with a heavy mailing operation must conduct regular audits of the mailroom.

Many firms have taken the view that the mailroom represents such a small exposure that close supervision is unnecessary. The head of the mailroom in a fair-sized Eastern firm got away with over $100,000 in less than three years from his manipulation of the postal meter. Only a firm that can afford to lose $100,000 in less than three years should think of its mailroom as inconsequential in its security plan.

Trash Removal

Trash removal has presented many problems. Employees have hidden office equipment or merchandise in trash cans and have then picked up the loot far from the premises in cooperation with the driver of the trash pick-up vehicle. Some firms have had a problem when they put out trash on the loading dock to facilitate pick-up. Trash collectors made their calls during the day and often picked up unattended merchandise along with the trash. On-premises trash compaction is one way to end the use of trash containers as a safe and convenient vehicle for removing loot from the premises.

Every firm has areas which are vulnerable to attack—what and where they are can only be determined by thorough surveys and regular reevaluation of the entire operation. There are no short cuts. The important thing is to locate the areas of risk and set up procedures to reduce or eliminate them.

WHEN CONTROLS FAIL

There are occasions when a company is so beset by internal theft that problems seem to have gotten totally out of hand. In such cases, it is often difficult to localize the problem sufficiently to set up specific countermeasures in those areas affected.

The company seems simply to "come up short." Management is at a loss to identify the weak link in its security, much less how theft is accomplished after security has been compromised.

Undercover Investigation

In such cases, many firms similarly at a loss, in every sense of the word, have found it advisable to engage the services of a security firm which can provide undercover agents to infiltrate the organization and observe the operation from within.

Such an agent may be asked to get into the organization on his own initiative. The fewer people who know of his presence, the greater his protection, and the more likely he is to succeed in his investigation. It is also true that when large scale thefts take place over a period of time, almost anyone in the company could be involved. Even one or more top executives could be involved in serious operations of this kind. Therefore secrecy is of great importance. Since several agents may be used in a single investigation, and since they may be required to find employment in the company at various levels, they must have, or very convincingly seem to have, proper qualifications for the level of employment they are seeking. Over- or under-qualification in pursuit of a specific area of employment can be a problem, so they must plan their entry carefully. Several agents may have to apply for the same job before one is accepted.

Having gotten into the firm's employ, the agent is on his own. He must conduct his investigation and make reports to his office with the greatest discretion to avoid discovery. But, he is in the best possible position to get to the center of the problem, and such agents have been successful in a number of cases of internal theft in the past.

These investigators are not inexpensive, but they earn their fee many times over in breaking up a clever ring of thieves.

It is important to remember, however, that these men are trained professionals. Most of them have had years of experience in undercover work of this type. Under no circumstances should a manager think of saving money by using employees or well-meaning amateurs for this work. Such a practice could be dangerous to the inexperienced investigator and would almost certainly warn the thieves, who would simply withdraw from their illegal operation temporarily until things had cooled down, after which they could return to the business of theft.

Prosecution

Every firm has been faced with the problem of establishing policy regarding the disposal of a case involving proven or admitted employee theft. They are faced with three alternatives: to prosecute, to discharge, or to retain the thief as an employee. The policy they have established has always been difficult to arrive at, because there is no ready answer. There are many proponents of each alternative as the solution to problems of internal theft.

However difficult it may be, every firm must establish a policy governing matters of this kind. And the decision as to that policy must be arrived at with a view to the greatest benefits to the employees, the company, and to society as a whole. An enlightened management would also consider the position of the as-yet-to-be-discovered thief in establishing such policy.

Discharging The Thief

Most firms have found that discharge of the offender is the simplest solution. Experts estimate that 90 percent of those employees discovered stealing are simply dismissed. Most of those are carried in the company records as having been discharged for "inefficiency" or "failure to perform duties adequately."

This policy is defended on many grounds, but the most common are:

1. Discharge is a severe punishment and the offender will learn from his punishment.
2. Prosecution is expensive.
3. Prosecution would create an unfavorable public relations atmosphere for us.
4. If we reinstate him in the company—no matter what conditions we place on his reinstatement—we will appear to condone theft.
5. If we prosecute and he is found not guilty, we will be open to civil action for false arrest, slander, libel, defamation of character, and other damages.

There is some validity in all of these views, but each one bears some scrutiny.

As to learning (and presumably reforming) as a result of discharge, experience does not bear out this contention. In a recent study, a security organization found that 80 percent of the known employee thieves they questioned with polygraphy substantiation admitted to thefts from previous employers. Now it might well be argued that, since they had not been caught and discharged as a result of these prior thefts, the proposition that discharge can be therapeutic still holds, or at least has not been refuted. That may be true and it should be considered.

Prosecution is unquestionably expensive. Personnel called as witnesses may spend days appearing in court. Additional funds may be expended investigating and establishing a case against the accused. Legal fees may be involved. But can a company afford to appear so indifferent to significant theft that it refuses to take strong action when it occurs?

As to public relations, many experienced managers have found that they have not suffered any decline in esteem. On the contrary, in cases where they have taken strong, positive action, they have been applauded by employees and public alike. This is not always the case, but apparently a positive reaction is usually the result of vigorous prosecution in the wake of substantial theft.

Reinstatement is sometimes justified by the circumstances. There is always, of course, a real danger of adverse reaction by the employees, but if reinstatement is to a position not vulnerable to theft, the message may get across. This is a most delicate matter that can be determined only on the scene.

As far as civil action is concerned, that possibility must be discussed with counsel. In any event, it is to be hoped that no responsible businessman would decide to prosecute unless the case was a very strong one.

Borderline Cases

Even beyond the difficulty of arriving at a satisfactory policy governing the disposition of cases involving employee theft, there are the cases which are particularly hard to adjudicate. Most of these involve the pilferer, the long-time employee, or the obviously upright employee who finds himself in financial difficulty and steals out of desperation. In each case the offender freely admits his guilt and pleads that he was overcome by the temptation suddenly thrust before him.

What should be done in such cases? Many companies continue to employ such men, provided they make restitution. They are often found to be grateful, and they continue to be effective in their jobs.

In the last analysis it is each individual businessman who must make the determination of policy in these matters. Only he can determine the mix of toughness and compassion that will guide the application of policy throughout.

Hopefully, every manager will determine to avoid the decision by making employee theft so difficult—so unthinkable—that it will never occur. That goal may never be reached, but it's a goal to strive for.

REVIEW QUESTIONS

1. What are some of the common 'danger signals' of employee dishonesty?
2. Polygraphing employees is illegal in some jurisdictions and opposed by unions in many. Discuss reasons for and against the use of the polygraph in background and internal theft investigations.
3. Discuss procedural controls for decreasing the incidence of employee theft in specific departments.
4. What should be management's role in effecting internal security?
5. Should employees be prosecuted for stealing? Why?

Chapter 12

INSURANCE

Many managers still cling to the notion that the most effective means of guarding against unforeseen business losses is insurance, and all too many still use insurance as a substitute for a comprehensive security program. The fallacy in this attitude is twofold.

In the first place, almost all casualty insurance companies have suffered losses in underwriting crime insurance over the last 20 years. In fact, payments for insured losses due to crime have regularly exceeded income from premiums since 1963. Obviously no company can continue to operate at a loss; casualty underwriters are no exception.

In the face of the rapid increase in crime against homes, persons and business establishments, most insurance companies have taken drastic steps to counter this trend. They have cancelled or refused to renew policies of insureds who have suffered losses from criminal activity. They have limited allowable coverage to a point well below replacement or even cash value of goods or property. They have limited the extent of coverage. They have set up limitations which exclude businesses in high crime areas or in high risk enterprises from any coverage at all. And, lastly, they have increased premium rates. The increase in burglary insurance has increased as much as 850 percent in some areas in the past 20 years. In New York City alone, burglary rates on liquor stores rose 400 percent between 1961 and 1971, and then increased over 100 percent in the next decade.

In the second place, it is virtually impossible to insure against all the losses that could be incurred. Hidden damages in loss of company morale, loss of customer confidence, interruption of vigorous participation in a highly competitive market—all are serious, if not fatal, blows to any business and can never be recompensed.

Clearly, insurance can never be a substitute for a security program. In many cases, the very fact that assets are insured to some degree tends to reduce the interest of the proprietor in instituting reasonable security procedures beyond those minimums specified in his policy. As an aspect of the overall picture, insurance tends also to reduce any interest the insured may have in capturing or prosecuting perpetrators of crimes, thus, in effect, encouraging the proliferation of like crimes.

Insurance is certainly important. It is clearly necessary for any businessman who wishes to protect himself against loss—to spread the risk—but it must be thought of as a supportive, rather than as the principal defense against losses due to crime.

INSURANCE AGAINST CRIME

Crime insurance covers the insured in the event of loss from robbery, theft, forgery, burglary, embezzlement, and other criminal acts. It is important, however, to know the specific coverage involved in any such policy and the circumstances under which recovery of losses is allowed.

For example, burglary is generally meant to refer to felonious entry and theft by force. In order to collect insurance after such an attack, there must be evidence of forced entry, such as broken locks or windows, tool marks, or other clear evidence that burglary was in fact committed. The mere fact that items are missing will in no way establish the fact that the insured has been the victim of a burglar. Theft or larceny coverage must be included in the policy in order to cover such incidents.

Robbery, too, must be specifically established according to contractual definition. Robbery can be loosely defined as the forcible taking of property by violence or the threat of violence aimed at a person or persons covered by the policy. Theft or purse snatching, for example, would not be covered under a provision covering robbery; neither would burglary.

It is, therefore, essential that the terms describing criminal activity be clearly understood so that the nature and the extent of the coverage conforms to the needs of the insured. This is particularly important with companies which are especially vulnerable to certain kinds of hazards.

It is also extremely important to check policies for exclusionary clauses which may exempt certain crimes from coverage or which simply do not include certain crimes in the contract. This will require a careful examination, since insurance contracts are notoriously long-winded of necessity to cover all of the possible contingencies within the area of their coverage. Certain of the absences of coverage or exclusion can get lost in the sea of verbiage.

Comprehensive Policies

Comprehensive policies covering dishonesty, destruction, and disappearance are designed to provide the widest possible coverage in cases of criminal attack of various kinds. The standard form is set up to offer five different kinds of coverage.

The insured has the option of selecting any or all of the insuring agreements offered and of specifying the amount of coverage on each one selected. In addition to the coverage options in the standard form, 12 endorsements are also available to the manager having a need for any or all of them.

The coverages available on the standard form consist of:

1. Employee dishonesty bond.
2. Money and securities coverage on the premises.
3. Money and securities coverage off the premises.
4. Money order and counterfeit paper currency coverage.
5. Depositors' forgery coverage.

Additional endorsements available are:

1. Incoming check forgery.
2. Burglary coverage on merchandise.
3. Paymaster robbery coverage on and off premises.
4. Paymaster robbery coverage on premises only.
5. Broad-form payroll on and off premises.
6. Broad-form payroll on premises only.
7. Burglary and theft coverage on merchandise.
8. Forgery of warehouse receipts.
9. Securities of lessees of safe-deposit box coverage.
10. Buglary coverage on office equipment.
11. Theft coverage on office equipment.
12. Credit card forgery.

Obviously, the premium on this coverage will vary according to the number of options selected and the amount of coverage desired for each.

Estimating Coverage

Figure 12-1 can prove useful in determining the amount of crime coverage a firm should carry depending upon its exposure.[51] Although this chart cannot specify the kind of coverage needed, since each company will have its own areas of vulnerability, it can indicate the potential dangers managers face, and the coverage recommended to protect against them.

To determine the suggested minimum amount of insurance required, compute the firm's dishonesty exposure index, and find the recommended coverage by referring to the list under suggested minimum amounts (Figure 12-2).

1.		Total current assets (cash, deposits, securities, receivables, and goods on hand).	$ ____	
	A.	Value of goods on hand (raw materials in process, finished merchandise or products).	$ ____	
	B.	5% of A.		$ ____
	C.	Total current assets less goods on hand. (The difference between 1 and 1A)	$ ____	
	D.	20% of C		$ ____
2.		Annual gross sales or income.	$ ____	
	A.	10% of 2		$ ____
	B.	Total of 1B, 1D and 2A—the firm's dishonesty exposure index.		$ ____

Figure 12-1. Dishonesty exposure index indicator.

Evaluating Risk

However the exposure is calculated, and however much coverage is then deemed necessary, it is essential that they reflect a realistic appraisal of the risk factors actually involved. Every manager dealing with protection factors must ask himself what the risk really is and what would happen to the company if any of the considered potential hazards came to pass.

In the last analysis, he is simply spending certain amounts of money on a gamble that certain potentially harmful events will or are likely to occur. In some cases he is betting that, although such events are not likely to occur, they would be so catastrophic if they did that the amount spent on insurance faces into insignificance compared to the possible losses. The balance is a delicate one and requires much thought and expertise to develop the most efficient insurance program.

Insurance rates, after all, based on actuarial tables which are, presumably, updated regularly to reflect the experience of various industries or types of business as a whole in losses from various sources or causes. Since most insurance companies are profit-oriented enterprises themselves, the rates also reflect that factor. They must, as part of their profit structure, charge for claims handling, sales commissions, administrative expense. This ultimately means that a manager must evaluate his own risk as against the insurance industry's evaluation of it.

Exposure Index		Amount of Coverage	
up to	25,000	$15,000 to	25,000
25,000 to	125,000	25,000	50,000
125,000	250,000	50,000	75,000
250,000	500,000	75,000	100,000
500,000	750,000	100,000	125,000
750,000	1,000,000	125,000	150,000
1,000,000	1,375,000	150,000	175,000
1,375,000	1,750,000	175,000	200,000
1,750,000	2,125,000	200,000	225,000
2,125,000	2,500,000	225,000	250,000
2,500,000	3,325,000	250,000	300,000
3,325,000	4,175,000	300,000	350,000
4,175,000	5,000,000	350,000	400,000
5,000,000	6,075,000	400,000	450,000
6,075,000	7,150,000	450,000	500,000
7,150,000	9,275,000	500,000	600,000
9,275,000	11,425,000	600,000	700,000
11,425,000	15,000,000	700,000	800,000
15,000,000	20,000,000	800,000	900,000
20,000,000	25,000,000	900,000	1,000,000
25,000,000	50,000,000	1,000,000	1,250,000
50,000,000	87,500,000	1,250,000	1,500,000
87,500,000	125,000,000	1,500,000	1,750,000
125,000,000	187,500,000	1,750,000	2,000,000
187,500,000	250,000,000	2,000,000	2,250,000
250,000,000	333,325,000	2,250,000	2,500,000
333,325,000	500,000,000	2,500,000	3,000,000
500,000,000	750,000,000	3,000,000	3,500,000

Figure 12-2. Suggested minimum amounts of honesty insurance.

More often than not, he will discover he is under-insured, but there are instances when he may well feel that his exposure to loss is well below that of his particular industry as a whole. In such a case, he might properly elect to save on his premium payments by insuring according to his appraisal of his risk (on the basis of the effectiveness of his security program, for example) instead of on the basis of his exposure.

There are never easy decisions to make. The under-insured are risking ruination, while the over-insured are spending substantial sums to no good purpose.

Whatever the evaluation, however, every manager must be thoroughly conversant with the risks he faces and make provisions for them accordingly. His first impulse must be to minimize the risk as much as possible by instituting security devices and procedures that will reduce the possibility of loss or at least will notify him promptly when a loss has taken place. He must then re-evaluate the risks in the light of this tightened or alerted security system and then re-evaluate his supportive system of insurance.

Fidelity Bonds

One important kind of insurance which is frequently badly underestimated is the fidelity bond—a form of insurance that will compensate for company losses from dishonest employees. These bonds are issued on the performance of specific, named employees after the bonding company has satisfied itself, through investigation, that such employees do not represent an evident threat to the assets of the company. In effect, the bonding company is guaranteeing the insured that bonded employees will perform in good faith—that they will not commit any dishonest acts against their employer. If any of such bonded employees violate this trust, the guarantor—the bonding company—will stand the loss up to the amount insured.

The investigation by the bonding company is valuable in that it provides a further check on the background of employees in sensitive positions as well as underwriting possible losses resulting from a violation of trust.

Most companies do require that employees handling cash or high value merchandise be bonded, but too many of these companies go along on a program calling for $5,000, $10,000, or $25,000 bonds, failing to consider that, if bonding is deemed necessary, it must provide for protection against potential damage that such an employee can cause. In situations where there is no system providing a regular, foolproof audit of cash and valuable merchandise, for example, an employee might steal enormous sums over a period of time, even if his daily handle is relatively small.

The Surety Association of America has published a list of losses from various kinds of businesses caused by bonded employees. They showed the extent of fidelity coverage and the actual loss. The net loss figures are dramatic. A small sample of their much larger list follows (Figure 12-3).

This sample clearly indicates the problem faced by business today. It also indicates that many businesses are not handling the problem with a coordinated systems approach. It may appear to be hindsight to point out that adequate bonding would have cost these companies the merest fraction of their ultimate losses. Yet we can assume that, in most of these cases, a more realistic evaluation of the exposure, risk and insurance costs would have prevented these substantial losses.

With losses attributable to internal theft estimated in the billions, it is easy to see why fidelity bonds are thought of as high priority insurance, especially since it provides particular protection in areas where exposure is generally the greatest. As

Business	Employee	Loss	Bond	Uninsured Loss
Whls. Produce	Bookkeeper	$185,820	$25,000	$160,820
Retail Dairy	Office Manager	11,000	2,500	8,500
Hospital	Chief Clerk	15,000	5,000	10,000
Machinery Mfr.	Sales Manager	96,940	50,000	46,940
Department Store	Several	81,000	15,000	66,000
Refrigerator Mfr.	Cashier	20,810	5,000	15,810
Rubber Mfr.	Bookkeeper	126,700	26,000	100,700
Advertising	Billing Clerk	90,875	10,000	80,875
Paper Products	Warehouseman	25,551	15,000	10,551

Figure 12-3. Losses caused by bonded employees in various businesses.

important as this form of insurance is, it is essential that it be handled properly to provide the full protection it is capable of providing.

It is important in cases involving bonded employees discovered in theft that no arrangement concerning restitution be made with them without consulting the bonding company. Case files are filled with situations where an employee agreed to pay back the value of stolen merchandise over a period of time. Typically, a few payments are made and the employee disappears. At that point, there is little or no likelihood that the bonding company will make good the loss. There may be good reasons to give an otherwise trusted employee a chance to make good his larceny, but if restitution is the sole or at least the prime consideration, the entire matter should be left to the insurer who is obliged to make good the loss to its extent or to the amount of the bond. Determination of the disposition of the perpetrator's case will be in his hands.

Another matter requiring some familiarity is the confusing exclusionary clause which states: "The insuring clause does not cover any loss, or that part of any loss, as the case may be, the proof of which, either to its factual existence or as to its amount, is dependent upon an inventory computation at a profit and loss computation." What insurance companies have maintained is that if an employee should confess to taking merchandise and selling it over a period of time, the amount of merchandise so appropriated may not be established simply by checking out what seems to be missing from stock. This, of course, assumes that the employee does not recall precisely how many items he stole over the period of time which is usually the case. The fact that records indicate there should be two hundred items in stock and an inventory can only find eighty does not, according to a strict interpretation of the exclusionary clause, serve to establish a claim for 120 items on the bond. In fact, courts, until fairly recently, held that inventory records were only useful to corroborate the fact of employee theft, not its extent. In the hypothetical case above, each of the 120 missing items would have to be accounted for with convincing independent proof of its theft by a bonded employee. As a result, full restitution for such losses would be difficult to obtain, in spite of confessions or other proof of employee dishonesty.

However, the 1970's ushered in a more liberal, consumer-oriented attitude in the courts, and this limiting exclusionary clause in bonding policies began to be viewed more flexibly than in the past. The Alabama Supreme Court allowed inventory records to be introduced. (*American Fire and Casualty Co. v. Burchfield,* 232 So.2d 606 (Ala., 1970)) An Ohio court, on the admission of theft by an employee, allowed the total of the inventory shortage to be insured. (*Sommer v. General Insurance Co. of America,* 259 N.E.2d 142 (Ohio, 1970).) An earlier case in New Jersey perhaps heralded this new attitude toward interpretation by carefully outlining the conditions of acceptance of inventory records as proof, not of the fact of loss, but of the extent of loss, provided that proof of employee theft be arrived at independently or as corroborated by any observed shortage in inventory. The judgment stated in part:

> "Such accommodation, in our judgment, should preclude recovery by the insureds under this bond: if they had no proof whatsoever of an employee-connected loss other than inventory profit and loss computations, no matter how reliable in the particular case. On the other hand, inventory records may by their very nature constitute inherently indispensable proof of an allowable claim under a fidelity bond in one or the other or both of two respects:
> (A) as the only available proof of the full amount of a loss, there being some appreciable proof from other facts or circumstances of a loss caused by employee dishonesty.
> (B) as corroboration sufficient to make a case for the fact finder of the fact of an employee-connected loss where independent proof whereof, considered alone, might be considered insubstantial." (*Hoboken Camera Center, Inc. v. Hartford Accident and Indemnity Co.,* 226 A2d 439, 448 (N.J. App., 1967)).

The court here states clearly that not only may inventory records be used to establish the extent of loss after independent evidence establishes employee theft, but it proceeds in (B) to allow such records to be part of the evidence establishing the fact of employee theft. Though there are few instances of other courts reiterating this viewpoint, many courts have since concurred in (A) of this decision.

In any claim in which inventory records play a role, it is very important that such records be well kept. The better they are maintained, the more powerful will be any case in which their accuracy, perhaps even their fail-safe quality, is at issue.

Generally speaking, few insurance carriers will allow use of inventory records in the initial stages of the claim, to describe the amount of loss. However, there is sufficient case law to establish such records as a valid part of the claim, in spite of the exclusionary clause. Insurance companies will ultimately deal with the issue and arrive at a mutually satisfactory settlement. They are not anxious to go to court in those states where precedent has established the use of inventory records as valid in establishing the extent of claims, and are reluctant to go to court and possibly establish such precedent in those states where it doesn't already exist.

PROPERTY INSURANCE

Insurance against criminal acts is only one of the many kinds of insurance that must be considered in protecting any business against unforeseen eventualities. The kinds of insurance that are necessary and the amount of coverage in each category will depend upon the nature of the business and the extent of its exposure to various hazards.

In a general way, however, there are certain considerations that must be taken into account before any program can be settled upon. For purposes of this discussion, we will omit considerations of liability insurance and confine ourselves to property insurance, which covers structures, goods, equipment, cash, papers, records, and negotiables.

The first consideration in evaluating any kind of coverage must be the nature and extent of the losses covered. These losses could be classified as *direct*—meaning loss of certain tangible benefits resulting from destruction or damage to the element concerned; or *extra expenses* losses—meaning those costs resulting from loss or damage to the element concerned, such as the rental of office space and/or equipment after a fire has damaged the insured office and equipment.

Loss of Use/Extra Expense Coverage

Most standard policies do not provide loss of use or extra expense coverage. Since both these matters can represent a very substantial loss to most companies, consideration must be given to expanding the provisions of the coverage to include them. Both of these losses can be covered either by endorsement or by additional policies which will provide that coverage on a broad basis.

Even a small fire in an office may render it inoperable from smoke and water damage, or damaged equipment for a substantial period of time. Even though all the damage is covered and will be cared for, the interim period when revenues may be lost and new facilities temporarily occupied may be as expensive as the fire itself.

Such a situation might be covered by a "business interruption" contract. Here, too, there are options. A business interruption policy can be drawn up on a comprehensive basis, which means that it will cover a broad base of situations that might create a stoppage. Such a policy must, of course, be examined for types of incidents which are specifically excluded from coverage. On the other hand, such a contract might be drawn up in which the incidents covered are specified and perhaps limited to just a few potential hazards.

The amount of coverage and the nature of recovery in business interruption contracts can be complicated. If recovery is on an actual loss basis, a careful audit of actual demonstrable losses must be presented to the insuror in order to collect. If the policy is drawn as a valued loss contract, an accountant must certify the daily amount that would be lost if an interruption were to occur. This amount is entered as part of the contract. The premium and recovery are based upon this amount, figured on the specified number of days to be covered for each interruption.

What Is Covered

Every manager must consider the property to be insured and check to make certain that all the property he has designated is covered by the policy issued. Because of the changing nature of casualty insurance in today's market, certain property may be excluded from coverage because of its location, or because the nature of the business creates or subjects it to special hazards that are not included under the basic coverage.

In such cases, special policies or endorsements may be required to fill out the insurance program. If the rates for such coverage are deemed excessive, it might be well to reconsider the original insurance program and to provide a sharply limited coverage in these special areas, while at the same time developing special programs to reduce exposure and vulnerability.

Persons Covered. Generally speaking, property insurance is recoverable only by the insured and his agents. This is extended to include heirs named in a will or receivers in a bankruptcy proceeding. If there are others who should be named, such as non-equity lenders or other interested parties, they must be entered into the contract by endorsement to make certain their interests are protected. If such an endorsement is not made, such persons have no rights under the policy and they must seek relief by other means if a loss should occur.

Time Covered. The period of time covered must be checked as well. Most policies are good for a year starting at a specified time of day on the effective date of the contract, and are in force until a specified time of day one year from that date. Some policies are effective for a longer period of time, so it is important to verify the precise period of coverage referred to in the policy.

Incidents Covered. Property insurance may be specific or comprehensive. It may also qualify its coverage in either kind of policy.

Specific policies will name the incident or incidents which are covered, and will further specify the degree to which they are covered. For example, in a policy in which the incidence of water damage is covered, there may be some kinds of water damage which are excluded from the coverage offered in a particular contract. It is especially important for a risk manager to be aware of these exclusions, and to avoid the common mistake of supposing that, simply because he is covered by a policy that names water damage as a specific covered incident, he is therefore covered for all water damage from every source.

Comprehensive policies cover a wide variety of incidents, except for those specifically exempted. Here again, they may limit the degree of their coverage of these incidents. They may also specify conditions which will invalidate the provisions of the contract. If, for example, a contract is drawn insuring a building against fire, the contract may state that the provisions of the contract are invalid during any period when the insured permits fire hazards to increase in the insured building.

How Much Insurance?

Since the options are essentially a choice between recovery of the cash value of the property or recovery of its replacement cost, there can be little hesitation in making a decision. Few experts disagree that insuring to the amount of replacement is clearly the wiser course to follow.

When property is insured for its cash value only, there will almost inevitably be a loss to the insured unless property values decline enough to make up for the extra costs involved in replacement. The latter might include demolition of the remaining structure in the event of fire, clearing the site for rebuilding, or the declining value of the dollar. History shows us that property values, or more specifically building costs, rarely decline in this manner. Protection should therefore be arranged on the basis of resuming business as it was before the damage took place. This can normally be done only by insuring for the replacement cost of the property.

Replacement cost coverage is more expensive than cash value coverage, since the insuror must set the premium sufficiently high not only to cover the estimated likelihood of a fire occurring, for example, but he must also try to anticipate the rate of increase in the cost of labor and materials in the reconstruction of the building in whole or in part.

Even insuring a replacement cost will not, as a rule, cover the full cost involved in a major disaster. Business interruption, extra expenses, site clearance, intervening passage of new and more exacting building codes—all add to the already inflated costs of replacing the existing structure so that business may resume as before as rapidly as possible.

There are many endorsements available to extend coverage to fully compensate for all expenses involved in replacement. They should all be considered in the light of individual needs. Obviously, the greater the coverage, the higher the cost of coverage; but this cost must be weighed against the risk and its consequences.

INSURANCE PROBLEMS

The Small Business Administration has shown that 25 percent of businesses in this country have had some kind of problem with property insurance in any 12-month period. These problems include cancelled policies, refusals to issue or renew insurance, prohibitive rates, and limiting coverage to well below the cash value of insured property.

In inner city locations or in certain types of business, policies, when issued, substantially limit the insuror's liability—frequently to the point where the policy is virtually useless as support protection.

Federal Crime Insurance Program

These actions by the insurance industry have created enough concern in the small business community to call for some kind of remedial action on the part of the federal government. Effective August 1, 1971, the Federal Crime Insurance Program came into being. This program, which requires the participation of individual states, provides for federally funded crime insurance at reasonable rates, based on the size and accepted risk of the insured property. Coverage is limited to a maximum of $15,000.

In order to qualify for protection under this program, however, a business must establish certain minimum protective devices and procedures. The businessman must, in short, recognize that he can get the supportive protection insurance offers provided he makes at least minimal efforts to protect himself.

The program prescribes locks, safes, alarm systems, and other protective devices, and establishes the kind of protection that various kinds of business must provide for themselves in order to qualify for this insurance. For example, gun stores, wholesale liquor and fur stores, jewelry firms and drugstores must all have a central station alarm system. Service stations must have a local alarm system, and so on. Small loan and finance companies, theaters and bars—businesses rated as high risk—are also eligible for insurance under the program.

The program still has a long way to go before it covers firms in every state, but its appearance on the scene is encouraging. Not only does it provide for insurance coverage of premises otherwise difficult or impossible to insure adequately or reasonably, but it also focuses attention on the very real need for the insured to take positive steps to provide protection of the premises to prevent loss, and to use insurance to defray those losses that do occur only when those security measures fail.

In short, it takes insurance from the front line of crime prevention— where it clearly cannot perform—and puts it into the reserve or back-up position where it can.

REVIEW QUESTIONS

1. Do you agree with the statement that "insurance must be thought of as a supportive rather than the principal defense against losses due to crime"? Why?
2. Why is it important for the insured to clearly understand the terms describing criminal activity (e.g. burglary, robbery)?
3. In terms of property insurance define the following: direct loss, loss of use, and extra expense losses.
4. What are the differences between SPECIFIC and COMPREHENSIVE property insurance policies? If you were a business executive, which would you prefer, *IF* you had to give priority to cost-effectiveness?
5. What were the problems that led to the establishment of the Federal Crime Insurance Program? What are some of the crime-preventive measures prescribed by the program?

PART IV
SECURITY APPLICATIONS

Chapter 13

RETAIL SECURITY

At the retail end of the distribution chain, the merchant is beset on all sides by assaults on his profit and loss position. The very nature of his business demands that quantities of merchandise be attractively displayed in easily accessible areas. The public can roam at will and handle much of the merchandise. Every effort is made to create a desire to possess the merchandise on hand, and every effort is made to make the possession as effortless as possible. Of course, the merchant expects payment for the goods. Others—customers and employees alike—sometimes overlook that aspect of the transaction, and the merchant takes a loss. Generally, each such loss is small, but the aggregate damage to his business from such erosion of his inventory can be enormous.

Most businesses are subject to problems of inventory shortages, but few feel the problem as acutely as the retailer. He necessarily deals in merchandise which must be received, stored, moved from warehouse or storage rooms to display areas on the selling floor. All of these operations pose a risk of loss from breakage or other damage, pilferage or quantity theft. Even inadequate or careless record keeping can effectively "lose" merchandise which fails to show up on proper inventory records. Merchandise on display is fair game for shoplifters during the day, and when the day is over the whole cycle begins again.

The three principal sources of loss to the retailer are the external losses from theft, the internal losses from employee dishonesty, and the losses that come from carelessness or mismanagement. Every area of loss must be counteracted in some way or the retailer may find that, while his gross business is booming, he is barely able to break even.

The arithmetic of it is as simple as it is familiar, but it bears repeating. If a supermarket operating at a net of one percent suffers inventory shortages from all sources amounting to $250 a week, it would have to have annual sales of $1.25 million worth of merchandise just to break even. Estimates of the Small Business Administration put the total loss to the retailing business from crime at $4.8 billion, in 1971. By 1975, according to the U.S. Department of Commerce, this figure was $6.5 billion. A 1978 report estimated annual retail losses due to shoplifting and employee theft at $10 billion, and another in 1979 put it at $12 billion.

Generally, shortage control procedures apply to retailers of every kind—from all-night restaurants to department stores. Legal considerations, surveillance techniques, cash register control, and the many other factors involved in a security program for a retailing facility are much the same. The details vary, and every establishment must ultimately make its own determination of what is best for its own application, but the basics are much the same throughout the trade.

SHOPLIFTING

Retailing today demands that merchandise be prominently displayed and exposed to enable customers to see it, touch it, pick it up and examine it. There is little likelihood that merchandise will ever go behind the counter again in the great majority of stores that display it so invitingly. In this mode it is hardly theft-proof, but its theft must be controlled if profit margins are to be maintained. Shortages from shoplifting are not likely to be eliminated, but they must and can be reduced by a thoughtful and energetic security program.

Extent of Shoplifting

It is difficult to accurately pinpoint the full dimension of the problem presented by shoplifting. Few shoplifters are apprehended, and of those who are, even fewer are referred to the police. In fact, "the most optimistic studies of retail effectiveness in spotting shoplifters indicate that stores who do a good security job apprehend no more than one out of every 35 shoplifters."[52] (Another report estimates that the figure is only one in 200).[53] Stores with vigilant personnel will report a relatively high level of shoplifting, while analogous concerns (in terms of store type, location, size) with indifferent anti-shoplifting programs report few such incidents.

A "West Coast study of 9,000 food store shoplifting cases indicated that more apprehensions are made in January than during any other month. Department stores, on the other hand, apprehend more shoplifters during the busy weeks before Christmas as well as during the week after the holiday. Unlike general larceny arrests, which according to national statistics peak in the month of August, a high incidence of shoplifting arrests is not reported during that time. However, most stores allow detectives to take vacations in August; those that retain a full detective staff during that month have found shoplifting apprehensions equal or exceed December arrests."[54] This experience has been frequently reported. It would appear that, where the vulnerability remains constant, the potential of shoplifting incident remains constant, ultimately varying only with the effectiveness of security measures. The unhappy lesson is simply that if you look for shoplifting, you can find it.

An interesting analysis based on the experience of 632 supermarkets in 18 Western supermarket chains suggests that shoplifting occurs at least six times per day per supermarket. The analysis further suggests that, since other reports have concluded that such incidence is considerably higher, by assuming only 10 thefts per store per day and extending the experience to all chain supermarkets, the national loss from shoplifting alone may be estimated at $450,000,000 annually.[55]

The American Management Association, in adjusting the Small Business Administration 1967 figure of $.5 billion in losses due to shoplifting, produced a figure of $1.6 billion for 1975. By using a different method of calculation, the AMA report placed 1975 shoplifting losses at $2 billion for all retailers. The precise dimensions of the problem can only be guessed at, but whatever figures are used as a starting point for such guesswork, the total loss to shoplifters is huge.

Interestingly enough, this analysis shows that, in spite of publicity to the contrary, adults comprise 60 percent of those apprehended and 75 percent of the loss in dollar value as a percentage of recovery. As between men and women, the division is almost exactly equal.

The same article reveals that, based on the only random study of department store shoplifting ever released, one out of each 15 customers entering a downtown department store will probably steal; each theft will average between $3.69 and $7.15; less than one percent will be apprehended; and age and race were not significant factors. It is interesting to note that men were involved in such incidents 25 percent less than women, and there seemed to be a correlation between frequency of incident and value of goods taken. The store with the highest frequency of observed shoplifting suffered losses averaging $7.15 per theft, whereas that with the lowest incidence lost $3.69. This test was confined to a study of downtown general merchandise department stores in major metropolitan areas and cannot be taken as representative of retailing in general; but the information may serve as a guide to more extensive conclusions.[56]

Methods of Shoplifting

Shoplifting is conducted in every imaginable way; the ingenuity of such thieves is legendary. By and large, the great majority of thefts are simple and direct, involving nothing more sophisticated than putting the stolen merchandise into a handbag or a pocket. But there are certain methods beyond the simple taking of items that are in general use and should be anticipated.

Bags and packages are frequently used to conceal stolen goods. Bags may be those taken from the store victimized. In some cases, items may be bagged by a clerk and stapled after the items have been paid for. The thief removes the staple, stuffs in stolen articles, and restaples the bag with a pocket stapler. Many items are packaged in such a way that they provide concealment for other stolen merchandise within them. Oversized cereal boxes, utensils, and unsealed boxes of bakery products are a few of the store-provided containers used to carry out small items.

Bulky clothing provides a number of possibilities for concealment. Small items can be held inside the clothing under the arms; larger ones can be carried between the legs under a skirt or overcoat.

Ticket switching is common in all stores where the price of merchandise is marked on the item in some way. Sometimes tags are exchanged with cheaper merchandise; sometimes prices are partially erased or altered. If prices are marked with erasable markers, they can be changed easily to indicate a lower price.

Who Shoplifts

The shoplifter comes in all sizes and ages, and is of either sex. Generally they are broken down by type into the professional, the amateur, the drug user, and the thrill seeker. Of these the amateur is by far the largest in number.

The amateur comes from every economic group and represents every level of education. His thefts are generally impulsive, although a significant number of them find some kind of economic or, more often, emotional satisfaction in their action, and they become virtually undistinguishable from the professional. The rest of them have no particular pattern of theft and may only steal once or, at most, a handful of times. Individually they do not represent a severe threat to the retailer, but the cumulative effect of such thefts, however motivated, can be very damaging.

The frequent repeaters ultimately become more methodical in their thefts and soon become a real problem. Among this group are those compulsive thieves known as kleptomaniacs—people who are unable to overcome their desire to steal. Such driven souls are very rare and do not make up a significant number or dollar loss to the retailing community.

The professional shoplifter is a very real danger. His methods are well-planned and practical. He appears in every way as an ordinary shopper, carefully fitting into the environment of the store he singles out. He selects merchandise with a high resale value and a ready market. He is well connected with fences and lawyers. He is in every sense himself a supplier in the sub-system of illegal merchandising. He is the first man in the chain of underworld retailing and his activities are damaging in a number of ways to the legal storekeeper. Not only does he create severe losses for the legitimate retailer, but he sets up a system whereby he is effectively in competition with his victim's own goods.

The drug user, trapped in his addiction, must find a regular source of funds to supply his needs. He turns to many sources for his insatiable demands, but shoplifting is often the easiest and most lucrative. Thefts of $500 a day may be required to supply his habit. Typically, merchandise is stolen for fencing at between 10 percent and 20 percent of its retail value. In other cases it may be stolen and returned for refund. Either way, the store suffers substantial losses.

Thrillseekers are, more often than not, teenagers who shoplift as a gesture of defiance or under peer pressure to do something daring. According to a 1979 survey by Central Service Systems of California (based on 20,111 apprehensions reported by clients), the 12 to 17 years age group apprehensions were 2.5 times its percentage of the population or 26.7 percent of all incidents. Sixty-six point six percent of all incidents were committed by shoplifters 29 years of age and younger.

Preventive Measures

Surveillance. The key to successful shoplifting is surveillance. Impulse theft, which comprises 95 percent of shoplifting incidents, is motivated by availability, desire and opportunity. Availability is a basic fact of life in modern retailing. Desire is not only a private matter, individual in character, but cultivated by the merchant aggressively selling his wares. Neither of these factors can be controlled to a significant degree, not should they be in retailing. But opportunity can be controlled.

The shoplifter characteristically snatches his loot when he thinks he is alone—when he is not being observed. An attentive sales staff occasionally asking if they can help, or rearranging merchandise in the vicinity of a customer acting in any unusual manner, can discourage most amateurs. Supervisors moving about the floor can also be effective in making known to the potential shoplifter that he may be observed at any time. Obviously such store personnel are primarily concerned with serving customers. But they can be an effective deterrent to shoplifting as well if they are aware of the problem and alert to any signs that might indicate a problem.

Mirrors. Convex mirrors, which are in wide use throughout the country, may be useful to avoid collisions of people or shopping carts rounding corners, but they have a limited use in the detection of shoplifters and may even hurt the program by creating an atmosphere of unwarranted confidence. Such mirrors distort the reflected scene in such a way that it is virtually impossible to see the details of action—and in shoplifting it is the detail of merchandise concealment that is of prime importance. Without a clear, precise image of what was taken and how it was concealed or where, it would be foolhardy, and possibly costly in legal fees, to confront a customer as a shoplifter.

On the other hand, flat mirrors of a decorative design might be built into the decor of the store at strategic spots which might otherwise be difficult to observe or keep under surveillance. Such mirrors, presenting a clear, undistorted image, can be very useful in the security effort.

Signs. There is considerable controversy over the use of signs warning of the results of shoplifting as a deterrent. Many merchants take the view that such signs are an insult to the great majority of honest shoppers who may become incensed and take their business elsewhere. Other merchants subscribe to the theory that such signs have no effect on honest people, since they clearly are not those to whom the message is addressed, but will remind those with larceny in mind of the gravity of their offense and thus deter them. There does not seem to be a clear-cut resolution of these viewpoints, except to say that there is no evidence anywhere that the posting of such warnings has ever had any effect on the incidence of shoplifitng.

Displays of Merchandise. Merchandise displays must be appealing to attract customers, but they can also be secure to prevent theft. Symmetry in certain kinds of items displayed can be important in enabling the clerk or floor personnel to tell at a glance if any of the items are missing. Thin, almost invisible wires that in no way detract from the display can secure small items to the display rack. Dummy items look exactly like the actual merchandise and should be used when possible. Fountain pens can be displayed in a closed case, with two or three different models on counter chains outside for handling and testing. There are thousands of items and countless ways to display them to catch the customer's attention. Each such display must accommodate some means to provide security for the goods it presents.

Check-out. Check-out clerks should check all merchandise for signs of switched or altered price tags.

Customers should never be permitted to carry out unwrapped or unbagged merchandise. All purchases should be wrapped or bagged, and if the additional precaution of stapling is taken, the receipt should be stapled to the package.

Checkers in supermarkets must check shopping carts for merchandise on the bottom shelf. They should further check all merchandise purchased for possible concealment of other goods. This includes paper bags holding purchases of produce. Other items may be concealed beneath the apples or potatoes they contain.

Refunds. Refunds should be issued on the return of merchandise with the original sales slip. Since these are frequently misplaced, if the customer insists on a cash refund, full particulars including the name and address of the customer should be entered on the appropriate refund form. The original of the form, which has been signed by the clerk making out the form, the customer, and the authorizing supervisor, should be presented to the customer for cashing at a refund window or other location handling such matters. Cashing of refunds should never be permitted at cash registers on the floor, since this practice could invite embezzlement by the operator of the register. A copy of the refund authorization should be turned in at the end of the day's business by the clerk who handled the transaction. All such refund forms should be numbered and accounted for, including damaged ones.

All customers must cash their own refunds, to avoid the possibility of forged slips supposedly being cashed by store personnel as a ''service'' to a nonexistent customer.

The refund system should be further checked by periodic audits of all refund forms used and unused and by contact with the customer so refunded.

Letters should be sent out regularly to a certain percentage of customers who have received refunds, asking if the service was adequate, and if their request for refund was promptly and courteously complied with. If such letters are returned as undeliverable, or if the customer denies having received a refund, an investigation of refund procedures is indicated.

Detecting the Shoplifter

Professional shoplifters will not be deterred by the normal means that would discourage the great bulk of amateurs from stealing. Only a well-trained security staff can apprehend them. "Most specialists agree that a store should be staffed with at least one floor detective for every $4 million it earns in gross sales."[57]

Such detectives must learn to blend in with the normal routine of the average customer in a given store at a given time of day. He must learn the different patterns of the secretary on her lunch hour, the bored matron who shops to kill time, the energetic early morning customer with a specific mission in mind. He must observe the difference in pace of customers in the 10 a.m. crowd and the hurry-to-get home 5:30 shoppers. After learning the techniques of anonymity, the detective must learn what to look for—how to spot a potential shoplifter.

A list of some of the signs to look for was developed by a large Mid-western store and includes the following:

1. Packages. A great many packages; empty or open paper bags; clumsy, crumpled, homemade, untidy, obviously-used-before, poorly-tied packages; unusual packages—freak boxes, knitting bags, hat boxes, zipper bags, newspapers, magazines, school books, folded tissue paper, briefcases, brown bags with no store name on them.
2. Clothing. A coat or cape worn over the shoulder or arm; coat with slit pockets; ill-fitting, loose, bulging, unreasonable, and unseasonable clothing.
3. Actions. Unusual actions of any kind; extreme nervousness; strained look; aimless walking up and down the aisles; leaving the store but returning in a few minutes; walking around holding merchandise; handling many articles in a short time; dropping articles on the floor; making rapid purchases; securing empty bags or boxes; entering elevators at the last moment or changing mind and letting elevator go; excessive inspection of packages; examining merchandise in nooks and corners; concealing merchandise behind purse or package; placing packages, coat or purse over merchandise; using stairways; loitering in vestibules.
4. Eyes. Glancing without moving the head; looking beneath the hat brim; studying customers instead of merchandise; looking in mirrors; quickly glancing up from merchandise from time to time; glancing from left to right in cross aisles.
5. Hands. Closing hands completely over merchandise; palming; removing ticket and concealing or destroying it; folding merchandise; holding identical pieces for comparison; working merchandise up sleeve and low-

ering arm in pocket; placing merchandise in pocket; stuffing hands in pocket; concealing ticket while trying on merchandise; trying on jewelry and leaving it on; crumpling merchandise (gloves and merchandise).

6. At counters. Taking merchandise from counter but returning repeatedly; taking merchandise to another counter or to a mirror; standing behind crowd and taking merchandise from counters; placing merchandise near exit counter; starting to examine merchandise, then leaving the counter and returning to it; holding merchandise below counter level; taking merchandise and turning back to counter; handling a lot of merchandise at different counters; standing along at counter.
7. In fitting rooms. Entering with merchandise but no salesperson; using room before it has been cleared; removing hangers before entering; entering with packages; taking in two or more identical items; taking in items of various sizes or of obviously wrong sizes; gathering merchandise hastily, without examining it, and going into fitting room.
8. In departments. Sending clerks away for more merchandise; standing too close to dress racks or cases; placing shopping bag on floor between racks; refusing a salesperson's help.
9. Miscellaneous. Offering questionable refunds; acting in concert —separating and meeting; setting up lookouts, interchanging packages, following companion into fitting room independently.[58]

Obviously, many of these are the actions of a perfectly well-intentioned person, but they may indicate a shoplifter, especially if several such indications appear in the actions of one person. In such cases surveillance is essential.

Arrest

There are many problems involved in making an arrest of a shoplifter. Chief among them, after considerations of justice and fairplay is the ever-present possibility of liability for false arrest or imprisonment, slander, or unreasonable detainment. Since the laws pertaining to shoplifting vary from state to state and since they are still subject to changes to conform to an equitable balance between the needs of the merchant and the protection of the general public, it is essential that legal advice be sought to guide company policy in such matters.

Every store management team should develop a specific set of instructions, and these instructions should be taught in such a way that adherence to them will be automatic in every covered instance. Any variation from approved procedure in dealing with shoplifting incidents can subject the store to severe financial reverses and damaging public relations consequences. All personnel must conform to the established policy. Since the owner is liable for all acts of his employees while they are on the job under the doctrine of *respondeat superior,* the term ''merchant'' will be used in this discussion even though most of the acts herein dealt with are usually performed by employees.

Under common law, the merchant discovers that he is operating in a most dangerous legal mine field. Although tort law allows a property owner the privilege of defending it against theft and even allows him to repossess his goods if they have been wrongfully removed from his premises, this privilege is not absolute but is limited in its scope. The privilege is entirely dependent upon the fact of wrongful taking. If the suspicion of theft turns out to be groundless, the privilege vanishes and the merchant is vulnerable to a tort action which can result in substantial damages—both compensatory and punitive.

The basis for such actions may be one or more of five torts.

1. False Imprisonment—Illegal restraint of a person against his will. Any exercise of force, or express or implied threat of force by which a person is deprived of his liberty or is compelled to remain where he does not wish to be is defined as ''imprisonment.'' False imprisonment may be committed by either words or acts alone or both together.

It is important to remember that the person need not be touched nor confined in any way in order for him to be imprisoned. He must feel he is being restrained by force or fear of force.

Of course, consent by such person is a defense to the action.

2. False Arrest—Any unlawful physical restraint of any person's liberty. This tort is very close to the tort of false imprisonment and often is seen in the same action with it. Though some authorities treat false arrest and false imprisonment as identical torts since both derive from the old common law action of trespass, they may both be filed in some situations. Although there is considerable variation in the statutory rules governing legal arrest from state to state, generally the right of a private citizen to make a legal arrest depends on a crime having, in fact, been committed and his having reasonable grounds for belief that the person arrested committed it. A police officer, however, may arrest without a warrant where he has reasonable grounds to believe a crime has been committed (whether or not a crime *was* committed) and that he has the criminal. Since most shoplifting involves only a misdemeanor, under common law neither private person nor officer may legally arrest without a warrant. But since most states allow both to arrest for a misdemeanor committed in their presence, such an arrest is usually, in fact, legal. In those jurisdictions where such misdemeanor arrests without a warrant are not authorized by statute, they are unlawful.

3. Malicious Prosecution—An action instituted with the specific intent to injure the defendant, knowing there is no proper basis for the charge. There are four elements, *all* of which must be proved in order to sustain the action.

1. A criminal proceeding was instituted by the defendant against the plaintiff. If the merchant (the defendant) simply files a report and the police, on their investigation, bring a charge, this element is not proven.
2. The proceedings against the plaintiff were either dismissed or the verdict, "not guilty."
3. Absence of probable cause to initiate the proceeding.
4. Malice must be proven. Malice need not be spite or ill will. If the purpose in instituting the action is other than bringing an offender to justice, even if it is to recover property or to collect a debt, that is adequate to establish malice.

4. Slander—Slander is the speaking of base and defamatory words which may prejudice another's reputation or means of livelihood.

A direct accusation is not necessary to constitute slander. Almost any question relating to payment for merchandise may be viewed as imputing theft.

It is necessary for the words to be spoken in the presence of another and loud enough so that one or more persons did, or could have, heard them.

In common law, truth of the words or their implication is a complete defense in a suit.

5. Assault—An assault is an intentional offer of force to do another injury. No contact is involved. Threatening words or gestures constitute assault, and can be the basis for an action.

Battery is actual contact whether it be a blow or simply an offensive touching. Searching a suspect's person may constitute battery. A merchant does have the privilege to recover his property promptly. However, no matter what the facts, he may not use unreasonable force and furthermore, if the subject does not have the property, the merchant will be liable.

It is plain that under the common law, merchants were beset with legal perils in the protection of their merchandise. Judgments against overly aggressive merchants were large enough and frequent enough to discourage many from taking direct action against shoplifters. Since individual thefts by shoplifters were usually of minor dollar value, it was only the minority who were willing to risk the possibility of damaging judgments to recover a few dollars worth or merchandise and, possibly, help to deter the continuing erosion of their profits by shoplifters.

State legislatures responded to the plight of merchants by enacting statutes which would provide some degree of protection to them in their efforts to recover their property and prosecute those guilty of larceny on their premises.

Generally speaking, these statutes provide the merchant protection from some of the hazards he faced under the common law. Although there are a variety of different wordings in the various state enactments, most of them aim to immunize the merchant from liability if he has *probable cause* to believe the person he detains has concealed his merchandise with an intention to steal it. This immunity is usually conditional upon the detention being for a reasonable time and conducted in a reasonable manner. Most of the statutes specify the purpose or purposes for which the detention is permitted. Most of them specify investigation, and many also specify the recovery of stolen goods. Only two states specify search as one of the protected purposes of the detention, but since recovery of property is mentioned by 15 states, it would seem reasonable to suppose that searching the suspect in order to regain property would be protected by probable cause if a mistake had been made. There seems to be no case history on this score, however, so the question remains unanswered as to how far the courts will go.

In many states, any effort to conceal merchandise is taken as presumption of intent to steal. Such enactments serve to aid the merchant in prosecutions. Such statutes by themselves, however, do little to protect the merchant from liability. And, in all but four of the 20 states with such statutes on their books, ''probable cause'' statutes are added.

The complexities of the legal climate governing the war against shoplifting require a sensitive understanding and the thorough briefing of an attorney experienced in the field.

Store policies on procedure of arrest and detainment must be predicated on legal advice, and they must then be followed to the letter. As a general rule, it should be noted that in cases of detainment where neither a confession of guilt nor a form releasing the store from any liability has been signed, it will be necessary to prosecute to avoid the suit for false arrest that would likely ensue.

Prosecution

The argument over when and whom to prosecute shows no sign of abating; certainly there appears to be little agreement among security professionals. The skew is probably more toward a tough attitude than otherwise, but every shade of opinion has its adherents.

Every study undertaken on the subject has shown that the rate of recidivism is enormous. The FBI has published figures indicating that 83.4 percent of those arrested for larceny in 1969 had been arrested before for the same crime. Such a figure suggests that neither arrest nor the fear of arrest is a very effective deterrent.

This notion was pursued further by a study with highly significant implications. ''A study of 400,000 case records of retail offenders who were either prosecuted or released by 30 leading stores in one city showed that one in four shoplifters repeated, some as often as 12 times. The discouraging findings induced a member of the local retailer's protective association to conduct a seven-year study of 400 shoplifters: 200 were selected at random as control subjects to be prosecuted or released according to the usual standards, and 200 were released on the condition that they would place themselves under the care of a psychologist or psychiatrist. Both groups were followed up over a period of seven years.

"The 200 control subjects followed the expected pattern, with one in four of this group repeating the offense and being arrested over and over again by member stores. But of the 200 who sought professional guidance, in seven years not a single one was picked up a second time by any of the association's member stores."[59]

The results of this study would seem to point a way to a solution which neither stern prosecution nor indiscriminate release has provided. It would, further, seem to bear out the contention of the eminent Dr. Karl Menninger, who said: "My answer is that we, the designated representatives of our society, should take over. It is *our* move and our move must be a constructive one, and an intelligent one, a purposeful one—not a primitive, retaliatory, offensive move. We the agents of society, must move to end the game of tit-for-tat and blow-for-blow in which the offender has foolishly and futilely engaged himself and us. We are not driven, as he is, to wild and impulsive actions. With knowledge comes power, and with power there is no need for the frightened vengeance of the old penology. In its place should go quiet, dignified, therapeutic programs for the rehabilitation of the disorganized person. If possible, the interest and help of a professional person should protect society during the offender's treatment, and should guide his return to useful citizenship as soon as this can be effective."[60]

CHECKS

The boom in private checking accounts which has been a part of the American economic scene for the past 25 years has led us to the point where business analysts and economists predict the cashless economy within the foreseeable future. At that time, they suggest, even checks will become passe, and all transactions will be based on debits and credits handled through instant recording of the exchange of goods or services, by centralized computers. Such computers would register a credit to an employee's account, credits where appropriate to various government agencies, and debit the account of the employer in the amount specified on payroll information. All obligations of the employee from mortgage payments to the expenses of a vacation trip would be handled in the same manner—by a simple series of ledger entries in an interlocking central computer system. Personal bank checks in use today are crude and primitive version of this predicted streamlined system.

With personal checking accounts approaching the 100 million mark, and with over half the adult population handling its cash through such bank checks, it would appear that the physical handling of cash has been delegated to the banks. That day has not, of course, arrived, but it seems close. Even today almost 90 percent of all business transactions are handled by check.

Nature of a Check

A check is nothing more than an authorization to the holder of funds or bank to debit the account of the authorizer and credit the named person or the bearer in the amount specified. Provided that sufficient funds are available in debited account, the exchange is made either in cash or by crediting the account of the payee. This system is really only a version of the most rudimentary kind of bookkeeping accomplished by the exchange of tons of paper in a never-ending chain of authorizations.

Checks can and have been written on almost anything, and in every conceivable form. Their legality as a negotiable instrument is in no way dependent on the usual check form in common use today. As a practical matter, many banks refuse checks other than those written on the forms they provide, simply because their procedure for processing the enormous volume they are called upon to record does not allow for personal eccentricities. The day when a Robert Benchley, a popular humorist of the 1930's and 1940's, could write checks in the form of risque poems on wrapping paper, is gone. Not because such a check duly signed and clear in intent is not a legal instrument for transferring funds, but because few people would cash it.

Checks are in such wide use today because they are safe and convenient. Since cash can almost never be identified if it lost or stolen, it is almost gone forever. Checks, on the other hand, have no negotiable value except as they are drawn and signed by the payor. Forgery is, of course, a problem, but that is considerably less of a risk than cash itself, which requires no criminal expertise of any level for its disposal.

Checks are invaluable as receipts of payments made and for recording an individual's accounts for tax and other purposes.

Checks can theoretically be used only by the designated payee. Therefore, if a drawn check is lost or mutiliated, payment can be stopped by notifiation to the bank and a new check can be drawn. Checks lost or stolen in the mail can be handled in the same way, whereas mailed cash is always at considerable risk of non-delivery because of theft, loss, or destruction.

Checks are a ready source of cash at any time. Many people are cautious about carrying substantial amounts of cash for fear of loss or theft. In the event a situation requiring funds arises, a check can provide the money necessary.

From the retailer's point of view, checks are a boon, since customers will buy what appeals to them at the moment and handle the transaction with the stroke of a pen. People who buy only with cash tend to be considerably more restrained in their buying habits. They are limited by the amount of cash they are carrying, and the psychological restraints imposed by the actual doling out of cash are considerably more persuasive than those involved in drawing a check. The operators of gambling casinos discovered many years ago that most players are cautious when handling cash, which has a reality that is automatically translated into food and doctor bills, car payments and the like. Chips in such establishments become simply tokens or scoring devices in a game and normal reticence is soon discarded. The check casher is usually a bigger buyer who allows his impulses greater sway in a retail situation.

Checks and the Retailer

It has been variously estimated that somewhere between 70 percent and 80 percent of all bank checks are cashed in retail stores. It has been further estimated that almost 30 billion checks are now written annually. With about three quarters of these checks handled initially by retailers, it can be seen that the scope of this traffic is tremendous. Retailers are standing in for banks as the major supplier of cash in the country. Banks are in no way relieved of their heavy burden of accounting or of their responsibility as the ultimate repository of the funds drawn on by these documents, but they are no longer the primary source of supply of funds to the consumer. They have, in effect, thousands of agents or branch offices who perform line functions for them.

This means that retailers must keep huge supplies of cash on hand to service check-cashing customers. This not only raises the risks of internal and external theft, but it serves to increase the cost of goods, since cash that might otherwise be invested in additional merchandise or other income-producing ventures must be held in reserve to handle the damands of customers.

Some economists take the view that this situation has its bright side, in that it forces many enterprises who might otherwise over-commit their funds to maintain a position of sufficient liquidity to protect against reverses. Nonetheless, many major retailers feel that their options are limited. Whatever the state of the economy, they must keep large supplies of cash available to provide what is traditionally a banking service.

Bad checks of various kinds add to the costs of doing business. Forgeries or fraudulent checks are a direct loss of both merchandise and cash. Checks which are ultimately collectable, but are returned by the bank for any number of reasons present a huge administrative headache in making the collection. Even though a majority of such checks are soon made good, often after a single letter or phone call, the costs in such follow-up and double handling, as well as the additional costs of a collection agency if such is required, may bring the costs of each such returned check to between $3 and $5. A charge for any returned check is usually added onto the eventual colleciton, but most retailers feel that they do not collect the full amount. Such returns actually cost them in direct and indirect losses.

Types of Checks

There are many kinds of checks representing different types of transactions that may be presented for cashing to a retailer. Since eventually the store must establish a check cashing policy which will involve the kinds of checks it is willing to cash, a brief list of check types is in order.

The *bank check* is that most commonly in use. It is normally a rectangular piece of tamper-resistant paper that will reveal any attempts at erasure or alteration. On it are lines for date, the amount (in both script and numerals), the name of the payee and the signature of the payor. Most checks are personalized by the imprint of the name and address of the account holder, and they carry the name and address of the bank in which the account is held. These checks also have a space in which the check can be numbered by the payee for his accounting convenience, or they may be prenumbered by the bank. They will usually, nowadays, also have a certain amount of computerized information (such as account number) for the bank's convenience in processing.

These checks are precisely as good as the credit or the account of the person drawing them, assuming that person is who he claims to be.

Blank checks are simply forms which can be filled out as above, but they contain no information such as bank name, payor name or account information. They must be filled out to designate the bank on which they are drawn as well as other pertinent information. They must be hand processed by the bank and are, therefore, unpopular or frequently unacceptable to many banks.

Travelers checks are checks drawn to a certain designated amount and purchased from various banks and other agencies. At the time of purchase each check is signed by the purchaser. When they are cashed, they must be signed again in the presence of the casher to allow him to compare the signatures. Since these checks are serialized, they may be recovered in the event of loss or destruction, provided a record of these numbers has been retained.

Payroll checks are simply bank checks drawn to an employee. Since they are issued on the account of a reputable firm, they are usually acceptable in retail establishments although they can be, and frequently are, forgeries.

A *third party check* is one issued by one person to another. The merchant may cash such checks provided the designated payee endorses it as having been paid.

Government checks, whether for social security, tax rebates, salary, or any of the many uses to which they are put, are essentially cash, except that the cash is assigned to a specific named person. Because the solvency of the payor is never a question, these checks have always had a lively traffic among thieves and forgers. It is particularly important that the identity of the payee be estblished and that these checks be examined for any signs of tampering.

Returned Checks

Generally speaking, checks are returned because of insufficient funds, closed account, or no account in the bank on which they were drawn. Other problems such as a damaged check, illegible signature or amount or improperly or incompletely drawn checks may cause the bank to return them for clarification, but this latter miscellaneous group represents a small minority of the returns.

''Insufficient funds'' returns are in the majority. Usually they indicate that the issuer is a bad bookkeeper, but he may also have in mind a loan from the store. He will cover the check eventually, but in the meantime he has the cash to tide him over.

''No account'' and ''closed account'' are almost invariably fraudulent checks. It is possible that such checks were issued in error (such as transferring a bank account before all outstanding checks have cleared), but it is unlikely.

Check Approval

The two most important and relatively simple steps that can and must be taken in approving checks are a thorough examination of the check itself and a positive identification of the check casher (see Figures 13-1 and 13-2). Neither of these techniques will screen out the accomplished forger or a skilled bank check artist; neither will they eliminate the problem of the checks issued by honest but careless people which are returned by reason of ''insufficient funds.'' But they will reduce or possibly eliminate a great deal of the most persistent losses.

Retailers are not obliged to cash checks. They do so in their own interests, as a service. If a check looks dubious or the casher appears in any way to be other than what he represents himself to be, the check should be refused.

Since, however, he will cash most checks presented, he must know what to look for. He must examine the check to verify the date, the amount of which is drawn both numerically and in script, the name and address of the customer, the name of the bank, the signature of the customer, and the endorsement if it is a third party check of any kind, including payroll or government checks.

He must also assure himself that the customer is properly identified. Generally, any official document which describes the holder and bears his signature can be accepted as adequate identification. One that also includes a suitably laminated (or otherwise affixed) photograph is even better. Such documents can be fabricated by an artful forger, but since forging of this kind requires some equipment and skill, it is not in wide or general use.

A driver's license, passport, national credit card, birth certificate, or motor vehicle registration are all acceptable identifications. Club or organization membership cards, Social Security cards, hunting licenses, and employee passes are not valid as dependable pieces of identification, since they are easily obtained or easily duplicated.

The best identification is by an authorization system established by the store itself. Credit cards or check cashing cards issued by the store provide a running record of the customer's account and serve as nearly positive identification for check cashing purposes. The cost of establishing such a system can be well worth it in stores suffering from substantial losses due to fraudulent or uncollectable checks.

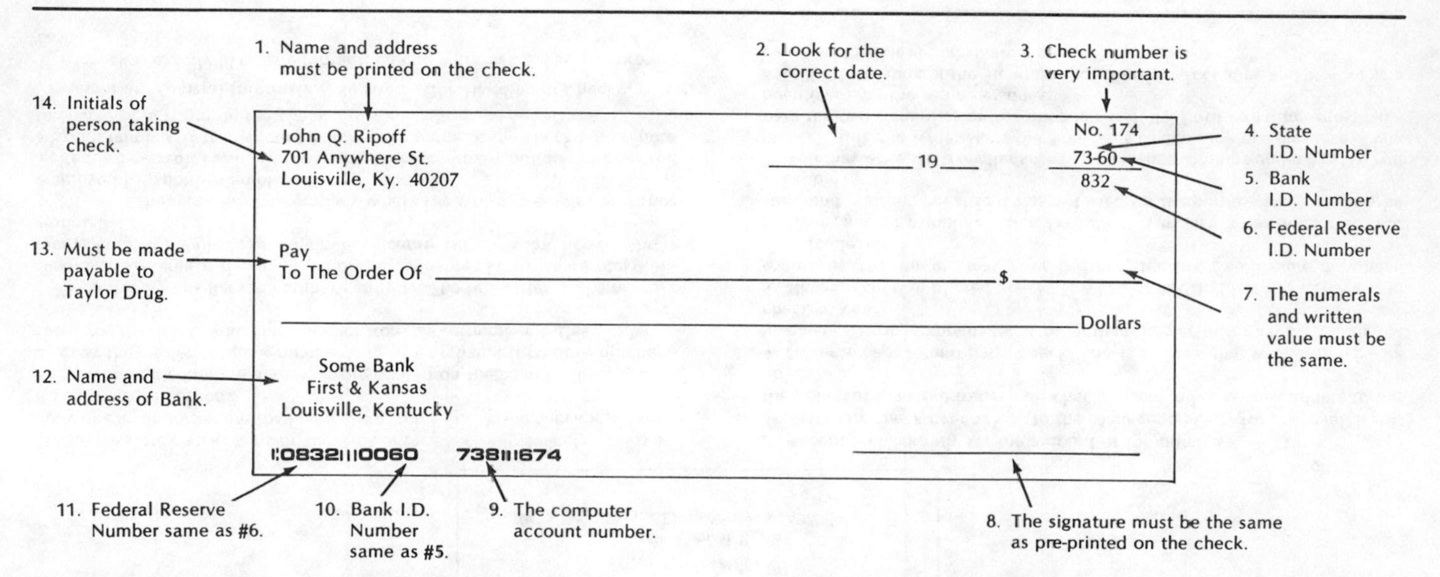

*1. Accept only those checks with the *name and address pre-printed* on the check. The address must be a street, *not a Post Office Box.*
2. *Accept only those checks with the correct date.* Checks over 30 days old are not covered by the bad check law. Never accept a post-dated check . . . a check that has a future date on it.
*3. Watch for low sequence numbers—Numbers 200 or less. It is a proven fact that a large number of "cold" checks have a check number of 200 or less.
4. Each state has an identification number. Kentucky is #73; Louisville #21.
5. Each bank has an identification number. *Some Bank* is #60. The bank number is shown in 2 places on the check—#5 and #10.
6. Identifies the geographical location within the Federal Reserve System.
*7. The amount of the check in *numerals and written value must be the same.* If they do not agree, the bank will favor the written amount.
*8. The signature should be the same as the pre-printed name—#1.
*9. Each account has its own computerized account number. Never accept a check without this pre-printed number on the check. A check without this number is called a Counter Check.
10. Bank I.D. number appears here and should be the same number as #5.
11. Federal Reserve number here should be the same number as #6.
*12. The name and address of the Bank is here. It must be local. *Do not* accept an out-of-town or out-of-state check.
*13. The check must be made payable to *Taylor Drugs.* A check made payable to anyone else is a third party check. *You must not accept a third party check* under any circumstances.
14. Place your initials in the upper left hand corner. If the check is returned you may be able to remember something that will help locate the writer.

Figure 13-1. Check procedure—accepting a check. (Courtesy of Taylor Drug Stores.)

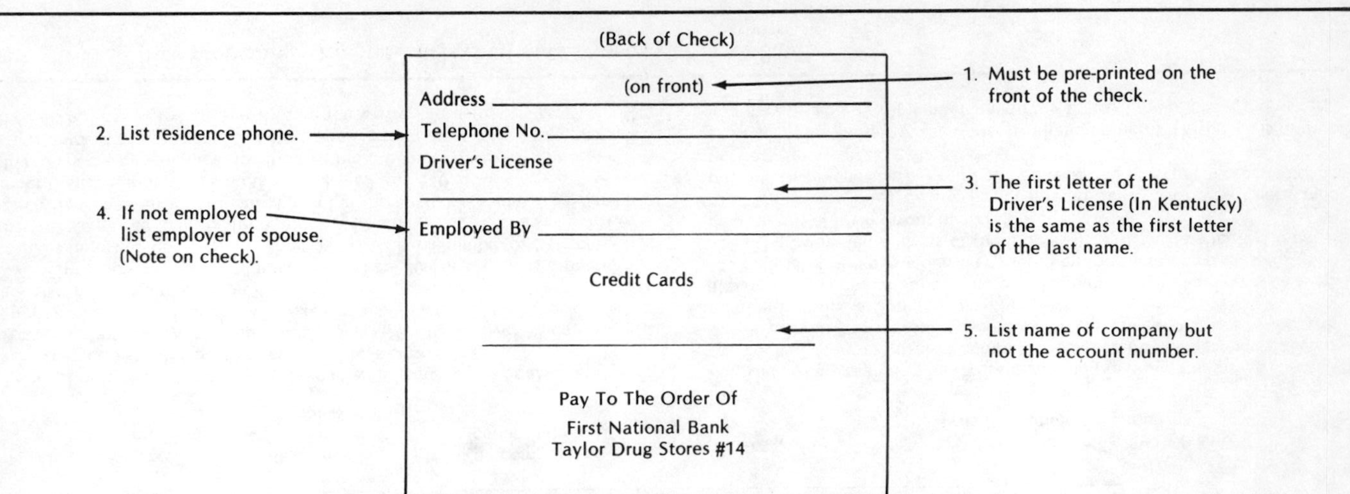
(Back of Check)
(on front)
Address
Telephone No.
Driver's License
Employed By
Credit Cards
Pay To The Order Of
First National Bank
Taylor Drug Stores #14
1. Must be pre-printed on the front of the check.
2. List residence phone.
3. The first letter of the Driver's License (In Kentucky) is the same as the first letter of the last name.
4. If not employed list employer of spouse. (Note on check).
5. List name of company but not the account number.

*Place the store check stamp on the back of the check *immediately*. This is exactly like an endorsement for *Taylor Drug Stores*, Inc. If the check was stolen it could not be cashed.

1. Be certain a complete address is pre-printed on the front of the check.
2. Copy the residence phone number here. The exchange (first three numbers) should be familiar to you. Think of your store phone number. Is this number in your area?
3. Copy the driver's license number carefully. The first letter of the number should be the same as the first letter of the last name of the check writer (Kentucky). Look at the photo on the driver's license. Is it the same person who is writing the check?
4. List here the name of the company where the writer is employed. If not employed list spouse's employer.
5. Request a credit card. List the name of the company but do not copy the account number. If the account is *not* in the name of the check writer, note the name of account on the check.

Important General Information Regarding the Handing of Checks

1. A check is *exactly like cash*. Protect the check the same way you protect cash.
2. Accept a check only for the amount of the purchase.
3. Refer any check over $25.00 to the Manager-on-Duty for approval. If it is the policy of the store to have all checks approved, you must adhere to this policy.
4. Never accept a third party check. This is a check that was written by someone (a person, company, etc.), other than the person who is presenting the check to you.
5. Never accept an altered check. This is a check that has been changed in some way, the amount may have been written over, an erasure may have been made, etc.
6. Never accept a money order. *Exception:* A money order bought at your store and returned because it was not used for the purpose intended may be accepted.
7. You may accept a Traveler's Cheque in denominations under $25.00. You must be sure the writer signs the Traveler's Cheque in your presence. Compare the two signatures on the cheque carefully. Both signatures must have been made by the same person.
8. Willful violation of the Standard Procedure for accepting and protecting a check will result in *immediate dismissal*.

Figure 13-2. Check cashing procedure—protecting a check. (Courtesy of Taylor Drug Stores.)

The information used to identify a customer should, in all cases, be entered on the back of the check for future reference. If, for example, a vehicle registration is accepted, its number should be entered on the check along with such other information as may be relevant.

ID Equipment and Systems

Equipment to record any check casher exists to help the retailer.

Camera devices which photograph the customer simultaneously with his check can be used to record the transactions.

Instruments to record his thumbprint on the check without the use of ink also serve in this capacity.

These devices are as effective as the system established for their use. If they are used carefully in accordance with a designated procedure, they can provide a useful record for the storekeeper. They do, however, tend to discourage check passing by thieves by their very presence, though they will by no means eliminate it.

Cooperative Systems to Control Check Fraud

Cooperative systems among a group of participating merchants seem to be the most effective way of controlling large scale check fraud. An article in *Successful Retail Security*[61] describes such a system set up in both Yakima and Spokane, Washington, which managed to reduce losses due to check cashing substantially.

This approach involved an alarm system which warns member firms when a team of check passers arrives in town or when there is a report of a theft of blank checks that might be used. When such information is received from a police agency or from one of the member stores, the organization notifies certain retailers on their list, these retailers in turn have other retailers they are to notify, and they pass the word on to other designated merchants. Through this information circuit, all participants can be notified of the threat in a very short time. During its first year in operation, no professionally written bad checks were passed in Yakima, and six bad check artists who tried were taken into custody.

Checks involving insufficient funds were also substantially reduced. Although the organization was not set up as a collection agency, it has been effective in recovery on such checks. Each member store reports on returned checks. The organization distributes weekly a list of all such outstanding accounts to every member, where it is posted by every cash register. No check will be cashed for customers with outstanding returned checks.

In order to effect recovery on returned checks, the customer is contacted by the organization and asked to set a date for taking care of the account. Ninety percent of the checks are taken care of after the first contact. Sometimes a second contact is necessary, and the Yakima experience shows that 75 percent of the remainder make restitution by this second contact. After that, legal action by the merchants may be necessary to collect the remaining two percent of returned checks. The result of these efforts was a report by a local bank that the incidence of NSF (not sufficient funds) checks was reduced by 50 percent after the second year of this operation.

The author's conclusion from his experience in this enterprise was that any area could reduce bad checks by 85 percent if 70 percent of the merchants in that area joined together in such an association. He suggests that such an organization be non-profit with a salaried executive, providing a valuable service at low service charges. In his experience, fees charged ranged from $7.50 per month for stores with a few bad checks up to about $25 per month for an average size supermarket, whose incidence of bad checks is necessarily higher.[62]

Store Policy

Every store must set its own policy for handling checks and maintain it. Certainly the policy should be reviewed and adjusted as necessary, but it must be strictly adhered to as long as it is in force. Employee indoctrination and continuing education in this problem is essential to the success of any program of check control.

As a guide, though certainly not as a rigid rule, certain limitations should be considered as follows:

1. No third party checks. Payment can be stopped on such checks and the retailer's recourse is only to the customer, not to the payor.
2. No checks on out-of-town banks. Such checks are difficult to verify, and the time involved in clearance is such that, if it is fraudulent, the customer has long disappeared before the check is returned.
3. No checks over a certain amount above purchase.
4. No checks cashed other than government or payroll checks.
5. No checks cashed without adequate pre-determined documents of identification.
6. No checks cashed drawn on other than personal checks imprinted with the name and address of the customer.

If such rules are followed, the loss from bad checks should be small, provided all checks are themselves carefully examined by the cashiers.

It is generally agreed that a retailer should never suffer losses greater than .005 percent of the face value of checks cashed, although some do, in fact, have losses as high as one percent. Such a drain on the resources of any company is clearly intolerable. It must be corrected by the establishment of reasonable controls and a program of employee education.

BURGLARY

According to the FBI's Uniform Crime Reports, burglary represented approximately 43 percent of all crime in the United States in 1978 and the incidence of burglary had increased by 208 percent in the years between 1970 and 1978. The report also shows that the incidence of burglary rose another 6 percent just for the year 1979. Less than 20 percent of the burglars of all occupancies are apprehended, and the rate of apprehension is dropping a few percentage points annually. So with burglary the most frequent crime, with an increasing rate of incidence and a decreasing rate of arrest, it is essential that every businessman—the retailer in particular—take most particular care to protect himself against this attack on his assets.

The Attack

A burglary attack on a retail establishment is similar to such attacks on other types of facilities, except that the job of the burglar is simplified by the ease with which he can inspect the premises before he makes his move. He can familiarize himself with the physical layout, store routines, police patrols and internal inspections, if any, simply by appearing as a customer. If he wishes to engage in a more exhaustive survey, he might even pose as a building or fire inspector and make a minute examination of alarm installations, safe location, interior lock construction, and every other detail of the store defenses that may interest him.

Assuming he has found a weak point, the burglar enters through a door or window, through the roof, or possibly from a neighboring occupancy. Most successful burglaries (and less than 10 percent of detectable attempts are unsuccessful) are made by forced entry. Curiously enough, the greatest number are made through the front door or main entrance.

A considerably smaller number of burglaries involve the stay-in who gathers his loot and then breaks *out*, and is gone before the guards or police can respond to an alarm.

Merchandise As Target

Most burglaries involve the theft of high value merchandise, although any goods will do in the absence of the ''big ticket'' items. Police reports repeatedly show the most astonishing variety of goods that have been targeted by enterprising thieves. Anything from unassembled cardboard cartons to bags of flour are fair game. Obviously such merchandise would hardly be considered as high risk assets, but they cannot be overlooked as possible loot. Burglars come in all shapes and sizes, and what may seem to one as cumbersome and unprofitable may be remarkably appealing to another.

Cash As Target

Cash is naturally the most sensitive asset and the most eagerly sought, but, since it is usually secured in some manner, it represents the greatest challenge to the burglar.

Stores keeping supplies of cash on hand are particularly susceptible, especially before payday, in areas in the vicinity of a company or companies with substantial payrolls. If such stores customarily cash payroll checks as a service to these employees, a burglar can assume that adequate cash must be on hand in anticipation of the next day's demands. Particular care must be taken in such cases if there is no way to handle cash needs other than to store it on the premises overnight. Every other avenue should be explored, however, before deciding to keep substantial supplies of cash on hand overnight.

Since any cash will normally be held in a safe, it requires some degree of expertise to get at it. Boring, jimmying, blasting, and even carrying away the entire safe are the methods most commonly used. Few burglars are sufficiently skilled to enter a safe by manipulation of the combination, so applied violence is the normal approach. Unfortunately, even in cases where an attack is unsuccessful, the damage to the container is likely to be severe enough to require its replacement.

This is equally true in all areas where entry is made or attempted. The high cost of repairs of damage to buildings and equipment can be almost as harmful as the loss in cash or merchandise. A survey of 4,800 company-owned and franchised stores in a nationwide chain of convenience stores indicated that 42 percent of the burglary loss was reported to be in building and equipment damage and theft.

Physical Defense Against Burglary

In order to reduce the threat of burglary, the premises must be secured against entry. The principles governing this hardening of the facility are essentially the same as those governing the protection of buildings of all types.

Doors must be of a heavy construction, hung in a frame sufficiently stong to avoid prying. Hinges must be either located within the doorway, or the screws or bolts must be set in such a way that they cannot be removed. If doors contain glass panels, they should be protected by grilles or screens, or they should be of burglar-resistant or polycarbonate materials.

Windows should be of these same materials, or they should be barred or screened. In many applications where windows serve only to admit light, they might well be replaced with glass brick.

Roofs should be protected by chain link fence or barbed wire if there is a possibility of gaining access to them from neighboring buildings, trees, poles, or by climbing drain pipes. Skylights should be protected by bars, grilles, or fencing materials.

All means of ingress of 64 square inches or larger must be screened, locked, or grilled in some manner. These would include air vents, manholes, loading chutes, and all other avenues through the outer walls.

Locks, adequate to the job, must be installed. And the installation must provide for adequate seating of the bolt.

Keys must be distributed sparingly and kept under tight control at all times. They should be inventoried regularly, including a physical inspection of all keys held by authorized holders.

Adequate lighting is as important to the protection of a retail outlet as it is to an industrial facility. In the case of supermarkets with large parking lots, lights should cover the area, for the safety and convenience of the customers as well as for the deterrent factor such lighting offers after closing hours.

Landscaping should be designed to avoid casting shadows or creating concealed approaches to the store.

In urban facilities, interior lighting inside the perimeter, especially in the area of entrances of any kind, will help to deter the would-be burglar, as well as to make his presence visible to passing patrols.

Fences and walls may be advisable in some situations, although their effectiveness must be reviewed in the light of the ease with which they can be scaled from nearby elevations.

Alarms

Alarms of some kind can make the difference between a really effective program and one that is only half safe. The type of alarm providing the best results is a matter of some disagreement, but there is little disagreement over the effectiveness of some kind of system, however simple.

Many stores report satisfaction with local alarm systems. They feel that the sound of the signaling device scares off the burglar in time to prevent his looting the premises. Many police and security experts feel such alarms are ineffective, because the response to the signal is only by chance and because such a device serves to warn the intruder rather than aid in his capture. Since most managers are less interested in apprehending thieves than in preventing theft, and since local systems are inexpensive and can be installed anywhere, they continue in wide use. Certainly they are preferable to no alarm system at all.

Whether the outer perimeter should be alarmed and whether space coverage alarms should be used is a matter peculiar to each facility. This must be carefully studied and determined by the manager. But it is always important to alarm the safe in some way.

Safes

The location of the safe within the premises will depend upon the layout and the location of the store. Many experts agree that the safe should be located in a prominent, well-lighted position readily visible from the street, where it can be seen easily by patrols or by passing city police. In some premises—especially where no surveillance by passing patrols can be expected—it is generally recommended that the safe be located in a well-secured and alarmed inner room which shares no walls with the exterior of the building. Floors and ceilings should also be reinforced. The safe should be further protected by a capacitance alarm. The classification of the safe and the complexity of its alarm protection will be dictated by the amount of cash it will be expected to hold. This should be computed on the maximum amount to be deposited in the safe and the frequency with which such maximums are stored therein.

Basic Burglary Protection

Obviously the amount of protection—and the investment involved in it—will depend on the results of a careful analysis of the risks involved in each facility. There is no single way—no magic solution—to the problem of burglary. Every system and each element of that system must be tailored to the individual premise.

The manager must evaluate the incidence of burglary in his neighborhood as well as the efficiency and response time of the police. He must consider the nature of the construction of his building and the type of traffic attracted to his area. He must consider the ease with which his merchandise can be carried off and by what probable routes. He must consider the reductions in insurance premiums provided by various security measures. He must consider the advice of city police in anti-burglary measures and their experience in analogous situations. In short, the retailer owes it to himself to learn as much as he can about coping with the problem of burglary and dealing with it as forcefully, economically, and energetically as possible.

ROBBERY

Robbery, next to larceny of $50 and over, is the fastest growing crime in the country. From 1970 to 1978 the number of incidents of robbery reported in the FBI's Uniform Crime Reports rose a staggering 272 percent. During the same period, this crime increased from 62.1 per 100,000 inhabitants to 189.8 per 100,000 inhabitants, an increase of 312 percent. In 1979 cities with a population of over 10,000 saw robbery increase by another 12 percent.

And retailers take the brunt of this attack. Since less than a third of the robbers are arrested and, as in the case of burglars, this percentage is declining annually, the retailer is obliged to take measures to protect himself from this most dangerous crime.

Nature of Robbery

An interesting profile of robbery, which might be projected into a larger, national view of the problem, can be found in the study of 4,800 company-owned and franchised stores of a national convenience store chain referred to earlier.

In this study we can see few differences in the experience with robbery due to geographical location, and the occurrence of such incidents is fairly constant throughout the week. The daily peak of robberies is 10:30 p.m. However, those stores remaining open all night suffered the majority of robberies between midnight and 3 a.m. Ninety-six percent of the money was taken from cash registers, and 81 percent of the robbers were armed with handguns. Eighty-seven percent of the assailants were under 30, and of these 97 percent were male. Although weapons were used (or the threat of harm inherent in a robbery), violence was actually carried out in only four percent of the cases. Eighty percent of the death cases in this study occurred in situations in which store personnel had done nothing to motivate the attack. And while 60 percent of the robberies were carried out by a lone robber, 75 percent of the death or injury occurrences took place in situations involving two or more robbers.

If we can take this microcosm as representative of the problem faced by the retailer, we find that the robber is most apt to be a young man working alone, carrying a pistol and threatening employees with bodily harm unless they hand over the money from the cash register. He rarely carries out his threats—probably because the employees wisely comply with his demands—unless he is accompanied by a partner, in which case he is very likely to cause death or injury without provocation.

Robbery Targets

The robber is a criminal making a direct assault on a person responsible for cash or jewelry. His target may be a messenger or store employee taking cash for deposit in the bank or bringing in cash for the day's business. Robbers frequently enter a store a few minutes before closing and hold up the manager for the day's receipts. They may break into the store before opening and wait for the first employees, who will be forced to open the safe or the cash room. In recent cases, managers or members of their families have been kidnapped or threatened in order to coerce access to company cash.

Delivery trucks and warehouses can also be targets for robbers. These latter cases require several hold-up men working together and, although they are a very real threat and do occur, the majority of cases are still carried out by the lone bandit attacking the cash register.

Businesses most frequently attacked are supermarkets, drugstores, jewelry stores, liquor stores, gas stations, and all-night restaurants or delicatessens.

Cash Handling

Probably the major cause of robbery is the accumulation of excessive amounts of cash. Such accumulations not only attract robbers who will have noticed the cash handled, but they are more damaging to business if a robbery does occur.

It is essential that only the cash needed to conduct the day's business be kept on hand, and most of that should be stored in a safe. Cash should never be allowed to build up in registers, and regular hourly checks should be conducted to audit the amounts each register has on hand.

Every store should make a careful, realistic study of actual cash needs under all predictable conditions, and the manager should then see to it that he has only that amount plus a small reserve on hand for the business of the day. Limits should also be set on the maximum amount permitted in each register, and cashiers should be instructed to keep cash only to that maximum. Overages should be turned in as often as necessary and as unobtrusively as possible.

In large volume stores, a three-way safe is an invaluable safety precaution. Such safes provide a locked section for storage of two or three hours' worth of money that may be called upon for check cashing or other cash outlay. The middle section has a time-lock which is not under the control of any of the store personnel and has a slot for the deposit of armored car deliveries or cash build-ups. Cash register trays are stored in the bottom compartment.

Movement of cash build-ups from registers to the safes should be accomplished one at a time, and every effort should be made to conceal the transfer. This is particularly important at closing time when registers are being emptied. The sight of large amounts of cash can prove irresistibly tempting to a potential robber.

Cashroom Protection

If the store has a cashroom, it must be protected from assault. Basic considerations as to location and alarming were outlined in the discussion on anti-burglary measures, but an additional precaution to protect against robbery should also be noted. The room, itself, must be secure against unauthorized entry during business hours. If possible, a list naming those persons authorized to enter the cashroom and under what conditions should be drawn up. It should be made clear that there are to be no deviations under any circumstances from these authorizations.

The door to the room must be secure and securely locked. If fire regulations indicate the need for additional doors, they should be equipped with panic locks and alarmed. The entrance should be under the control of a designated person who, through a peephole or other viewing device, can check persons wishing to enter. In larger facilities, entrance can be controlled by an employee at a desk outside the room, although this arrangement should be studied carefully before implementation, since a robber could force such an employee to permit entrance unless his position is protected from the threat of attack.

Opening Routine

Since the potential for robbery is always present when opening or closing the store, it is important that a carefully prescribed routine be established to protect against the possibility.

At opening, the employee responsible for this routine should arrive accompanied by at least one other employee. The second stands well away from the entrance while the manager prepares to enter. The manager checks the burglar alarm, turns it off and enters the store. He checks the interior, including washrooms, offices, backrooms, and other spaces that might offer concealment to robbers. After a specific, pre-established time, he reappears in the main entrance and signals his assistant in a pre-determined code.

This code, which should be changed periodically, should be given both in voice and in some hand signal or gesture such as adjusting his tie or smoothing his hair. One coded reply and gesture should indicate that all is clear; another should indicate trouble. It is important that this code be innocent and reasonable sounding to allay any suspicions on the part of the robbers if any are present.

If the assistant gets the danger sign, he should reply with something to the effect that he will be along as soon as he gets a paper or checks his car or gets a cup of coffee. He should then proceed slowly and deliberately to a predetermined phone and make a call to the police. Experienced security men all stress the value of a card with the number of the police station typed on one side and coins to make the call taped to the back. For lack of a dime, stores have been robbed.

Transporting Cash

Since a basic principle in robbery protection is minimizing cash on hand, money will necessarily have to be moved off the premises to a bank from time to time. Ideally, this should be done by an armored car service. If this cannot be done because of cost or non-availability of such service, efforts should be made to get a police escort for a store employee acting as messenger.

In any event, the messenger so assigned must be instructed to change his route; he should be sent out at different times of day and, if possible, on different days of the week. He should regularly change the carrier in which the money is stored. It might be anything from a brown paper bag to a tool box. He should stay on well-populated streets and he should move at a predetermined speed. The bank should be notified of his departure and given a close estimate of his time of arrival.

Closing Routine

Shortly before closing time, the manager makes a check of all spaces, much as at opening. An assistant unobtrusively watches for any unusual activity on the part of departing customers. When all registers have been emptied and the money locked up, he stations himself in the parking lot and watches as the manager completes his closing routine. When the manager has checked out the alarm and locks up, the assistant is free to go. If the manager signals trouble, during any of this routine, the assistant will follow the same routine as at opening.

Other Routines

Since any entry into the store leads to a potential for robbery, it is important that the manager never try to handle it alone. There have been a number of instances where hold-up men have called the manager and reported damage or a faulty alarm ringing in order to lure him into opening the store. He is then in a position to be coerced into opening the safe or, at least, providing a bypass of the alarm system. If he is so notified, the manager should phone the police and/or the appropriate repair people and wait for them to arrive before getting out of his car. He should never try to handle the matter without some back-up present.

Employee Training

A full cooperative effort by all employees is essential to any robbery prevention program. A properly indoctrinated employee will know how to use a silent, "hands off" alarm if such an installation is deemed advisable. He will learn to question persons loitering in unauthorized areas. He will be alert to suspicious movements by customers, and will know what to do and how to do it if he is aware that a hold-up is in progress in some other part of the store. He will remain cool, make mental notes of the robbers' appearance for later identification. He will cooperate with the robbers' demands to an acceptable minimum. He will not, under any circumstances, try to fight back or resist. If possible, he will hand over the lesser of two packs of money if that choice is open to him, but he will never do so if there is the slightest chance that the robbers will become aware of what he has done. He will remain alert to the possibility of robbery, and, if he should become involved in one, he will make careful observations for assistance in apprehending the criminals.

INTERNAL THEFT

Methods of Theft

The list of ingenious techniques employed by dishonest employees to steal from the stores that hire them is endless. Whatever systems are installed to control inventory shrinkage from internal theft, some clever employee finds his way around them.

Although the report of the Small Business Administration conservatively estimates that employee theft represents about 12 percent of the losses suffered by all small business, the 1972 study by the Department of Commerce previously referred to (*The Economic Impact of Crime Against Business*) states that: "While shoplifting appears to be the most serious problem for retail establishments, most observers believe that because of the reluctance of businessmen to admit the magnitude of their employee theft problem, that figure is seriously understated. Some believe that employee theft accounts for substantially more loss than shoplifting by customers.[63] How much more is indicated by a number of knowledgeable observers who estimate that, as between shoplifting and employee theft, the latter accounts for anywhere from 60 to 80 percent of the loss.

It is important for every manager to familiarize himself with some of the ways employees steal in order to better understand the scope and nature of the problem. Familiarity alone will not serve to stamp out this problem, but it will aid in focusing on those areas of greatest danger and help to establish some countermeasures.

Cash Registers. Perhaps the most widely employed method of theft involves some kind of juggling with the cash register. Since cashiers and managers both have access to customer cash to be deposited in registers which record the transaction, there are literally thousands of opportunities to manipulate the accounting before individual sales or even the receipts of the day are posted. Cash never recorded as received is clearly much harder to locate or identify as missing than cash or merchandise which has been entered and later stolen.

Methods used with the register usually involve the regular theft of small sums by under-ringing the amount received. This involves ringing $9 for a $10 purchase, for example, and pocketing the difference at the end of the day before checking in. An even cruder method is to ring up "no sale" or "void" instead of the amount paid from time to time, and pocket the cash received at the sale.

From time to time, employees are found who have an opportunity to remove the tape from their register shortly before the end of the day. They put in a fresh tape and ring up all sales accurately. When they check in, they pocket all the proceeds on the new tape, destroy it and hand in the prematurely removed old tape as their record of the day's receipts. They have effectively gone into business for themselves on company time.

Several flagrant cases of such private enterprises have involved managers buying their own register and, either alone or in collusion with another employee, setting up an additional check-out lane during a few hours of heavy traffic. Such cases are unusual but not unheard of.

The Giveaway. Checkers in supermarkets have frequently been discovered giving away large amounts of merchandise to fellow employees when they check out with purchases by ringing up only a small fraction of the value of the merchandise. In other cases, they have similarly accommodated friends or family members.

Many cases are reported where various store clerks set themselves up as traders, exchanging stockings for another clerk's shoes, or blouses for costume jewelry. In a somewhat similar approach, employees have been found who, after selling an item for its regular price, enter it as having been sold to an employee at the regular employee discount price. The seller then pockets the difference.

Price Changing. Every kind of store must be alert to price changing by employees and customers alike. Employees have an opportunity to alter price tags and buy the merchandise on their own or in collusion with an outside confederate.

Vendor Kickbacks. Collusion with vendor in receipting more merchandise than is delivered is common in every business, and cannot be overlooked in retail establishments particularly.

Refunds. In cases where refund controls are inadequate, it is a simple matter for employees to write up refund tickets and submit them for cash, either in person or through a confederate.

Merchandise Theft. It is common for employees to transport items of merchandise to their cars in the course of the day. This is often done in several trips or perhaps by accumulating items in one package to be removed just after the store opens or before it closes. If package control procedures are in operation, such employees may purchase an inexpensive item or two and pass out the package with the legitimate sales slip for authorization.

Stocking. Stocking crews have an excellent opportunity to steal enormous amounts of merchandise during the off-hours when they are typically in operation.

Embezzlement. Embezzlement in retail establishments takes the same forms as those discussed in an earlier chapter on internal theft, and the same precautions must be exercised to prevent it.

Screening and Supervision

The first and most important countermeasure that every store should establish is a firm employment policy involving a careful screening of every job applicant. These procedures, which have been discussed in another chapter, may seem burdensome and even more costly than the more *pro forma* checking used by too many stores, but in the long run they will pay many times over.

The next basic is enlightened supervision. No system of controls can be effective if it is not adhered to, and unless it is supervised it may soon fall into disuse. Supervision to confirm that all established procedures are being followed, as well as regular audits of their effectiveness, are vital to the success of any retail security program.

Shopping Services

Auditing of the efficiency, effectiveness and honesty of sales people by shopping service investigation has long been used by retail firms as one of the most accurate methods of determining the conduct of their operation. Such device is not inexpensive but has established itself as sufficiently effective in the reduction of theft and the improvement in the performance of personnel to pay for itself many times over.

The tests conducted by such services or shoppers usually take note of the employee's appearance, manner, helpfulness and salesmanship, as well as checking for any signs of dishonesty.

Dishonesty testing consists of the creation of situations where the employee could easily steal or fail to ring up cash. The employee is then observed or audited to check on performance under the circumstances. The situations created are in no way construed as entrapment or enticement but represent re-creations of normal situations which could be anticipated in the regular course of business. Since shoppers making these tests are unknown to the employees, their reactions to the tests can be taken as indications of their performance in similar situations with any customer. A full report is submitted after each such test for the manager's reference and review.

Most stores contract for such services on a yearly basis at a set fee, with the understanding that inspections will be made with a certain prescribed frequency. In this way, store management has some confidence that an ongoing audit made by objective outside investigators will help to uncover inadequate or dishonest performance by sales employees and cashiers.

Shipping-Receiving Controls

As in other types of business, the receiving and shipping area of any retail operation is a particularly sensitive one. Since the life blood of the business flows across these areas many times in the course of a year, it is essential that there will be full accountability for all movement of merchandise, that a perpetual inventory be an integral part of the system, and that the area be restricted to those persons specifically authorized to be there.

All merchandise loaded or unloaded should be subject to periodic spot checks, including a complete unloading and recounting of merchandise already loaded from time to time. Cargo seals should be secured and inventoried regularly. All broken shipments should be investigated and secured. All loading and unloading procedures should be supervised.

Maintain a rest room and lounge area for drivers which is separate from the facilities used by stockroom, warehouse or dock personnel, and insist on this separation. Do not permit drivers to enter storage or merchandise handling areas.

The effectiveness of procedures should be audited from time to time by introducing errors into various operations. For example, the number of cases to be shipped might be invoiced incorrectly to see if the checker catches the error; truck seals might be logged and noted on invoices incorrectly to verify the alertness of personnel involved.

Trash Removal

Trash removal has always been a problem because it provides an efficient means of removing merchandise from the premises without detection. It is important to have a supervisor on hand when trash is loaded for removal, and it is especially important that trash collection for such removal be separated from loading or receiving dock areas.

Package Control

It is important that some kind of control procedure be established to inspect packages removed from the premises by employees. Retail outlets have a particularly difficult time with this problem, since employees have regular access to large amounts of merchandise—much of it small enough for easy portability. Receipts must accompany every purchase removed and all packages should be subject to inspection as desired by security personnel.

Employee Morale

The state of employee morale is the key factor in any store security program. If employees are totally familiar with all store rules and policies; if they feel that they are appreciated as human beings as well as store employees; if they feel they are an important functioning part of the organization, they will respond with increased efficiency and the problem of dishonesty will diminish. It is essential that management bear in mind that, along with its desire to receive reports up the ladder, it must reciprocate by communicating back down the structure. Communication must be a two-way path.

Employees should be motivated to perform, not compelled to. This can be accomplished only by clear statements of policy, firm supervision insisting on compliance, and intelligent leadership. In this atmosphere morale should grow and the company will prosper.

REVIEW QUESTIONS

1. What are common preventive measures that can be used in a retail store to help reduce the incidence of shoplifting? How effective are each of these measures?
2. What are the legal implications that accompany the arrest of a shoplifter?
3. Identify check-cashing policies a retail store might set up to reduce its losses from bad checks.
4. How can a retail store's vulnerability to burglary (or robbery) be reduced?
5. Discuss the statement: ''The state of employee morale is the key factor in any store's security program''.

Chapter 14

CARGO SECURITY

DIMENSIONS OF THE PROBLEM

During the three years of hearings by the Senate Select Committee on Small Business (1970-1972), testimony of witnesses and staff investigation led to the conclusion that the direct losses from theft in cargo carried by the air, truck, rail, and maritime carriers amounted to almost $1.5 billion in 1970.

The indirect costs resulting from claims processing, capital tied up in claims and litigation, and market losses due both to non-delivery and underground competition from stolen goods were estimated at between two to seven dollars for every dollar of direct loss—an 8 to 10 billion annual loss in the national economy.

Since then, Executive Order 11836 formalized the National Cargo Security Progam. This program was designed to reduce cargo losses through voluntary action within the shipping industry with support and assistance from the government at all levels. Two years later (March, 1977) "A Report to the President on the National Cargo Security Program" was issued from which it could be interpreted that the problem has not grown and may even have shrunk slightly to approximately $1 billion in direct cost. The problem is still enormous.

Inadequacy of Security

The shipment of goods is vital to the economy and, ultimately, to the survival of the country. Yet the Committee found that law enforcement efforts, conducted by local and state police and the FBI, were totally inadequate to prevent this enormous theft of cargo and to apprehend the criminals involved.

Testimony additionally indicated that, although some trucking concerns were making an effort to establish effective security programs, a considerably greater part of the industry, from shippers to receivers, from warehouse managers to drivers, were bogged down in confusion and ineffective half-measures that served more to nurture and encourage crime than to prevent it. Control systems were frequently misunderstood, paperwork was complicated and often unchecked, physical security was inadequate, supervision was inadequate, traffic control was poor, and security policies were often misdirected or unenforced.

The Committee called for a program directed at crime *prevention* and law enforcement *coordination* rather than an expansion of police or reactive forces. The Committee concluded that security measures do exist which can halt the increase in theft in the shipping industry, and it encouraged the industry, along with law enforcement officials, to set up coordinated and cooperative programs that will apply the most advanced technology to the solution of these problems.

The Role of Private Security

Since, according to an analysis made by the Department of Transportation (DoT), 85 percent of goods and materials stolen go out the front gates on persons and vehicles which are authorized to be in cargo handling areas of transportation facilities, it would appear that by far the greater burden falls upon the security apparatus of the various private concerns involved. It is true that public law enforcement agencies must make a greater effort to break up organized fencing and hijacking operations, and they must find a way to cut through their jurisdictional confusions and establish more effective means of exchanging information. But the bulk of the problem lies in the systems now employed to secure goods in transit.

There is no universally applicable solution to this problem. Every warehouse, terminal, and means of shipment has its own particular peculiarity. Each one has weaknesses somewhere, but certain principles of cargo security, when thoughtfully applied and vigorously administered, can substantially reduce the enormous losses which are so prevalent in today's beleaguered transport industry.

ACCOUNTABILITY PROCEDURES

The paramount principle is accountability. Every shipment, whatever its nature, must be identified, accounted for, and accounted to some responsible person at every step in its movement. This is difficult in that the goods are in motion and there are frequent changes in accountability, but this is the essence of the problem. Techniques must be developed to refine the process of accountability of all merchandise in shipment.

Invoice System

This accountability must start from the moment an order is received by the shipper. As an example of a typical controlled situation, we might refer to a firm that supplies its salesmen with sales slips or invoices that are numbered in order. This is very important if any control is to be maintained over these important forms. Without numbering, sales slips could be destroyed and the cash, if any, could be pocketed; or they could be lost so that the customer might never be billed. When these invoice forms are numbered, they can be charged out to the salesman, in which case every invoice can and should be accounted for. Even those forms which have been spoiled by erasures or physical damage should be voided and returned to billing.

Merchandise should only be authorized for shipment to a customer on the basis of the regular invoice form. This form is filled out by the salesman receiving the order and sent to the warehouse or shipping department. The shipping clerk signs one copy, signifying that the order has been complied with, and sends it to accounts receivable for billing purposes. The customer signs a copy of the invoice indicating receipt of the merchandise, and this copy is returned to accounts receivable. It is further advisable to have the driver sign the shipping clerk's copy as a receipt for his load. In some systems a copy of the invoice is also sent directly to an inventory file for purposes of on-going inventory and audit. Returned merchandise is handled in the same way, but in reverse.

In this simplified system there is a continuing accountability for the merchandise. If anything is missing or unaccounted for at any point, the means exist whereby the responsibility can be located. Such a system can only be effective if all numbered invoices are strictly accounted for, and where merchandise is assembled and shipped only on the basis of such an invoice. The temptation to circumvent the paper work in the name of "rush order" or "emergency" frequently arises, but if the company succumbs, losses in embezzlement, theft, or lost billing can be substantial.

Similarly, transport companies and freight terminal operators must insist on full and uninterrupted accounting for the goods in their care at every phase of the operation from shipper to customer.

Separation of Functions

Such a system can be comprised by collusion between the various people who constitute the links in the chain leading from order to delivery, unless efforts are made to establish a routine of regular, unscheduled inspections of the operations down the line, and, depending on the nature of the operation, regular inventories and audits.

It is advisable, for example, for the shipper to separate the functions of selecting the merchandise from stock from the packing and loading function. This will not, in itself, eliminate the possibility of collusion, but it will provide an extra check on the accuracy of the shipment. And as a general rule, the more people (up to a point) charged with the responsibility and held accountable for merchandise, the more difficult and complex a collusive effort becomes.

In order to clearly fix accountability, it is advisable to require each person who is at any time responsible for the selecting, handling, loading, or checking of goods to sign or initial the shipping ticket which is passed along with the consignment. In this way errors, which are unfortunately inevitable, can be assigned to the man responsible. Obviously, any disproportionate number of errors traced to any one person or to any one aspect of the shipping operation should be investigated and dealt with promptly.

Similarly, at the receiving end, the ticket should be delivered to a receiving clerk. The truck will then be unloaded by appropriate personnel who will verify the count of merchandise received without having seen the ticket in advance. In this way each shipment can be verified without the carelessness that so frequently accompanies a perfunctory count that comes with the expectation of receiving a certain amount as specified by the ticket. It will also tend to eliminate theft in the case of an accidental or even intentional overage. Since the checker doesn't know what the shipment is supposed to contain, he cannot rig the count.

Driver Loading

Many companies, with otherwise adequate accountability procedures, permit drivers to load their own trucks when taking a shipment. This practice can defeat any system of theft prevention, since the driver is accountable only to himself when such a method of loading is in effect. The practice should never be condoned. In the long run it would not be an economy in extra time spent in checking the load or extra personnel needed to do it. The potential losses in theft of merchandise by overloading or future claims of short deliveries would almost certainly exceed the costs involved in the time or personnel involved in instituting sensible supervisory procedures.

THEFT AND PILFERAGE

Targets of Theft

An analysis of claims data from the transportation industry shows that there are a few very specific commodities that attract the attention of pilferers and thieves.

Nine commodities (clothing, electrical appliances, automotive parts, food products, hardware, jewelry, tobacco products, scientific instruments, and alcoholic beverages) make up about 80 percent of total national losses. Broken down by industry, clothing represents 35.6 percent of the losses in trucking and 45.5 percent in air shipments; electrical machinery and appliances make up 11.8 percent of the losses in trucking and 15.4 percent of the losses in air transport. These two products alone represent approximately half of all the losses in both air and truck transportation. Obviously a company, carrier, or terminal operator would be well advised to take special precautions when handling such merchandise.

The complete list of principal target merchandise follows:[64]

Truck

1.	Clothing, textiles	35.6%
2.	Electrical machinery, appliances	11.8%
3.	Metal products, hardward	8.9%
4.	Transportation equipment, motor vehicles	5.9%
5.	Food, food products	5.6%
6.	Chemicals, petroleum, rubber, plastics	5.0%
7.	Alcoholic beverages	4.4%
8.	Tobacco products	3.7%
9.	Wood products, furniture	3.6%
10.	Medicine, drugs, cosmetics	3.6%
		88.1%

Air

1.	Clothing, textiles	45.5%
2.	Jewelry, coins	21.0%
3.	Electrical machinery, appliances	15.4%
4.	Machinery, except electrical	3.6%
5.	Instruments	3.5%
		89.0%

Rail/Maritime

1.	Transportation equipment, motor vehicles	28.0%
2.	Food, food products	22.0%
3.	Chemicals, petroleum, rubber, plastics	11.0%
4.	Metal products, hardware	8.1%
5.	Electrical machinery, appliances	8.1%
6.	Wood products, furniture	4.5%
7.	Machinery, except electrical	4.3%
8.	Alcoholic beverages	4.0%
9.	Tobacco products	4.0%
		94.0%

Such a list can be a guide to security managers handling merchandise that falls into one of these categories. This tabulation of industry-wide losses provides a forewarning of the relative dangers.

But the kinds of things stolen vary considerably, depending on the locality and the nature of the merchandise available. A thief might prefer to steal part of a shipment of clothing but, if none is available to him, he will just as vigorously pursue a truckload of dog food, as long as he has a means of disposing of it. Anything can be stolen if the thief is given the opportunity. Anything that has a market—and it wouldn't be shipped unless a market existed—can be considered attractive to a thief.

It is important for transportation industry managers to recognize that all merchandise is susceptible to theft, but at the same time to know the high-loss items at their locations, so that they may exert extra efforts to secure such goods. It is also useful to know that motor carriers are the victims of 74 percent of the theft-related losses, rail and maritime about 24 percent, and air cargo losses are below one percent of the total.

Pilferage

According to the Department of Transportation analysis referred to earlier in this chapter, 25 percent of the cargo theft in this country is the result of pilferage or thefts of less than a case. Thefts of this nature are generally held to be impulsive acts, committed by persons operating alone and who pick up an item or two of merchandise which is readily available when there is small risk of detection. Typically, the pilferer takes such items for his personal use rather than for resale, since most of those who are termed pilferers are unsystematic, not of a committed criminal nature, and are unfamiliar with the highly organized fencing operations that could readily dispose of such loot.

Such pilferage is always difficult to detect. Because it is a crime of opportunity, it is rarely committed under controlled circumstances which can either pinpoint the culprit or gather evidence that would later lead to his discovery. Generally, the items taken are small and readily concealed on the person or easy to transport and conceal in a car.

Pilferage is usually aimed at items in a freight terminal which are awaiting transshipment. In such instances merchandise may be left unprotected on pallets, hand carts, or dollies awaiting the arrival of the next transport. In this mode it is highly susceptible to pilferage as well as to a more organized plan of theft.

Broken or damaged cases offer an open invitation to pilferage if supervisory or security personnel fail to take immediate action. Accidental dropping of cases to break them open is a common device used to get at the merchandise they contain. If each such case is carefully logged, listing the name of the man responsible for the damage, as well as the names of those who instantly gather at the scene, a pattern may emerge that will enable management to take appropriate action.

Whatever form pilferage takes, it can be extremely costly. Although each individual instance of such theft may be relatively unimportant, the cumulative effect can be enormous. Twenty-five percent of an estimated total direct loss of $1.5 billion adds up to a staggering bill for petty theft.

Deterring Pilferage

Here again, accountability controls can provide an important deterrent to pilferage loss. If some one person is responsible for merchandise at every stage of movement or storage, the feasibility of this kind of theft can be substantially reduced.

In those cases where they do occur, a rapid and accurate account of the nature and extent of such losses can be an invaluable tool in indicating the corrective action to be taken. Properly supervised accountability controls can locate the point along the handling process where losses occurred and, even in those cases where they will not identify the culprit, they will underscore any weaknesses in the system and indicate trouble areas which may need more or different security application.

Movement of personnel in cargo areas must be strictly controlled. All parcels must be subject to inspection at a gate or control point at the entrance to the facility. Private automobiles must be parked outside the area immediately encompassing the facility and beyond the check point. All automobiles should be subject to inspection upon departure if a parking area is provided.

Every effort should be made to keep employee morale high in the face of such security efforts. Though some managements have expressed an uneasiness about inspections and strict accountability procedures, fearing that they might damage company morale, it should be pointed out that educational programs aimed at acquainting employees with the problems of theft and stressing everyone's role in successful security have resulted in boosting morale and in enlisting the aid of all employees in the effort.

Here, as in other areas of security, employees should be encouraged to report losses immediately. They must never be encouraged to act as informers or asked to report on their co-workers. If they simply report the circumstances of loss, it is the job of security to carry forward such investigation or to take such action as may seem indicated.

Large-Quantity Theft

Referring again to the DOT report, we find that 60 percent of the theft of cargo consists of merchandise in quantities of one or more cases but less than a full load. Thefts in such amounts are no longer in the category of pilferage. This becomes thievery, usually engaged in by one or more persons who are in it for profit—for resale through traffickers in stolen goods.

The thief may or may not be an employee, but since in either case he needs information about the nature of the merchandise on hand or expected, he will usually find an accomplice inside who is in a position to have the information. He is interested in knowing what kinds of cargo are available in order to make a decision as to what merchandise to hit, depending on its value, and on the demands of the fencing organization with which he deals.

Dealers in stolen goods are subject to the vagaries of the marketplace in the same way legitimate businessmen are. Whereas a certain kind of merchandise may find a ready market today, it may move slowly tomorrow. Such dealers are anxious to move their goods rapidly, not so much because of fear of detection (since even if they are found, mass produced goods of any nature are difficult, if not impossible, to identify as stolen once they find their way into other hands), but because they generally want to avoid the overhead and the attention created by a large warehousing operation.

Removal of Goods

Once the thief has the information, he needs to arrange to take over the merchandise and remove it from the premises. To do this, he will usually try to work with some employee of the warehouse or freight terminal he has singled out.

In cases where accountability procedures are weak or inadequately supervised, he has few problems. In tighter operations he may try to bribe a guard, or he might forge papers which represent him as an employee of a customer firm. He might even create confusion, such as a fire in a waste bin or a broken water pipe, in order to divert attention from his actions in those few moments he needs to accomplish the actual theft. Generally, the stolen goods are taken from the facility in an authorized vehicle driven either by the thief, who has supplied himself with false identification papers and forged shipping documents, or by an authorized driver who is, more often than not, working with him.

Disposal of Stolen Goods

Disposal of the goods usually presents no problem, since it is customary for him to steal from an order placed for any one of certain kinds of merchandise. His loot is normally pre-sold.

It might be noted at this point that it is for this reason, principally, that cooperation between private security and public law enforcement is so vital in the war against cargo thefts. No thief will continue in his efforts unless he has a ready market for the goods he steals. In the hearings of the Select Committee on Small Business, one of the most inescapable conclusions to be drawn was that the fences were the kingpins behind the majority of the thefts taking place. It is they who dictate the nature of the merchandise to be taken, the price to be paid, and the amount wanted. They direct the thefts without ever becoming involved—in many cases never even seeing the merchandise they have bought, warehoused and sold. Without the fences—many of whom are otherwise legitimate businessmen—the markets would shrink, the distribution networks would disappear, and losses would be dramatically reduced.

Unfortunately, only sporadic and ineffective efforts have been made to break up the big fencing operations, and the business of thievery thrives. Any assistance that private security can give to public law enforcement by way of instant and full reports on thefts could help to combat these shady operations, and all private industry will benefit immeasurably in the long run.

Terminal Operations

Terminal operations are probably more vulnerable to theft than other elements of the shipping system. Truck drivers mingle freely with personnel of the facility, and associations can readily develop that often lead to collusion. Receiving clerks can receipt goods that never arrive; shipping clerks can falsify invoices; checkers can overload trucks, leaving a substantial percentage of the load unaccounted for and therefore disposable at the driver's discretion.

Here again, these thefts can be controlled with a tight accountability system, but too often such facilities fail to install such a procedure or to follow up on it after it is in effect. This is poor economy, even under the most difficult situations when seasonal pressures are at their highest.

Railway employees on switching duty at a freight terminal can also divert huge amounts of goods. They can easily divert a car to a siding accessible to thieves who will unload it at an opportune time. This same device can be employed to loot trucks which have been loaded for departure the next morning. Unless these trucks are securely locked and parked where they can be under surveillance by security personnel, they can be looted with ease. Drivers can park their trucks unlocked near a perimeter fence for later unloading, unless the positioning and securing of the vehicle is properly supervised.

In all cases a professional thief is a man with a mission. He cannot be deterred by the threat of possible detection. He recognizes the possibility, accepts the risk, and makes it his business to circumvent detection. Only alert and active countermeasures will serve to reduce losses from his efforts.

Surveillance

There must be a strict guard surveillance at entrances and exits, and there must be patrol activity at perimeters, and through yards, docks, and buildings. Key control must be tight and painstakingly supervised. Cargo should be stored in controlled security areas which are enclosed, alarmed, and burglar-resistant. High value cargo should be stored in high security areas within the cargo area. Special locks, alarms, and procedures governing access should be employed to provide the highest possible security for these sensitive goods.

Employee Screening and Controls

Employees must be carefully screened. Screening is a prudent practice in any operation, but particularly so in a terminal where the thief is in such constant, intimate contact with such a tempting variety of merchandise. Employment in such a facility is the goal of many thieves, but a reasonably thorough screening procedure will weed out many of them.

Every effort must be made to frustrate the thief who may have gotten through the employee screening procedures in order to work out a schedule of theft from the inside. His efforts can be counteracted to some degree by denying him the information he needs to plan for major attacks. Shipments of unusual value should be confidential, and only those employees who are directly concerned with loading or unloading or transporting such shipments should be aware of their schedules. Teletype information about such movements should be restricted, and trailer numbers should be covered while the vehicle is in the terminal. Employees involved in any way with such shipments should be specially selected and further indoctrinated in the need for discretion and confidentiality.

A Total Program

Adequate physical security installations supported by guard and alarm surveillance will go a long way toward protecting the facility from the thief, but these measures must be backed up by proper personnel and cargo movement systems, strict accountability procedures, and a continuing management supervision and presence to insure that all systems are carried out to the letter. Regular inspections of all facets of the operation, followed by prompt remedial action if necessary, are essential to the success of the security effort.

PLANNING FOR SECURITY

It is important to the security of any transportation company, shipper or freight terminal operation to draw up an effective plan of action to provide for overall protection of assets. The plan must be an integrated whole, wherein all the various aspects are mutually supportive.

In the same sense, in large terminal facilities occupied by a number of different companies, all individual security plans must be integrated to provide for overall security as well as for protection of individual enterprises. Without full cooperation and coordination among participating companies, much effort will be expended uselessly and the security of the entire operation could be threatened.

Such a plan should establish area security classifications. Designated parts of the building or the total yard area of the facility should be broken down into controlled areas, limited areas and exclusion areas. These designations are useful in defining the use of specific areas and the mounting security classifications of each of them.

Controlled Areas

Controlled areas are those areas whose access is restricted as to entrance or movement by all but authorized personnel and vehicles. Only part of a facility will be designated a controlled area, since general offices, freight receiving, personnel, rest rooms, cafeterias, and locker rooms may be used by all personnel, some of whom would be excluded if these facilities were located within an area where traffic was limited. Within this area itself, all movement should be controlled and placed under surveillance at all times. It should, additionally, be marked by a fence or other barrier, and access to it should be limited to as few gates as possible.

Limited areas are those within the controlled area where a greater degree of security is required. Sorting, handling of broken lots, storage, and recoopering of cases might be vulnerable functions handled in these areas.

Exclusion areas are used only for the handling and storage of high value cargo. They normally consist of a crib, vault, cage, or room within the limited area. The number of people authorized to enter this area should be strictly limited, and the area should be under surveillance at all times.

Since such areas should be locked whenever they are not actually in use, careful key control is of extreme importance.

Pass Systems

All employees entering or leaving should be identified and checked for their authorization to be there. Each employee should be identified by a badge or pass, using one of several systems:

- The single pass system, wherein the badge or pass coded for authorization to enter specific areas is issued to an employee who keeps it in his possession until his authorization is changed or until he terminates.
- The pass exchange system, in which he exchanges one color-coded pass at the entrance to the controlled area for another, which carries a different color code specifying the limitations of his authorization. Upon leaving, he surrenders his controlled area badge in exchange for his basic authorization identification. (In this system the second badge never leaves the controlled area, thus reducing the possibility of switching, forging or altering.)
- The multiple pass system, which provides an extra measure of security by requiring that an exchange take place at the entrance to each restricted area within the controlled area.

Vehicle Control

The control of the movement and the contents of all vehicles entering or leaving a controlled area is essential to the security plan. All facility vehicles should be logged in an out on those relatively rare occasions when it is necessary for them to leave the controlled area. They should be inspected for load and authorization.

All vehicles entering the controlled area should be logged and checked for proper documents. The fastest and most efficient means of recording necessary data is by a camera such as a Regiscope, which will record all of the required information on a single photograph. This should include the driver's license, truck registration, trailer or container number, company name, way bill number, delivery notice, a document used to authorize pickup or delivery, time of check, and the driver's picture if it is not on his license.

The seal on inbound loaded trailers should be checked and the driver issued a pass which will be time-stamped on entering and leaving the area and which will designate the place for pickup or delivery.

All vehicles leaving the area should surrender their pass at the gate where their seals will be checked against shipping documents. Unsealed vehicles will be inspected, as will the cabs of all carriers leaving the facility. Partial load vehicles should be returned to the dock from time to time on a random basis for unloading and checking cargo under security supervision.

Loading and unloading must be carefully and constantly supervised, since it is generally agreed that the greater part of cargo loss occurs during this operation and during the daylight hours.

Other Security Planning

The security plan must also specify those persons having access to security areas.

It must specify the various components necessary for physical security such as barriers, lighting, alarm systems, fire protection systems, locks, and communications.

The plan must detail full instructions for the guard force. These must contain general orders applicable to all guards and special orders pertaining to specific posts, patrols and areas.

There must be provision for emergency situations. Specific plans for fire, flood, storm, or power failure should be a part of the overall plan of action.

After the security plan has been formulated and implemented, it must be reexamined periodically for flaws, for ways to improve it and to keep it current with existing needs. Circulation of the plan should be limited and controlled. It must be remembered that such a plan, however well conceived, is doomed at the outset unless it is constantly and carefully supervised.

Security Surveys

The security manager of freight terminals or companies engaged in shipping will find himself continually occupied with surveys of the facility under his security supervision.

An initial survey must be made to formulate the security plan governing the premises. It should be thorough enough to detect the smallest weakness in the operation and provide the information needed to prepare adequate defenses. Further surveys will be necessary to evaluate the effectiveness of the program established, and follow-up surveys should determine whether all regulations and procedures are being followed. Additional surveys may be necessary to reevaluate the security picture following changes in operational procedures in the facility, or to make special studies of particular features of the security plan.

Inspections

In addition to these surveys, which are essentially designed to evaluate the security operation as a whole or to re-evaluate it in the light of changing conditions, the security manager should make regular inspections of the facility to check on the performance of security personnel and to check the operating condition of the facility. Such inspections should include potential trouble areas and should not overlook a check of fire equipment and alarm systems.

Education

If the security plan is to succeed, it must have the full cooperation and support of all employees in the facility. This can only be achieved by a continuing program of education in the meaning and the importance of effective security in every phase of the business.

All personnel should be indoctrinated at the time of their employment, and a continuing program should be instituted to update the staff on current and anticipated problems. More advanced courses on procedures might be instituted for management personnel.

Part of this program should be devoted to educating employees in the importance of security to each individual and his job.

Security reminders are also important to keep the subject of security constantly alive in the minds of everyone. Posters, placards, and notices prominently posted are all effective devices for getting the message across. Leaflets or pamphlets covering more detail can be distributed to each employee in his pay envelope.

CARGO IN TRANSIT

The Threat of Hijacking

Although the crime itself is dramatic and receives much publicity, armed hijacking of an entire tractor-trailer with a full load of merchandise represents only 10 percent of the losses suffered by the shipping industry as a whole. This is clearly not to say that it is a minor matter. On the contrary, such a crime is of extreme importance to the carrier taking such a loss because the enormity of the theft represents a huge financial blow in one stroke, whatever its significance in the overall percentages.

There is, however, little that can be done by a driver who is forced over by a car carrying armed and threatening hijackers. When it comes to that point, he has little choice but to comply. The load is lost but, hopefully, the driver is unhurt. There is little that private security can do in such cases, and it is indeed fortunate that the incidence of such crimes is as low as it is. The matter is in the hands of public law enforcement agencies and must be handled by them.

If there is cause to believe that hijacking of a load is an imminent possibility, trucks should be scheduled for non-stop hauls and re-routed around high risk areas. Schedules should be adjusted so that carriers do not pass through high risk areas at night. In extreme cases, trucks might be assigned to travel in pairs or in larger convoys. Very high value loads that are deemed especially vulnerable can be followed by company cars. Such procedures constitute selective protection at best, since they can be used only infrequently and are impractical for general application.

Other aspects of theft on a line haul can be dealt with, however, and it is important that procedures be established that will serve to protect the cargo.

Personnel Qualifications

The cardinal rule in the management of a trucking concern is that those assigned to line haul duties must be of the highest integrity. Drivers and helpers must be carefully screened before they are hired, and they must be carefully evaluated by personnel and security managers before they are given this critical responsibility. An irresponsible or criminally inclined driver can cost the company all or part of his load, and whether the loss is unintentional or the result of the driver's carelessness, the cost to the company will be the same.

Many cases have been reported where the driver has set himself up as a "victim" of a hijack. This is difficult, if not impossible, to prevent unless the honesty of the driver is unquestionable—an attitude which can only be determined by a careful screening process and regular analysis of the man's behavior and performance. Any changes in demeanor or in his life style should be noted as a possible deterioration of morale or a basic change in attitude toward his job that might lead to future problems.

Procedures on the Road

All employees should receive specific instructions about procedures to be followed in every predictable situation on the road. The vehicle should be parked in well-lighted areas where it can be observed. It should be locked at all times, even if the driver sleeps in the cab. Trailers should be padlocked as well as sealed.

Drivers must be instructed never to discuss the nature of their cargo with anyone. Thieves frequently hang out at truck stops hoping to pick up information about the nature of loads passing through. All too often, the most innocent conversations among truckers can lead to the identification of a trailer containing high value merchandise which, when spotted, becomes a target.

The driver should never deviate from the preplanned route. In case he is forced to take a detour or his rig breaks down, or if in any way his schedule cannot be met, he must notify the nearest terminal immediately.

Trucks should be painted on the top and sides to facilitate identification by helicopter in the event of theft.

Seals

Among the many seals available today, the one in most common use is the metal railroad type. This is a thin band of metal which is placed and secured on the trailer in such a manner that the door or doors cannot be opened without breaking it, thus revealing that the doors have been opened and a theft has or may have taken place. They are easy to break and must be broken when the destination has been reached and the merchandise is unloaded. They in no way secure the doors, but are placed there simply as a device to indicate whether the doors have been opened at any point between terminal stops. Each seal is numbered and should identify the organization which placed the seal (See Figure 14-1).

All doors on a trailer must be sealed. Those trucks or trailers with multiple doors may habitually load by only one, in which case those doors not in regular use may carry the same seal for months at a time with only the rear being regularly sealed and unsealed.

Seal numbers must be recorded in a permanent log as well as in the shipping papers. The dock superintendent or security personnel should be responsible for recording seal numbers and affixing seals on all trucks. The seals should be positioned in such a way that locking handles securing the door cannot be operated without actually breaking the seal. Some truck or railway car locking devices are so large that several seals may have to be used in a chain to properly seal the carrier. In this case all seal numbers must be recorded.

Resealing. Trailers loaded to make several deliveries along the route must be resealed after each stop. To accomplish this, enough seals must be issued to be placed on the truck after each delivery is made. In this case the truck is sealed at the point or origination and the seal number logged and entered in shipping documents. The

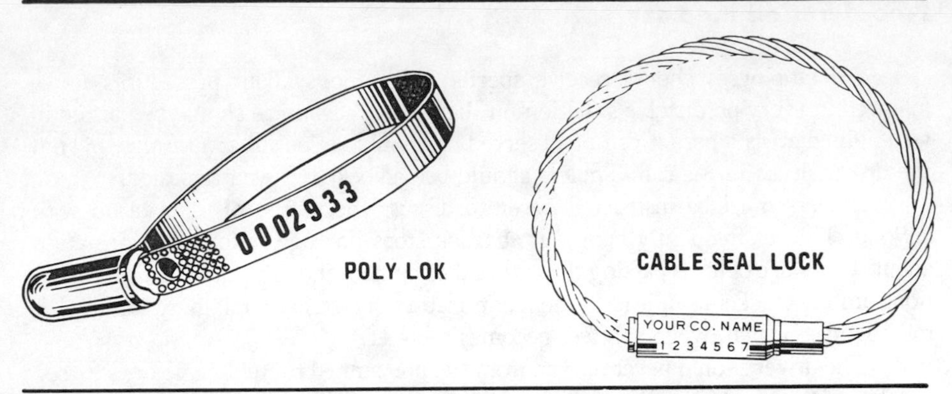

Figure 14-1. Two security seals in common use. (Courtesy of E.J. Brooks Company.)

additional, as yet unsealed, seals are also logged, and the numbers to be used at designated points of delivery are entered in the shipping documents.

When the first point is reached, the receiving clerk verifies the seal number of the truck on arrival. After his merchandise is unloaded, he affixes the next seal as directed by the shipping documents. This procedure is followed at all stops, including the last one. There, a seal is affixed to the now empty vehicle, which may return to the point of origin, where the seal will again be checked against the shipping documents.

Empty trailers should also be sealed immediately after being unloaded. This practice will discourage the use of such empties to remove unauthorized material from dock areas, and it precludes the necessity of physical inspection of all vehicles leaving the controlled areas, providing the seal and the papers check out.

Seal Security. All seals must be held under the tightest possible security at all times. Unissued seals should be logged and secured upon receipt. They should then be issued in numerical order, and the assignment of each should be duly noted in the log. The seal supply should be audited daily; a careful check to account for each one, issued or not, should be made at the same time. Without such regular inventories the entire system can be seriously compromised. As a further part of such audit, all seals which are damaged and cannot be used, as well as seals taken from incoming vehicles, should be logged and secured until they can be destroyed.

REVIEW QUESTIONS

1. Why is it said that the greater burden in preventing cargo thefts falls upon private security rather than on public law enforcement?

2. Describe the operation of an invoice system that would establish accountability for all merchandise in shipment.
3. What are some policies and procedures that would be effective in deterring pilferage?
4. Define CONTROLLED AREAS; LIMITED AREAS; EXCLUSION AREAS.
5. Offer procedures for the control of the movement of and contents in all vehicles entering or leaving a controlled area.

Chapter 15

COMPUTER SECURITY

In the past 30 years American business, in the process of almost explosive evolution, has leaped light years ahead in its capacity to gather, store, and deal with vast amounts of computerized information. This ability has made it possible to effect great improvements in the efficiency and flexibility of all financial matters as well as to create systems providing instant playback describing inventories, market position, and potential revenues at any given time.

There seems to be no end to the benefits to business to be derived from computerization, and the more benefits it creates, the more dependent upon its magic we become. In the past decade or so, the computer has become the nerve center of our commercial activity. Indeed, many modern companies are frank to admit that they could not operate without the computer.

A report by the Stanford Research Institute points out that ''recent studies indicate that about 7 percent of the present U.S. work force of 84 million people work directly with computers, and another 15 percent work indirectly with them.[65] This same report goes on to say, ''All but the smallest business and government agencies own, lease or use computer service. Most large organizations are discovering that they can function for only a few hours or few days at most without correct functioning of their computers. At least 60 percent of all banks are automated and would be unable to function unless their demand deposit accounts were successfully processed on computers.

''It is estimated that computer manufacturing, data communications, and operation of computer systems will together represent 14 percent of the gross national product by 1980. A consensus of experts recently indicated that losses, injuries, and damage directly associated with computers will exceed $2 billion annually by 1982.[66]

Computer Vulnerabilities

There can be little question of the importance of the computer in the business world of today, and its increasing significance in the future.

Yet, in spite of the benefits to be derived from a computerized operation, it creates a great potential for danger. Probably no one element in business, including catastrophic fire and the ravages of non-computerized internal theft, presents a greater potential to wipe out an entire business so quickly and so effectively.

The dangers that can befall a computer center or can be created by it encompass virtually the entire operation of the business it serves. Embezzlement, programming fraud, program penetration, operator error, input error, program error, theft of confidential information, and plain carelessness are a few of the problems that can arise in routine operation. Add to this the potential for fire, riot, flood and sabotage.

The Need for Security

In the face of the risks to this sensitive machinery and to the enormous accumulation of data concentrated in a limited area, as well as business's ever-increasing reliance on the computer, many firms continue to ignore the dangers. Many of them confess an uneasiness about security for the computer but claim they cannot afford an effective protective program.

There is no question that computer security can be costly, but the stakes are too high to try to effect economies in this area. The fact is that any company involved with a computer—and that includes most companies in this country—cannot afford not to have a comprehensive security program that will protect their computer operation. Since such installations can represent an investment running into the millions of dollars, a security program costing from $50,000 to $250,000 is not an unreasonable expense.

Ideally the program designed to provide computer security will be an integrated part of the company's overall loss prevention effort. It should be administered by a specialist in computer operation, but it will report to, and be coordinated with, the larger program. It must encompass physical security, procedural or operating security, access and traffic controls, disaster planning and contingency procedures, employee education programs and a painstaking screening procedure in the employment of computer personnel.

An effective security program is essential to any electronic data processing (EDP) operation, and it must have the vigorous support of an informed management. It is not always easy to convince the chief financial officer of the need to expend substantial funds for security, but when the risks are as high as they are in computer operations, he must somehow be convinced.

PHYSICAL SECURITY

The physical security needs of a computer installation are, in most respects, the same as any business or industrial establishment. Its goal is essentially to deny access to the computer area to all unauthorized persons.

In many respects, this is easier to apply to a computer installation than to other facilities, since only a limited number of people have reason to be in the area in the first place. Whereas an industrial facility may have to make provision for the movement of employees of various trades, truckers, supervisory personnel and others in great numbers, a computer center is not obliged to put up with any appreciable traffic of this kind to operate. It can, and should, be closed to all but those specifically assigned to it.

Structural Barriers

To prevent surreptitious or forced entry, the facility should be housed within an area which is already secured against intrusion or, if possible, in a separate building far enough from other structures to be free from the danger of fire spreading from them. In any event, it should be protected by appropriate fencing, whether standing alone or housed within the perimeter protection of another building. It should be adequately lighted and it should not be obstructed by ornamental plantings and landscaping which could provide concealment from an intruder.

Entrances should be limited. The best design would incorporate one entrance only and such fire doors fitted with panic bars as are consistent with local fire laws and good safety procedure. Fire doors should be locked, allowing exit but preventing entrance, and they should be alarmed to signal their use.

Any other means of entrance through vents, trash chutes, air conditioning ducts or drains over 64 square inches should be covered with bars or heavy wire mesh. If the roof can be reached from poles or neighboring buildings, it should be appropriately fenced. If the building is only one or two stories, the roof can be assumed to be accessible and should be reinforced to prevent forced entry by chopping through it. As many windows as possible should be sealed by bricking them up. Those that, for any reason, cannot be sealed should be covered with a heavy wire mesh of a close enough weave to prevent the introduction of devices such as Molotov cocktails or bombs.

Alarms. Ideally, the entire facility should be alarmed, although, if it operates on a 24-hour basis, a full alarm system may not be necessary. If it is not that active, a central station alarm system is essential. The system to be used will depend upon the layout of the facility, but it should certainly cover the entrance, the entrance to the computer itself, and the tape library. In the case of a one or two-shift operation switch locks which lock off power to the machines should be installed.

Adjoining Spaces. In cases where the computer is located in building serving other purposes as well, care must be taken to prevent entry into the area by drilling through the floor from areas below or by moving through the crawlspace in the ceiling. Modern construction typically includes space for air conditioning and other services above the false ceiling. This space is accessible simply by pushing up the suspended ceiling tiles and crawling from office to office. Such a space could provide easy egress to the computer or the computer library unless it is blocked by a bearing wall or by other means. Cables generally run under a raised floor; this space must also be made inaccessible to intruders.

In this discussion of physical security, we have alternated between assumption of a facility located in a detached building and one located in an existing structure. It might be in either. It would be unrealistic to expect it to be located as sole occupant in a separate structure in an urban setting, but it might be well to consider placing it outside the city. Proximity to the computer is rarely necessary, and the advantages of removing it to a setting where it can be isolated from any but its own problems are considerable. A sizable computer operation represents an enormous capital outlay. It would seem to be poor economy to locate it in a "space available" site in an existing building which is probably poorly equipped to handle the needs of this vital (and expensive) installation.

Site Selection

In considering location, it is important to evaluate the environmental and human factors that might affect the site as well as the services available in the area. Some of the factors that must be considered are:

Fire and Police Protection. Check response capability and general efficiency. In this same regard, the availability of central station alarm protection and its reliability must be investigated.

Surrounding Neighborhood. High crime areas are undesirable—not so much because of computer vulnerability as because it is a problem in employee morale. Such a neighborhood could have an adverse effect on employee retention and could also serve to increase insurance premiums in various areas of coverage.

Access. The site must be easily accessible to its personnel. Ideally, it should be served by adequate public transportation—remembering that virtually every such installation spends part of all of its time on a three-shift or 24-hour schedule.

Maintenance. The time of response of manufacturer maintenence and repair personnel must be considered.

Space Requirements. The equipment required for a specific operation may take more space than the headquarters building can provide. The weight of it also may be such that extensive and expensive remodeling may be needed in the existing building.

Natural Phenomena. Consideration must be given to the possibility of floods, earthquakes, hurricanes, and tornadoes. Obviously such disasters cannot in themselves be anticipated, but their incidence in the planned site can be. Earthquakes can be anticipated on or near existing faults, floods can be considered a possibility in certain known areas. Some areas suffer from tornadoes or hurricanes more than others. All of these factors must be considered before a site is finally determined upon.

This evaluation should also take into account the presence of radar installations. In the past several years, there have been several cases of tapes being erased by radar radiation as much as 1/8 of a mile away.

Environment. A tape library suffered a loss of more than $75,000 from acid fumes given off by a neighboring manufacturing facility. This is as much a problem of internal climate control as of location, but the presence of such pollutants as well as a high incidence of particulate matter must be known and dealt with.

Power Source. Check the power company's record of reliability and its reputation for efficient and speedy response to power failures. Any fluctuation in line voltage can cause inaccuracies in data transfers. While some operations are not seriously hampered by fluctuations, with others the problem can be a serious one. Where this is the case, the practice is to isolate the computer from the power from the local utility by the use of batteries, constantly recharged by line voltage, which supply the computer through solid-state interrupters.

Site location must also take into account the presence of water mains and proper drainage in the computer room and library if an already existing building is under consideration. Much damage has resulted from flooding caused by broken water mains or pipes, further complicated by the lack of adequate drainage.

Access Control

Since one of the principal means to computer security is access control, it must be planned so that those employees authorized to be in the EDP center can get in with a minimum of restrictions and red tape, while unauthorized persons are positively excluded. There are a number of systems currently in use that accomplish this with a high degree of success.

Entrances. There should be an absolute minimum of entrances to the facility. The entrance should be locked at all times. If at all feasible, this entrance should be under the supervision and control of a security employee.

Some installations use a double door arrangement with an electrical control allowing entrance and exit. In this construction, an outside door leads into a small anteroom large enough to accommodate two or three people. Beyond that is another door leading into the facility. When employees arrive, they are identified by the guard and allowed through the first door. They remain in the anteroom until the outside door closes and locks before they are allowed in the second door. This arrangement protects against unwelcome persons crowding into the facility along with authorized employees. In larger EDP centers this entire operation can be supervised by a guard observing on CCTV. He activates the electrically controlled doors remotely from his control position.

Locks and Keys. In smaller installations where such an elaborate set-up is not practical, authorized employees may be issued keys to the entrance. This system is generally unsatisfactory, since keys can be lost or compromised.

A better system might be the installation of one of several kinds of combination locks. Combinations can be changed periodically to prevent entry by former employees or other persons who may have discovered the combination. Other locks are available that are opened by the insertion of cards magnetically coded both to identify the user and to open the door. Systems using both the coded card and a combination lock are in effect to avoid the possible compromise of either system above.

There are also locks that identify personnel by their fingerprints. The index finger is placed in a slot and an identifying card is placed in another slot at the same time. A film clip containing the employee's fingerprint is automatically selected from a file, compared with his finger and entrance is effected. There are even electronic keys that are, in essence, mini-computers, which can be used to gain entrance.

Each system has its strengths and its weaknesses. No system is infallible, but they should be studied to determine which is the most suitable for a particular installation. Cost is always a consideration, as is the efficiency and dependability of the mechanism.

If any system provides high security at the expense of reasonable convenience to authorized personnel, it may ultimately prove to be totally unsatisfactory. As with any other control, care must be taken in the selection of an access system that it does not create more problems than it solves. If the efficiency and morale of employees are substantially reduced by the installation of any system, it almost certainly should be changed, no matter how security effective it may be.

FIRE PROTECTION

EDP centers are particularly vulnerable to fire. The maze of wires and equipment under the floor and the accumulation of paper from printouts, in addition to the normal fuel load characteristic of any office, make for substantial quantities of combustibles to feed any fire that might get started. The losses in such a fire usually are tremendous. The equipment involved is extremely costly, and the loss of data on destroyed tapes can be many times more harmful and expensive. RCA recently estimated it would take 9.2 man-years and 478 computer hours to reconstruct the master files of its New York City EDP center.

Causes of Fire

The cause of fires in EDP centers is a matter of some dispute. Potentially, it would appear that the principal cause should be from electrical malfunctions. The miles of wiring and the possibilities of short circuits are powerful persuasions to exercise great care and to conduct frequent inspections in those areas accessible to inspection.

Statistically, however, the major cause of fire is from adjacent occupancies—either in the same or neighboring buildings. Most EDP centers have been constructed and equipped to resist fire very effectively, but they are often located in a building which is not so well protected. Obviously, the computer will go down with the rest of the building in case of fire no matter how well it is, itself, protected. There have been cases where the floor has collapsed in a fire and dropped the computer and its attendant equipment into the fire below.

Careless use of cigarettes around accumulated waste paper is the second most common cause of fires in EDP centers. Electrical fires are the third most common, but, although it is third on the list, the threat of fire from the power system is always present.

Sensors

Since some polyester-backed tape used in computers can be rendered unreadable at temperatures of 150°, it is essential that a sensor system providing the earliest possible warning of a fire be installed. The most economical installation of such a system is during the construction of the facility, but it must be done whether the facility is already built or not. Sensors should be located on ceilings, under raised floors, inside false ceilings, in air ducts and within the equipment itself. This same protection must be extended to the tape library as well. The earliest warning comes from ionization or products-of-combustion detectors, which respond to the earliest stages of combustion before the development of smoke or flame. These sensors must be checked regularly to be certain they are in working condition.

Sprinkler System

Perhaps the most effective automatic fire extinguishing system in general use is the sprinkler system, but its use in computer installations is a matter of considerable

How Safe is Your Computer Room?

- Tape and other magnetic media should be kept in a secure library separate from the computer room.
- Smoking, eating, horseplay, or socializing should not be allowed in the computer rooms.
- Communications lines and equipment should be aggregated within locked telephone closets.
- Equipment cabinets should be kept closed and locked unless access is needed for authorized servicing.
- There should be no accumulation of paper trash in computer rooms.
- Computer rooms should be windowless restricted areas with access controlled by the senior operator on duty.
- There should be an easily accessible portable CO_2 fire extinguisher of at least 15 lb. capacity within 50 ft. of any point in a computer room.
- Emergency "stop" buttons should be covered by spring-loaded housings to prevent accidental actuation.
- Logged-in terminals should never be left unattended.
- There should be no directional signs posted outside the computer room that might give assistance to possible intruders.
- Computer print-out should be handled in a secure manner until delivered to the intended user.
- Documentation, spare parts, and removable media should be locked away when not in use.
- Dust sources such as keypunches, chalkboards, and outer clothing should be kept out of computer rooms.
- Data-preparation areas should be located in secure zones separated from the computer room.
- Customer-service counters should have limited access and be separated from the computer room.
- No keypunches, photocopiers, cameras or other reproduction equipment should be allowed in computer rooms.

Figure 15-1. How safe is your computer room? A Checklist. Reprinted with permission of Cahners Publishing Co., from *State and County Administrator*, May 1978, 14.

controversy. The Factory Mutual group of insurance companies favors the use of sprinklers in computer installations—and they are experts in the field of fire protection. IBM, the world's largest manufacturer of computers, has said that they are harmful to the equipment and should never be used. There are compelling arguments for both views.

Water from a sprinkler system will run indefinitely after it has been triggered by an alarm or by heat, until it is turned off manually. This continuous stream of water cools the burning material below the point of combustion so that it will not start again after the water is turned off. This cooling effect can be important in cases of smoldering combustion (as opposed to flaming combustion).

Unfortunately, there are serious side effects in a computer application. If the power to the computer is not turned off before the water hits the equipment, the effect on electronic and electro-mechanical parts can be ruinous. Then, too, tapes can be destroyed. Tapes are not at all affected by a short immersion in water, nor are they damaged by temperatures as high as 250° in a dry environment, but some can be ruined by a combination of heat and humidity. As noted above, they will deteriorate to unusability at 150° in a humid environment. Thus, the problem with sprinklers. When they discharge into a fire, they produce steam—which may quickly wipe out the tapes subjected to it.

There are answers to these problems. One lies in the installation of a system which alarms first and provides enough of a delay before water is discharged to allow an employee time to turn off the equipment before water is discharged from the sprinklers. This procedure could also be automatic, whereby the equipment was inactivated by the alarm prior to the flow of water. In such a case, the equipment would have to be dried out before it could be used again.

In cases where the facility is shut down, all equipment should be covered to protect it from the possibility of such water damage. If a fire started inside one of the machines, it would burn away the cover and become accessible to the extinguishing efforts of the sprinkler, but the other unaffected machines would still be protected by covers.

Other Extinguishing Systems

Halon 1301, an inert, odorless gas, is an efficient extinguishing agent recommended by many authorities in the field. This gas is non-conductive, leaves no residue, and is highly effective in fighting fires. It presents problems, however. It has no cooling effect, and although it is extremely effective when used against flaming combustion in solid fuels, it is considerably less effective against smoldering combustion. Every automatic halon system is limited by the quantity of halon contained in its storage receptacle. Whereas water will continue to flow until turned off, the halon will be discharged only until it is exhausted. Usually, this will be enough, but in some cases of stubborn smoldering fires, it may not be.

Finally, whereas halon in its normal use presents no hazard to humans, it is a toxic chemical. Although the National Fire Protection Association has stated it has no effect on humans in concentrations up to seven percent, it does have harmful effects in concentrations beyond that. They further suggest that concentrations of from 15 to 20 percent might even cause death if exposure to such concentration is prolonged.

Carbon dioxide systems are as effective as halon, but CO_2 is very dangerous to use since it is deadly in the concentrations needed. All personnel must be evacuated from the immediate premises before it can be discharged. If a CO_2 system is installed, all personnel must be trained in evacuation procedures and indoctrinated in the dangers of this gas.

Dry chemicals are effective in smothering a fire, but they leave behind a residue of powder on the equipment and in the air which may be more harmful to the machines than the fire itself.

CO_2 extinguishers should be placed throughout the facility and all personnel should be schooled in their use. These extinguishers are not a substitute for a total extinguishing system, but they are valuable for bringing small fires under control before they become big enough to require disruptive and expensive discharge of the main system.

General Fire Prevention

Perhaps the most important element in the fire program, is the creation of a carefully supervised fire-resistant environment. This involves a trash disposal system which will prevent the otherwise inevitable build-up of waste paper in the computer area; the establishment of no smoking rules in and around the computer room and the tape library; frequent inspections and clean-ups under the raised floor; fireproofing all rugs, drapes, wallpaper and furniture; sealing off all holes and passages that violate the fire-resistant integrity of floors, walls and ceilings; installing multiple manual cut-offs of air conditioning.

SYSTEM PROTECTION

Back-Up Systems

Because when a company changes over from manual to computer operation it becomes dependent on the computer, it is vital that some contingency plan be developed for emergencies of any kind that would render the computer inoperable. Such a breakdown could come from any of a number of sources—machine malfunctions, power failure, natural disaster, fire, malicious destruction, or even building renovation.

It is customary in setting up such a plan to locate a computer of the same model or conformation, and to enter into some kind of mutual assistance pact, in which it is agreed that, if either party should suffer from some kind of breakdown, he can continue to operate in the other's facilities.

This is sensible pre-planning, but it must be pursued much farther than this. It is important, in the first place, to remember that not all machines—even of the same model—are compatible. Systems vary according to ancillary equipment and memory size. Even in cases where the equipment is identical in every detail, differences in programs may create incompatibility.

There is no real way of knowing whether one system will operate with another's software without a trial run. Before any mutual agreement is entered into, it is essential that each test his needs on the other's equipment. If it is determined that compatibility does exist, periodic test runs should still be made to verify the continuing compatibility of the systems. Since equipment and program changes are common in computer operations, there is always the distinct possibility that what is a suitable back-up system today will be unusable when it is needed most. Only regular testing can establish this. Agreements to notify the other of any changes in the system cannot be depended upon as reliable.

Another problem that arises regularly in such agreements is the change in each party's requirements for machine time. If both concerns are on a 16-hour machine use schedule, they have a basic problem to begin with. Even if they are each on an eight- or a 12-hour schedule at the time the agreement is entered into, there is no assurance that their needs will not increase to such an extent that neither can fully accommodate the other. Ideally, a similar agreement should be made with several companies to assure continued operation in the face of an emergency. This complicates the problem, but it is sufficiently important to make the effort worthwhile.

Since most manufacturers can now deliver an exact copy of a customer's machine promptly, an alternative to shared backup systems is an alternate site where a duplicate system can be installed. Site preparation, however, can still require a long time. One economical approach to providing such an alternate processing site for emergency use is for several companies to share a site that affords all facilities except the computer.

Off-Premises Protection

However effective security measures may be, there is always the possibility the defenses may be penetrated. If such penetration should result in the loss or destruction of stored data, the results could be catastrophic.

Here, as elsewhere, it is important to rememeber that a company, once it has undergone the difficult process of converting its operation to computers, cannot operate manually. Therefore, the loss of much of its carefully accumulated data would be ruinous. The destruction of a database carrying perpetual inventory information, or a record of accounts receivable, could create problems from which the company might never recover. It is, therefore, essential that duplicates of such data be stored in some remote location where their protection can be assured.

Data stored in this manner must, of course, be regularly updated to reflect the current status of the company. How duplicate data are created, as well as the method and frequency of updating, are matters to be determined by the cost and other factors involved. In some instances duplicate tapes of all programs have been run as they are made. In the past it has been common to provide off-site storage for as many as five generations of data, or to use a three-generation system consisting of a duplicate of data up to the time of storage and journal tapes to provide the information needed to update the latest version. Today most companies use database management systems that store data on disk units and not necessarily in the form of distinct files. These databases are usually updated in real time. Therefore, it is no longer possible to rely for backup on rotating the generations of tape files. Arrangements must be made for off-site storage of update tapes and for periodic dumps of the database.

As a final note, it must be pointed out that the destruction of computer data is less often the result of criminal activity than of natural disaster, fire or, even more often, program error, machine malfunction, or input error. Whatever the cause, however, the results are the same, and some carefully supervised system of duplication must be installed to protect against these possibilities of loss.

Protecting Against Unauthorized Use

One of the particularly disturbing problems faced by a company which has converted to a computer operation is the possibility that unauthorized persons might operate the machine either at the installation or through remote terminals, thus gaining access to privileged data which they can then convert to their own use. This is a very real and valid concern, since all the company's files and records have, in fact, been transferred to an instrument that is accessible to many people within the company or, in the case of time-sharing operations, to many people outside the company.

Passwords. The machine contains a complete and accurate data source essential to the operation of the business it serves. It will, upon command, produce any or all of that information. It will do precisely what it is told to do. Security of data lies in concealing what the machine is told to do. This usually consists of a file system with the establishment of a password when the file is created. The machine is, in effect, told to release such files only after the programmed password is used when the information is requested. It is further instructed never to reveal the password in readouts or system reviews.

Graduated Access. Some files are further protected by a preprogrammed permissions system in which the names of authorized system users are entered into the file on graduated access basis. In this way, a file can be used to its fullest extent by allowing full and restricted use simultaneously to users at various levels of authority. Machines whose use is restricted to only a few special employees are seriously hampered in their effectiveness. A system of graduated access, however, opens up its potential usefulness and at the same time restricts its use to a need-to-know or use basis.

Remote Terminal Security. In the early stages of computer technology, information was produced only at the site of the machine itself. Protection of its information was simpler, since denial of physical access to the machine denied access to its information. This is not to say that unauthorized personnel did not obtain sensitive information directly from the computer, but when this occurred it was because of a failure in the personnel access procedures.

Today, however, thousands of concerns operate on what are referred to as management information systems. Such systems, much like airline reservation systems, locate consoles, teletypes and other readout devices in convenient locations within the company building or in a network of locations tied in with telephone lines across the country. Many of these devices enable a user to access the computer. In the same way, smaller companies using a computer on a time-sharing basis have similar terminals with which they can access the computer operated by a service bureau.

In both cases the security problem is the same, though its focus is somewhat different. In the case of the proprietary computer, it is necessary to restrict information to those people and those locations where such information is authorized. The time-sharing customer must be restricted to access to his own data base. He must be prevented from entering the files of other customers. How he restricts that information within his own company is for him to decide and program accordingly, but he, as well as all other customers, must be assured that his file is secure from access by any but authorized personnel.

In either system, both personnel and location authorization must be considered as part of the security system. For example, the personnel manager may be authorized to retrieve payroll information at a location within the confines of the personnel department. Although he is authorized to receive the data, it would be denied him if he asked for it from any other location. This is an important precaution in areas where a visual display might well expose confidential information to any number of unauthorized people.

Checks and Audits. Examining the total process of information retrieval in a security situation, we would find that an initial check and ongoing audit of the online user's authority is basic to the system. The user would activate a remote station, at which time he would be asked for his name, organization and password. The software would identify the individual, verify his authority, acknowledge the password, and combine that verification with the status of the terminal in use and the access level of the user. After this initial access, the user would ask for a particular file, using such other passwords as may be indicated. Upon further verification, the computer would provide him with the data he needed. Any information sought from the machine would be validated before delivery of such information. If, during this time, at any stage, the user attempted to retrieve unauthorized information, appropriate security alarm procedures would be initiated.

Encryption. The best protection for data, either when stored within a computer system or while in transit from one terminal to another, is encryption. The National Bureau of Standards Data Encryption Standard has been implemented in both hardware and software and is commercially available.

Protecting Against Errors

It is important to review the operation of a computer system in order to grasp the significance of even the smallest error, whether caused by carelessness or by the intent of someone to compromise the machine.

In the first place, computers store information in their own language and there are many different languages available for machine use, depending on its application. Print outs are frequently in direct statements easily comprehended by the uninitiated, but the stored information is in the form of digits. Everything has been turned into a number and, in order to retrieve it, that number must be called for. If anything has in any way compromised the integrity of that number, the information cannot be retrieved. If, for example, some item has been designated a three but, by error, has been entered into the machine as a two, it may never be found. Since the program clearly carries it as a three, there is no way of knowing that it has somehow found its way into the data bank as a two.

It is essential, therefore, that procedures of operation as well as programming include many methods of cross-checking, and the computer must be further programmed to notify the operator if the cross-checks fail.

An insufficiently trained or unqualified operator may not recognized such problems when they arise, and may continue to damage the stored information in such a way that he compounds the original error and creates a chain of lost information. Such a situation can be extremely damaging to an otherwise efficient operation.

This kind of loss of control over the information stored is one of the costliest and most frustrating problems facing EDP managers today. The damage potential from unqualified personnel who, though sincere, are nonetheless quite capable of bungling a job, is a greater security hazard in most companies than all the criminal conspiracies combined.

It is essential that only highly qualified personnel be permitted to function in areas affecting the accuracy and efficiency of the computer operation. To operate otherwise is to invite disaster.

It is equally essential that systems be established for every part of this invaluable function. A computer is very costly to install and operate. The demands for its use grow as its usefulness becomes more evident throughout the company. Economy dictates that it be used as efficiently as possible.

Obviously, priorities will be established governing the programs to be run, and explicit systems and procedures regarding the operation of these programs must be developed, communicated to all employees involved, and prominently posted in an appropriate place within the computer complex. These instructions should be reviewed regularly for update.

It is frequently the case that procedural rules are developed, distributed, posted, and then not pursued. Inevitably, operators work out shortcuts which may or may not be beneficial. It is management's responsibility, however, to see to it that all procedures are followed as published.

It must also be remembered that some procedures can be burdensome and inefficient. If such appears to be the case after reasonable re-evaluation, they must be changed. Operators will frequently find simpler and safer ways to achieve an end. Their suggestions can never be ignored. If they come up with a new way to accomplish a job, it should be examined and, if it is feasible, adopted as a change in operating procedures. The important thing is to insist that prescribed routines be followed to the letter, but at the same time, to keep an open mind to suggestions for better ways.

Computer Crime Measures

Most operating routines are designed to avoid mistakes that will result in a loss of control over the retrieval of information, or to avoid operational missteps that could damage or destroy data already accumulated. These are the computer professional's nightmare. But it has become increasingly evident that controls must be established to prevent the use the computer for criminal acts.

The computer presents a formidable challenge to the security man. Its very efficiency makes its use as an instrument of embezzlement almost impossible to detect—most of the instances of such use that have been uncovered have been as a result of some unforeseen accident, not from routine or even special audits. The revelations yet to come may be astonishing. In fact, Joseph J. Wasserman, who heads a Bell Telephone Laboratories task force seeking to devise methods of auditing computers used by the Bell System suggests that many companies already have been hit by heavy losses but don't know it. He predicts that within a few years someone will uncover a computerized embezzlement that will make even the $150 million salad oil swindle seem puny.

Computers are operating faster and faster and producing fewer and fewer of the printouts that auditors and financial officers need to follow the flow of dollars processed by the machines. "If auditing staffs don't get involved in designing computer systems soon, they might as well climb up on their stools, pull down their green eyeshades and pray for early retirement to come," says Mr. Wasserman.[67]

Sophistication in computer disciplines is necessary to use the machine for one's own purposes. Fortunately, this knowledge is not possessed by every thief, but the conditions are ideal for the thief educated in the ways of this technology. Computer records present a different world from the old-fashioned ledger. There are no erasures—only removals which leave no trace. No numbers or entries need to be doctored. They are simply changed, and it can be done in seconds. With most systems such a change can never be detected.

Almost none of the major embezzlements uncovered were found by audits of programs, but were discovered only when a check was accidentally returned by the post office, or an accomplice informed the police for revenge, or when an account was audited off schedule. In other words, the result of the rigged program instruction was stumbled upon—in many cases years after systematic embezzlement had begun—

but neither the machine nor its software was set up to frustrate these schemes or to detect them after they were established.

Remember—a computer can do only what it is told, and it does that incredibly well. It makes no judgments and asks no questions. It simply performs as directed.

There is not, at this time, a method which is totally embezzler-proof. That day may come, but the state of the art today is such that, if it could be done at all, the machine would be essentially inoperable. For all practical purposes, it is a hazard to which every computer operation is heir.

There are, however, some procedures which, if developed and supervised, will substantially reduce the danger and make a really determined thief work much harder for his loot.

First—Reduce or eliminate contacts between the four principal categories of computer personnel—the programmers, the operators, the test and auditing personnel, and the maintenance crew. C.F. Hemphill and J.M. Hemphill, in their informative book on computer security, *Security Procedures for Computer Systems,* refer to a major Eastern banking chain that has discouraged contact between operators and programmers (whose collusion would be the most likely and the most effective from a criminal standpoint) by "requiring these two classifications of employees to enter widely separate entrance gates, to park in segregated lots, and to enter work areas through entrances on opposite sides of the building. Continuing this separation, cafeteria access, coffee breaks, and toilet facilities are provided in different areas of the building. Passing from one area to the other with a written pass is possible only during hours of controlled supervision. At other times, an alarm system and uniformed guards prevent traffic between programmer areas and operating areas."[68]

These are extreme measures, to be sure, and perhaps more stringent than is necessary or even possible from a labor relations view in other applications. But the point is clear. Programmers must not be permitted to be involved in machine operation—not personally nor through collusion with an operator. If such a situation were permitted, any program man who was criminally inclined could build any loophole he liked into the system and then feed it any information necessary to accomplish his purpose.

For much the same reason, though to a lesser degree, contact between programmers, operators, test and audit personnel, and maintenance personnel should be minimized. It must be remembered, however, that employees in all of these categories are professionals in a complex and exacting field. They cannot be treated arbitrarily, simply because they probably won't put up with it; and they cannot be hampered or tied up in excessive red tape, or their usefulness and efficiency will be substantially reduced.

Second—Insist that all programs be documented as the program is being developed. Frequently, programmers fail to record their progress or even record the details of the program after it has been completed. This can lead to some confusion and certainly loss of time when errors need correction or changes need to be made in the program. This oversight is also a handicap to new men assigned to the project, who have no way of knowing what the program consists of, either up to the stage when they are assigned or, in the case of a completed program, in its final form.

Third—Insist that programmers stay out of the computer room. They rarely, if ever, have a function to perform there, and their presence could lead to problems for the company.

Fourth—Transfer programmers and operators to different programs and different machines from time to time. This may serve to discourage any plans to set up long term programs of embezzlement. More often than not, the schemes that have been discovered involved the regular issuance of checks or a regular transfer of funds rather than a single raid on company assets. Such transfers could minimize the risk of such steady drains on company funds and increase the likelihood of discovery of any plot by the newly assigned operator or programmer.

Fifth—Control all aspects of machine operation. If an operator has free access and control over input operations, he has effectively become a programmer-operator and can, on his own, use the machine in a number of unauthorized, and possibly criminal, ways. The status of any program run should always be known and its progress supervised.

Sixth—Insist on careful logging of all aspects of the operation. If trouble develops in the operation, reference to such a log, if well and accurately kept, can serve to locate the source of the difficulty.

Seventh—Separate computerized check writing from the source authorizing their issuance. This is only reasonable business practice which would be demanded in any company—computerized or not.

Eight—Insist on careful audits and evaluation of machine usage. Such an examination will keep management abreast of computer needs and activities for consideration in assigning priorities of use. It will also help to discover unauthorized or unnecessary usage of this valuable instrument.

Ninth—Establish a routine for the regular disposal of all output information, punch cards, and program information. This material should be shredded or carried off under supervision, to be incinerated.

Tenth—Check on programs periodically after adoption. Such checks can be made by comparing the original copy of the program against the copy being used by the operator. If any changes have been introduced, they should be immediately apparent. No changes should be permitted without supervisor approval.

Eleventh—Test-run all new programs thoroughly before allowing them to become operational. Without such pre-testing undertaken under a prescribed routine, some programs can create havoc with customer relations or they can damage existing data.

Personnel Considerations

In the last analysis, no system of safeguards can absolutely prevent theft by computer. There are too many people who must have access to programs and the machine itself in order to function. Any one or group of them could turn their expertise and virtually private knowledge of the company's systems into a criminal scheme. No company can ultimately protect itself against EDP personnel by guards, access systems, or even validation systems. Such protective devices can keep out strangers or dabblers, but not the authorized personnel. They create and operate the systems, and they can destroy them.

This is by no means to say that they are, as a group, inclined to damage the company. Quite the contrary, they are usually dedicated specialists whose instinct is to always improve existing operations by creating foolproof and tamper-proof programs. And they should be recognized as such. Security systems which materially reduce their efficiency will be resented, although they, as a group, are inclined to be the first to recognize the need for reasonable security procedures.

Screening and Evaluation. Careful screening of personnel is of the highest priority in any computer security program. Exhaustive evaluation of background, previous employment experience, if any, and level of training and competence is a must. Having satisfied this basic requirement, every new employee must be thoroughly indoctrinated in the peculiarities of the company's operation. He is then phased into an operational function, but without undue haste. This phase-in period will permit management to further evaluate his ability and attitude.

It is also important to remember that the evaluation at this phase is mutual. The new man will be weighing the acceptability of his new position to him with just as much thought and concentration as the company will be determining his suitability to its operation. Since recent studies have indicated that the average turnover in computer personnel amounts to approximately 15 percent, and competent trained people are in short supply in this exploding market, it is important to keep this mutuality of satisfaction in mind.

A Special Breed. At this point, it would be well to consider that computer people are a special breed of men and women. They speak a different language and, generally speaking, their dedication is to the demands of the machine, not to the company that owns it. They pursue challenges wherever they may arise. Encouraging company loyalty is an important consideration and should be pursued in a mature and intelligent manner. The computer room is the last place for locker room pep talks.

Management must recognize that the demands of this technology introduce different work routines, and any attempt to force non-applicable company policies in this regard on computer personnel will be non-productive and probably damaging. This is not to say that they need to be coddled or accorded any special privileges; it is simply to point out that they are specialists in a specialized field, and many of the standard rules and routines are irrelevant and don't apply.

Salaries. They must be adequately paid. This will probably adjust itself, since the demands for good people in this field are so great that offers below the going market will be ignored. But pay scales must be regularly reviewed. If the salary range in a company freezes at a point which is later exceeded by the computer business as a whole, large scale resignations can be expected. Excessive turnover of such personnel, beyond the already considerable rate, is expensive and counter-productive.

If all this seems excessively conciliatory, that's unfortunate. Those are, generally, the facts. Computer people today are, in some respects, the glamor element—the stars of the business world. They are now, and for the immediate future, in a position to demand respect and appreciation. Any other attitude by management would be to deny the realities in today's commercial world.

The best security lies in enlightened employee relations combined with conscientious leadership—and the computer staff can be the most important weapon in the fight against computerized crime.

Insurance

Since the potential for loss in a computer operation is so enormous, and since no existing system can provide a guarantee against such losses, insurance coverage is essential to back up the operation after all other reasonable precautions have been taken.

Because this kind of coverage is relatively new in the field of casualty insurance, new policies are still being developed. And since data on liabilities and claims is necessarily limited at this stage, rates may vary considerably from company to company, depending on exposure and operational techniques.

At the moment, more claims are presented for damage from water damage than any other type of loss, but this may not continue as coverage becomes more comprehensive and auditing techniques become more sophisticated.

Equipment insurance is, perhaps, the first consideration. It covers all equipment for a wide variety of risks from fire to accidental damage. Since such insurance may be offered with a deductible feature, savings may be effected in certain instances. In the case of leased equipment, the contract must be studied to determine the liabilities of lessee and lessor, since some agreements do not hold the lessee liable for damage, while others do in certain instances.

Software can be covered under the company standard fire insurance contract, but such insurance covers only the physical material such as tape. A separate policy or an endorsement on the standard policy will be required to cover the cost of reconstructing records destroyed by various hazards. Such a policy probably will not cover operator errors which result in damage to the data base.

Business interruption and extra cost insurance covering computer operations is available and recommended.

Accounts receivable insurance is available, but may not be necessary if a good program of off-site storage is followed.

Errors and omissions insurance could be very important to the operator of a data processing service. This covers liability incurred by such a service making honest mistakes in the performance of a job for a client firm. Recent court cases, in which defense disclaimed responsibility for damages because "it was the computer that did it," seem to indicate that the courts do not consider computer error to be a mitigating circumstance or a valid defense. Insurance coverage in this area is advised.

Many of the above may be covered by an all-risk policy which provides comprehensive coverage of most phases of the computer operation, but only a very careful study will determine its practical application to any given business.

REVIEW QUESTIONS

1. Justify the statement that, ideally, a computer center probably should be located 'by itself out in the country.'
2. What significance do the terms 'man-trap' and 'password' have in the area of computer security?
3. In computer center fire control, cite the advantages and disadvantages in using: 1) sprinkler systems, 2) Halon 1301, 3) carbon dioxide and 4) dry chemicals.
4. What is meant by the term 'remote terminal'? How do remote terminals complicate the security of computer systems?
5. What special risk is involved in having a computer programmer also function as an operator?

Chapter 16

INSTITUTIONAL SECURITY

The development of a security system adheres to certain principles that remain essentially constant no matter what its application may be. Basics are always present. The hardening of the target to physical attack, physical security in the interior, accountability transfer in internal transactions—these and other basics, discussed in previous chapters, appear in systems covering facilities from steel mills to stereo stores; from employment offices to import-export businesses. Yet, for all the similarities, there are differences. The shift in emphasis in different applications seems to alter a great deal. The basics remain, the principles are the same, but each system has its own needs and its own personality. There is much that is the same and much that is different.

In this section, we are examining certain specific areas that have special needs. The common thread runs through all of them, but each one's uniqueness is examined to broaden the view of this complex industry.

BANK SECURITY

Of all the institutions that have a high appeal for the criminal, the banks must be rated at or near the top of the list. In spite of the reputation as a hard front to criminal attacks, the bank, with all of its ready money, exerts a powerful influence on the acquisitive instincts of all classes of the underworld element. Banks have never been unaware of the target they represent, and yet many of them had made little or no provision for an overall security program until some measure of standardization was effected by the Bank Protection Act, passed by Congress is 1968 to become effective in January, 1969.

The Act covered all financial institutions which were chartered by the federal government, or were members of the Federal Reserve System, or whose deposits were insured by either the Federal Deposit Insurance Corporation (FDIC) or the Federal Savings and Loan Insurance Corporation (FSLIC). This represents a very high percentage of the total financial institutions serving the public.

The Act established minimum standards for the protection of assets of these institutions principally against robbery, burglary, and larceny. This protection was directed toward assaults from outside and makes no effort to direct its attention to problems of internal theft, collusion, computer fraud, carelessness or bad management. This is not to say that the Act is short-sighted. It is rather to emphasize that it has a specific mission of establishing minimum standards of protection against the external attacks on the attractive prize that financial institutions represent. It is not an omnibus act designed to provide the total security program which these largely private enterprises may or may not elect to initiate.

Bank Protection Act

The full text of the Bank Protection Act is available in libraries to any student who wishes to inspect its provisions in their entirety. For the purposes of this discussion, some excerpts may be useful. After preliminary discussion defining federal agencies supervising the Act, timetables of compliance, and insurance coverage, the Act provides:

> "On or before February 15, 1969, the board of directors of each financial institution shall designate an officer or other employee of the bank, who shall be charged, subject to supervision by the board of directors, with responsibility for the installation, maintenance, and operation of security devices and for the development and administration of a security program which equals or exceeds the standards prescribed by the Part.
>
> **Section 216.3—Security Devices**
>
> (a) *Installation, maintenance, and operation of appropriate security devices.* Before January 1, 1970, the security officer of each financial institution, under such directions as shall be given him by the board of directors, shall survey the need for security devices in each of the institution's offices and shall provide for the installation, maintenance, and operation, in each office, of:
>
> (1) a lighting system for illuminating, during the hours of darkness, the area around the vault, if the vault is visible from outside the banking office;
>
> (2) tamper-resistant locks on exterior doors and exterior windows designed to be opened;
>
> (3) an alarm system or other appropriate device for promptly notifying the nearest responsible law enforcement officers of an attempted or perpetrated robbery or burglary; and
>
> (4) such other devices as the security officer, after seeking the advice of law enforcement officers, shall determine to be appropriate for discouraging robberies, burglaries, and larcenies and for assisting in the identification and apprehension of persons who commit such acts.
>
> (b) *Considerations relevant to determining appropriateness.* For the purpose of subparagraph (4) of paragraph (a) of this section, considerations relevant to determining appropriateness include, but are not limited to:
>
> (1) the incidence of crimes against the particular banking office and/or against financial institutions in the area in which the banking office is or will be located;

(2) the amount of currency or other valuables exposed to robbery, burglary, or larceny;

(3) the distance of the banking office from the nearest responsible law enforcement officers and the time required for such law enforcement officers ordinarily to arrive at the banking office;

(4) the cost of the security devices;

(5) other security measures in effect at the banking office, and;

(6) the physical characteristics of the banking office structure and its surroundings.

(c) *Implementation.* It is appropriate for banking offices in areas with a high incidence of crime to install many devices which would not be practicable because of costs for small banking offices in areas substantially free of crimes against financial institutions. Each institution shall consider the appropriateness of installing, maintaining and operating security devices which are expected to give a general level of bank protection at least equivalent to the standards described in Appendix A of this Part. In any case in which (on the basis of the factors listed in paragraph (b) or similar ones, the use of other measures, or the decision that a technological change allows the use of other measures judged to give equivalent protection) it is decided not to install, maintain and operate devices at least equivalent to these standards, the bank shall preserve in its records a statement of the reasons for such decisions and forward a copy of that statement to the regulatory agency.

Security Procedures

(a) *Development and administration.* On or before July 15, 1969, each bank shall develop and provide for the administration of a security program to protect each of its banking offices from robberies, burglaries, and larcenies and to assist in the identification and apprehension of persons who commit such acts. This security program shall be reduced to writing, approved by the bank's board of directors, and retained by the bank in such form as will readily permit determination of its adequacy and effectiveness, and a copy shall be filed with the regulatory agency.

(b) *Contents of security programs.* Such security program shall:

(1) provide for establishing a schedule for the inspection, testing, and servicing of all security devices installed in each banking office; provide for designating the officer or other employee who shall be responsible for seeing that such devices are inspected, tested, serviced, and kept in good working order; and require such officer or other employee to keep a record of such inspections, testings and servicings;

(2) require that each banking office's currency be kept at a reasonable minimum and provide procedures for safely removing excess currency;

(3) require that the currency at each teller's station or window be kept at a reasonable minimum and provide procedures for safely removing excess currency and other valuables to a locked safe, vault, or other protected place;

(4) require that the currency at each teller's station or window include "bait" money, i.e., used Federal Reserve notes, the denominations, banks of issue, serial numbers and series years of which are recorded, verified by a second officer or employee, and kept in a safe place;

(5) require that all currency, negotiable securities, and similar valuables be kept in a locked vault or safe during non-business hours, that the vault or safe be opened at the latest time practicable before banking hours, and that the vault or safe be locked at the earliest time practicable after banking hours;

(6) provide, where practicable, for designation of a person or persons to open each banking office and require him or them to inspect the premises, to ascertain that no unauthorized persons are present, and to signal other employees that the premises are safe before permitting them to enter;

(7) provide for designation of a person or persons who will assume that all security devices are turned on and are operating during the periods in which such devices are intended to be used.

(8) provide for designation of a person or persons to inspect after the closing hour, all areas of each banking office where currency, negotiable securities, or similar valuables are normally handled or stored in order to assure that such currency, securities, and valuables have been put away, that no unauthorized persons are present in such areas, and that the vault or safe and all doors and windows are securely locked; and

(9) provide for training and periodic retraining of employees in responsibilities under the security program, including the proper use of security devices and proper employee conduct during and after a robbery, in accordance with the procedures listed in Appendix B of this Part.

Filing of Reports

(a) *Compliance reports.* As of the last business day in June of 1970, and as of the last business day in June of each calendar year thereafter, each institution shall file with the regulatory agency a statement certifying to its compliance with the requirements of this Part.

(b) *Reports on security devices.* On or before March 15, 1969, and upon such other occasions as the regulatory agency may specify, each bank shall file with the regulatory agency a report on Form P-1 (in duplicate) for each of its offices that is subject to this Part.

(c) *External crime reports.* Each time a robbery, burglary or nonbank employee larceny is perpetrated or attempted at a banking office, the bank shall within a reasonable time, file a report in conformity with the requirements of Form P-2.''

Up to this point the Act could be read as a general description of a security program adapted by almost any sizeable retail operation. There is little in it (with the possible exception of the on-premise overnight storage of cash) that would identify it as a program tailored for banks. The Act has great value, however, in that it commands the attention of every covered institution. It calls on certain action by the boards of directors and any security program that has the attention as well as the participation of top management has an excellent chance of success. The program herein envisaged is clearly one that goes beyond the outline presented in the Act but if every industry were so motivated to consider its own best interests, whether participation is reluctant or enthusiastic, crime against business could be cut substantially.

On November 1, 1973, Appendix A of Part 326 was amended by the following rules which remain in effect. Security officers are guided in their selection of security equipment, devices, and procedures simply by acquainting themselves with the Act. This kind of guidance is virtually unprecedented in the security profession. An examination of these provisions, though lengthy, is valuable to any student seeking to find those elements common to most security applications.

Minimum Security Devices and Procedures (Amended)
Appendix A: Minimum Standards for Security Devices

Effective November 1, 1973, Appendix A is *amended* to read as follows:

In order to assure realization of maximum performance capabilities, all security devices utilized by a bank should be regularly inspected, tested, and serviced by competent persons. Actuating devices for surveillance systems and robbery alarms should be operable with the least risk of detection by unauthorized persons that can be practicably achieved.

(1) *Surveillance systems.* (i) *General surveillance systems should be:*

(A) equipped with one or more photographic, recording, monitoring, or like devices capable of reproducing images of persons in the banking office with sufficient clarity to facilitate (through photographs capable of being enlarged to produce a one-inch vertical head-size of persons whose images have been reproduced) the identification and apprehension of robbers or other suspicious persons.

(B) reasonably silent in operation; and

(C) so designed and constructed that necessary services, repairs or inspections can readily be made. Any camera used in such a system should be capable of taking at least one picture every 2 seconds and, if it uses film, should contain enough unexposed film at all times to be capable of operating for not less than three minutes, and the film should be at least 16mm.

(ii) *Installation and operation of surveillance systems providing surveillance of other than walk-up or drive-in teller's stations or windows.* Surveillance devices for other than walk-up or drive-in teller's stations or windows should be:

(A) located so as to reproduce identifiable images of persons either leaving the banking office or in a position to transact business at each such station or window; and

(B) capable of actuation by initiating devices located at each teller's station or window.

(iii) *Installation and operation of surveillance systems providing surveillance of walk-up or drive-in teller's stations or windows.* Surveillance devices for walk-up or drive-in teller's stations or windows should be located in such a manner as to reproduce identifiable images of persons in a position to transact business at each such station or window and areas of such station or window that are vulnerable to robbery or larceny. Such devices should be capable of actuation by one or more initiating devices located within or in close proximity to such station or window. Such devices may be omitted in the case of a walk-up or drive-in teller's station or window in which the teller is effectively protected by a bullet-resistant barrier from persons outside the station or window. However, if the teller is vulnerable to larceny or robbery by members of the public who enter the banking office, the teller should have access to a device to actuate a surveillance system that covers the area of vulnerability or the exits to the banking office.

(2) *Robbery and burglary alarm systems. (i) Robbery alarm systems.* A robbery alarm system should be provided for each banking office at which the police ordinarily can arrive within 5 minutes after an alarm is actuated; all other banking offices should be provided with appropriate devices for promptly notifying the police that a robbery has occurred or is in progress. Robbery alarm systems should be:

(A) designed to transmit to the police, either directly or through an intermediary, a signal (not detectable by unauthorized persons) indicating that a crime against the banking office has occurred or is in progress;

(B) capable of actuation by initiating devices located at each teller's station or window (except walk-up or drive-in teller's stations or windows in which the teller is effectively protected by a bullet-resistant barrier and effectively isolated from persons, other than fellow employees, inside a banking office of which such station or window may be a part);

(C) safeguarded against accidental transmission of an alarm;

(D) equipped with a visual and audible signal capable of indicating improper functioning of or tampering with the system; and

(E) equipped with an independent source of power (such as a battery) sufficient to assure continuously reliable operation of the system for at least 24 hours in the event of failure of the usual source of power.

(iii) *Burglary alarm systems.* A burglary alarm system should be provided for each banking office. Burglary alarm systems should be:

(A) capable of detecting promptly an attack on the outer door, walls, floor, or ceiling of each vault, and each safe not stored in a vault, in which currency, negotiable securities, or similar valuables are stored when the office is closed, and any attempt to move any such safe;

(B) designed to transmit to the police, either directly or through an intermediary, a signal indicating that any such attempt is in progress; and for banking offices at which the police ordinarily cannot arrive within 5 minutes after an alarm is actuated, designed to actuate a loud sounding bell or other device that is audible inside the banking office and for a distance or approximately 500 feet outside the banking office;

(C) safeguarded against accidental transmission of an alarm;

(D) equipped with a visual and audible signal capable of indicating improper functioning of or tampering with the system; and

(E) equipped with an independent source of power (such as a battery) sufficient to assure continuously reliable operation of the system for at least 80 hours in the event of failure of the usual source of power.

(3) *Walk-up and drive-in teller's stations or windows.* Walk-up and drive-in teller's stations or windows contracted for after February 15, 1969, should be constructed in such a manner that tellers are effectively protected by bullet-resistant barriers from robbery or larceny by persons outside such stations or windows. Such barriers should be of glass at least 1-3/16 inches in thickness,[1] or of material of at least equivalent bullet-resistance. Pass-through devices should be so designed and constructed as not to afford a person outside the station or window a direct line of fire at a person inside the station.

(4) *Vaults, safes, safe deposit boxes, night depositories, and automated paying or receiving machines.* Vaults, safes (if not to be stored in a vault), safe deposit boxes, night depositories, and automated paying or receiving machines, in any of which currency, negotiable securities, or similar valuables are to be stored when banking offices are closed, should meet or exceed the standards expressed in this section.

(i) *Vaults.* A vault is defined as a room or compartment that is designed for the storage and safekeeping of valuables and which has a size and shape which permits entrance and movement within by one or more persons. Other asset storage units which do not meet this definition of a vault will be considered as safes. Vaults contracted for after November 1, 1973,[2] should have walls, floor, and ceiling of reinforced concrete at

---

[1]It should be emphasized that this thickness is merely bullet-resistant and not bulletproof.

[2]Vaults contracted for previous to this date should be constructed in conformance with all applicable specifications then in effect.

least 12 inches in thickness.[3] The vault door should be made of steel at least 3½ inches in thickness, or other drill and torch resistant material, and be equipped with a dial combination lock, a time lock, and a substantial lockable day-gate. Electrical conduits into the vault should not exceed 1-½ inches in diameter and should be offset within the walls, floor, or ceiling at least once so as not to form a direct path of entry. A vault ventilator, if provided, should be designed with consideration of safety to life without significant reduction of the strength of the vault wall to burglary attack. Alternatively, vaults should be so designed and constructed as to afford at least equivalent burglary resistance.[4]

(ii) *Safes.* Safes contracted for after February 15, 1969, should weigh at least 750 pounds empty, or be securely anchored to the premises where located. The body should consist of steel, at least 1 inch in thickness, either cast or fabricated, with an ultimate tensile strength of 50,000 pounds per square inch and be fastened in a manner equal to a continuous ¼ inch penetration weld having an ultimate tensile strength of 50,000 pounds per square inch. The door should be made of steel that is at least 1½ inch in thickness, and at least equivalent in strength to that specified for the body; and the door should be equipped with a combination lock, or time lock, and with a relocking device that will effectively lock the door if the combination lock or time lock is punched. One hole not exceeding ½ inch diameter may be provided in the body to permit insertion of electrical conductors, but should be located so as not to permit a direct view of the door or locking mechanism. Alternatively, safes should be constructed of materials that will afford at least equivalent burglary resistance.

(iii) *Safe deposit boxes.* Safe deposit boxes used to safeguard customer valuables should be enclosed in a vault or safe meeting at least the above-specified minimum protection standards.

(iv) *Night depositories.* Night depositories (excluding envelope drops not used to receive substantial amounts of currency) contracted for after February 15, 1969, should consist of a receptacle chest having cast or welded steel walls, top, and bottom, at least 1 inch in thickness; a steel door at least 1½ inches in thickness, with a combination lock; and a chute, made of steel that is at least 1 inch in thickness, securely bolted or welded to the receptacle and to a depository entrance of strength similar to the chute.

---

[3]The reinforced concrete should have: two grids of #5 (5/8" diameter) deformed steel bars located in horizontal and vertical rows in each direction to form grids not more than 4 inches on center; or two grids of expanded steel bank vault mesh placed parallel to the face of the walls, weighing at least 6 pounds per square foot to each grid, having a diamond pattern not more than 3" x 8"; or two grids of any other fabricated steel placed parallel to the face of the walls, weighing at least 6 pounds per square foot to each grid and having an open area not exceeding 4 inches on center. Grids are to be located not less than 6 inches apart and staggered into each direction. The concrete should develop an ultimate compression strength of at least 3,000 pounds per square inch.

[4]Equivalent burglary-resistant materials for vaults do *not* include the use of a steel lining, either inside or outside a vault wall, in lieu of the specified reinforcement and thickness of concrete. Nonetheless, there may be instances, particularly where the construction of a vault of the specified reinforcement and thickness of concrete would require substantial structural modification of an existing building, where compliance with the specified standards would be unreasonable in cost. In those instances, the bank should comply with the procedure set forth in section 216.3(c) of Regulation P.

Alternatively, night depositories should be so designed and constructed as to afford at least equivalent burglary resistance.[5] Each depository entrance (other than an envelope drop slot) should be equipped with a lock. Night depositores should be equipped with a burglar alarm and be designed to protect against the ''fishing'' of a deposit from the deposit receptacle, and to protect against the ''trapping'' of a deposit for extraction.

(v) *Automated paying or receiving machines.* Except as hereinafter provided, cash dispensing machines (automated paying machines) including those machines which also accept deposits (automated receiving machines) contracted for after November 1, 1973, should weight at least 750 pounds empty, or be securely anchored to the premises where located. Cash dispensing machines should contain, among other features, a storage chest having cast or welded steel walls, top, and bottom, at least one inch in thickness, with a tensile strength of at least 50,000 pounds per square inch. Any doors should be constructed of steel at least equivalent in strength to the storage chest and be equipped with a combination lock and with a relocking device that will effectively lock the door if the combination lock is punched. The housing covering the cash dispensing opening in the storage chest and the housing covering the mechanism for removing the cash from the storage chest, should be so designed as to provide burglary resistance at least equivalent to the storage chest and should also be designed to protect against the ''fishing'' of cash from the storage chest. The cash dispensing control and delivering mechanism (and, when applicable, cash deposit receipt mechanism) should be protected by steel, at least ½ inch in thickness, securely attached to the storage chest. A cash dispensing machine which also receives deposits should have a receptacle chest having the same burglary resistant characteristics as that of cash dispensing storage chest and should be designed to protect against the fishing and trapping of deposits. Necessary ventilation for the automated machines should be designed so as to avoid significantly reducing the burglary resistance of the machines. The cash dispensing machine should also be designed so as to be protected against actuation by unauthorized persons, should protected by a burglar alarm, and should be located in a well-lighted area. Alternatively, cash dispensing machines should be so designed and constructed as to afford at least equivalent burglary resistance.[6] A cash dispensing machine which is used inside a bank's premises only during bank business hours, and which is empty of currency and coin at all other times, should at least provide safeguards against ''jimmying,'' unauthorized opening of the storage chest door, and against actuation by unauthorized persons.

Thus the Bank Protection Act was the first legislation creating a comprehensive security program for a large and significant private commercial enterprise. Whatever its shortcomings as a total program, they pale beside the contribution made by its enactment. For the first time, an entire industry has had mandated the need for an

---

[5] Equivalent burglary-resistant materials for night depositories include the use of one-fourth inch steel plate encased in 6 inches or more of concrete or masonry building wall.

[6] Equivalent burglary-resistant materials for cash dispensing machines include the use of 3/8 inch thick nickel stainless steel meeting American Society of Testing Materials (ASTM) Designation A 167-70, Type 304, in place of 1 inch thick steel, if other criteria are satisfied.

effective security effort, and for the first time, minimum standards have been laid down to apply throughout a segment of a business community. In view of the wave of bank holdups in 1979 and early 1980, some additional standards with respect to the screening of tellers might be considered, though taken all in all the Act has probably been very effective. Some errors and failures must be attributed to the careless or indifferent management of this legislated program in individual applications. There will always be an "it can't happen here" attitude in any extensive security program. But, for better or worse, this act is as significant in its own way as OSHA is.

Personnel

There is, of course, a need to exercise great care in the selection of employees for a bank. They must not only be able to withstand the temptations created by money and negotiables in evidence, but they must have a level head in providing the customer service expected of them and at the same time be alert to the innumerable frauds to which banks are subjected. They must also be able to work quickly, efficiently, and accurately.

A careful pre-employment screening is vital. Security must work in close cooperation with the Personnel Department as in other applications previously discussed. A review of salary levels at the earlier stage of employment may be in order since banks are frequently at the lower end of the pay scale in lower echelon job categories. This is a matter requiring the greatest tact and diplomacy in dealing with policies of this delicacy with top management. It may have to be handled, however, to avoid the danger of staffing the bank with undependable and potentially dishonest employees.

Because the Securities and Exchange Commission requires that all employees handling securities be fingerprinted and since the F.B.I. is authorized to check prints of employees of federally chartered or insured financial institutions, many pre-employment screenings may be more extensive than most of those previously discussed.

Alarm Systems

All financial institutions are required to have an alarm system "or other appropriate device" for notifying law enforcement of an attempted or perpetrated robbery or burglary. Since many communities do not permit the installation of alarm systems terminating in a public facility, the alternative is usually a central station. Even in communities where police monitoring of alarm boards is permitted, it may not be advisable to use such a facility. Police dispatchers are usually busy with a number of duties and may not see the alarm signal for some time. Police dispatchers also frequently fail to distinguish betwen equipment problem conditions and dispatch units to the scene in either case. Customers and police alike are disturbed, and the police soon become apathetic about responding to what has frequently proven to be a false alarm.

Central station installations, though costlier, are generally more satisfactory. Line service, as well as alarm monitoring, is in the hands of one company. Service is faster and more thorough.

Alarm Transmitters

Buttons at the teller stations are cheap to install and are commonly used although they are the least satisfactory robbery alarms. Such alarms are accidentally tripped from time to time and too visible to a robber who is apt to see the movement of the teller's hand. At least the fear of being observed deters many tellers from taking the chance, and the alarm is not sent. Footrails are somewhat more expensive to install and they cause more false alarms. Their activation in the course of a robbery cannot be seen, however, and thus the alarm is very likely to be transmitted.

Bill traps are expensive but effective. They are, as the name indicates, alarms that are triggered by the removal of bills from an activating device. There are virtually no false alarms and there is no hesitation to use it. If the robber is emptying the drawers, he will activate the alarm himself.

Surveillance Systems

A vital aspect of the Act is the requirement of photographic, recording or monitoring devices capable of reproducing clear images for identification and apprehension of robbers or other suspicious persons. Certainly a good picture of a robber or a fraud suspect in the act is a great aid to police, both in identifying and in apprehending the perpetrator.

The market is filled with products to handle this legislated need. The most widely used film format is the 16mm movie camera which is set to photograph two to four frames per minute continuously during banking hours. A "hold-up alarm override" shifts the camera into rapid or movie sequence when the alarm is activated. In this mode, the camera takes continuous pictures of the robbery until the film is exhausted or the alarm is reset.

The photographs from such a camera are useful for identification in many of the frauds that plague banks. For example, a digital clock/calendar positioned in the camera' s field will identify the date and time of any transaction. If all documents are stamped with the date and time when they are handled by the teller, the recording of that transaction can easily be found in the film files. For this reason, files should be made, and since frauds, particularly those involving government checks, may take many months to surface, films should be retained for two years or more.

Photographic surveillance service organizations are available to assist with advice and service. A good service will not only process the film but will inspect the rolls regularly for focus, effectiveness, coverage, and usefulness in achieving the agreed upon end. The organization will also provide prints when necessary, store negatives and offer expert testimony in court.

Other formats in use in film surveillance rigs are 35mm or 70mm, both of which are more expensive but which give clearer, wider pictures and may be necessary in certain cases. Eight mm or Super 8 cameras are sometimes used for internal information only since they do not qualify as adequate surveillance gear under the Act.

Closed circuit television is useful in those relatively few situations where cameras are continually monitored by a guard who can summon help in the event of a problem. The cost of such an operation in bank surveillance is very high. The more normal application is much the same as with film, the difference being that tape is too expensive to file for long periods of time and without that file one of the significant uses of surveillance systems is lost.

Robbery

Banks have always been subjected to robbery attempts either by the lone note passer or the armed gang. The latter group has been increasing recently as the barriers between tellers and customers have come down. In an effort to create a warmer, more personal atmosphere and thus stimulate business, banks have moved away from the barriers and grills that have characterized them for many years. This desire to eliminate the obstacle has, as in the case of supermarkets and other self-service retail outlets, increased customer activity and increased security problems many times over. Some banks have started a revisionary movement back to barrier, usually of a highly technical design dominated by bullet resistant acrylics and brushed stainless stell.

Except in the case of an armed gang, bank holdups are usually conducted in such a way that only the teller approached is even aware that a holdup is taking place. The important thing for the teller to remember is to activate the holdup alarm. If this is done at the outset, the bandit will never notice any movement involved. He does not, after all, know the work area that well, no matter how he has cased the bank before he goes for the money. After activating the alrm, the teller should begin doling out the money as slowly as possible, being certain to include ''bait money'' in the first handful passed over to the thief. It is important that the teller look closely and carefully at the bandit. He must look for characteristics that will not change. The clothes will be discarded; the moustache will be shaved; the hair will be cut, re-styled, and dyed; but scars will remain, and the shape of the eyes, tattoos, hand shape, height, face shape, and body structure will remain much the same.

After the robber turns to go, the teller (who is still probably the only person who knows about the robbery) should activate the holdup alarm again, and when he thinks it is prudent to do so, look out the door to determine where the bandit has gone and by what transportation. More bank robbers are caught by a license number reported by an alert witness than by any other means.

As soon as the robber disappears, the teller should proceed to the phone, call the authorities, and report the facts. After the call, he should try to piece together all the details of the robbery. The smallest ones count. He should write down as much as he can recall.

It is advisable to shout "ROBBERY" after the bandit has gone in order to activate whatever plan the bank has for dealing with such situations. Such a plan is essential for preserving evidence as is until the police arrive, handling nervous customers who are in the bank, monitoring traffic, closing cash drawers and shutting the front door to further business. Any contract with the media should be based on a news brief prepared by a senior officer of the bank.

If the bank is hit by an armed gang, the main thing to remember is to do as they say. If possible, hit the alarm—it might be possible to do so in the first five or ten seconds of their entrance, before they are in absolute control. Beyond that, try to observe. Visual observation may not be possible. Such gangs will frequently demand that everyone lie face down on the floor. In such a situation, prior instruction should be to do as ordered. Do not sneak a peek or try to be cute with the bandits. Observation of voice or speech patterns may be possible and a brief glimpse when the bandits first enter may prove valuable to police later.

Although holdups lag far behind credit card and check frauds, forgeries and embezzlements as a financial drain on the bank, they are extremely dangerous, disruptive and destructive of customer confidence and employee morale.

Kidnap/Extortion

A gunman or even a gang of them that enters a bank demanding money is limited in his potential from the outset. The time is very limited, the chances of being identified are substantial, the cash take is limited to the amount held in one or perhaps a few drawers, and his picture is taken for later study and identification. The kidnapper/extortionist is in a different game. His victim can be the president of the bank; there is virtually no limit to his demand. He need not fear identification and time is a matter of his own choosing.

In order to combat this threat, it is essential that a plan be drawn up to detail the bank's response to such an assault. Since the Federal Bank Robbery Act (Title 18, Sec. 2113) does not recognize the extortion of money by holding a bank officer or a member of his family hostage as an attack against the bank, there is some question as to whether the removal of bank funds to pay a ransom demand is permitted. One solution to this problem has been to cause the board of directors to pass a resolution that an attack against any employee or his family involving a demand for bank funds would constitute an attack against the bank. Obviously an officer of the bank would have to be notified of the situation and of the intention of the threatened employee to withdraw funds for such payment. In such an event, the bank would sustain the loss unless it could be covered by a blanket bond. This question of recovery from the bonding company is highly controversial. Most bonding companies take the view that they are not liable for payments of kidnap/extortion claims unless such claims are specifically included in an endorsement. Two cases have been decided in favor of the plaintiff banks in this issue, although the matter is still very much open.

The situations which may arise are:

1. Kidnapping an officer outside the bank.
2. Hide-in holding branch manager when he enters to open up.
3. Holding a member of an employee's family.

There are many variations of these basic scenarios but these are where the bandit's leverage lies.

It is essential that a plan covering all contingencies be drawn and given minimum necessary circulation for compliance. The FBI must be called in at the earliest safe moment. The FBI has experience in these matters, and local police, however competent, generally have not. This whole area of extortion by hostage requires the closest study of the local conditions as well as a broad knowledge of the industry's experience with such crimes. Only this way can a reasonable and effective plan be developed to counter-attack and protect in this area.

MUSEUM SECURITY

Since banks are attractive targets for criminal assault because of the product they handle, they are prepared to withstand the anticipated attacks by limiting cash and negotiables in public access areas and by constructing their premises in such a way as to reduce public proximity to cash handling activities.

Museums, on the other hand, are constructed specifically to display their treasures to the widest possible audience. The public is encouraged to attend frequently to view the exhibits. Nothing may be done to detract from the exhibit on display.

Since the 1960's, art thefts have increased at a tremendous rate all over the world. The market in fine art is international, and although the finest pieces are usually highly recognizable as to their previous place of display, avid collectors are not always too circumspect when acquisition is imminent. Museum directors have reacted sharply to the mounting losses worldwide. They have looked to increasingly sophisticated devices to protect their priceless art objects. The trend in museum security is more toward the use of electronic equipment rather than on increasing the guard force protecting the premises.

The installation of electronic equipment complements any existing guard force. Exhibited items can, and should be monitored at all times while the facility is open to the public. In this way, the guard force can intervene in any attempted theft or interference with an object in response to an automatic signal from properly designed protective gear.

When the museum is open to the public, every item on display such as paintings, sculptures, smaller items in display cases, and furniture from settings, must be supervised to prevent theft or vandalism.

The job of the security system is to protect each object in the most appropriate way. Some items invite tactile exploration. These are usually sculptures, usually large enough to discourage moving them or firmly fixed in a pedestal or other display piece. Obviously, such a display cannot be alarmed to warn that it is being handled since that is one purpose of its display. Such displays would normally be alarmed if its size made it appropriate to place contacts underneath. Significant movement or removal of the statue would be instantly signaled. Paintings must never be handled or touched. They would be alarmed to signal that they had been disturbed or, in some instances of especially important pieces not under the surveillance of a guard, the alarm would warn of an approach to the painting past an indicated point.

However they are protected, there are certain conditions that must be met. To begin with, any system must be sufficiently flexible to permit frequent modification of the displays. Most museums have permanent collections always on hand, but there are special exhibition salons in which temporary shows are arranged for varying periods of time. Sometimes even the permanent collection is displaced to provide room for some particularly exciting exhibition for a period of time. The security system must be designed with this possibility in mind and must be able to adapt to the vagaries of the art world of providing total protection for the irreplaceable pieces.

In the second place, no matter what the item or the manner in which it is being displayed, it must be shown as the artist or museum director determines without any interference from any security equipment or procedure. The security system must adapt to the aesthetics of the exhibit and cannot be allowed to intrude upon or mar the effectiveness of the work or its manner of display. In museum security, the security will have no hand in influencing display to accomodate his responsibilities. In an industrial context the security director should be involved in input for design so that efficiency, ambience, sales appeal and security aspects all find expression. In a museum, the security director plays a significant role in determining overall needs, but he is silent and uninvited when it comes to exhibits. Only after the pieces have been selected and their display determined does the security director become involved. His job is simply to see that the integrity of the exhibit is maintained.

Security Rails

For the display of wall hangings such as paintings, carpets, weapons, masks, and carvings, support rails replace mouldings or standard picture rails. They are mounted high enough on the walls to provide a semi-permanent point to which hangings will be fixed. These rails will hold sensors and cables connecting to the alarm system. For the greatest flexibility, these rails should extend around all walls (and the length of them) where items might conceivably be hung. The rails themselves are neutral and after installation appear to be mouldings. They can be painted any way and hung in any position that would be appropriate for a moulding in that decor. The wall hung items are suspended from sensors that can be moved along the rails to any position required, and they are easily replaced if they become inoperable or damaged.

These sensors can be obtained in either an electro-mechanical or an electronic variety. The electro-mechanical type has a contact of the variety used frequently in door or window applications. The weight of the object holds the contacts open (or closed) and removal of the item will be alarmed. The electronic unit is more sophisticated, and may be more useful in that the slightest weight change—not simply removal—will be transmitted via the suspension wires. This unit can be adjusted to a very high sensitivity so that the slightest touch on the protected object will be recorded, and the alarm will be activated. Choice between the two will be dictated by the level of security required. The sensors are located inside the security rail, and when the cover lid has been fitted they cannot be detected. Only the hooks for hanging are visible below the rail.

Showcase Protection

Many items, usually smaller ones, are displayed in glass cases. The items so arrayed can frequently be viewed from all sides to provide full visibility of the items. The integrity of the glazing material is of great importance in this application. This means that display cases may be constructed of burglary-resistant acrylics in some situations, but not all of them. Alarming the glass itself with tape or with the use of alarm glass threaded with copper wire is not acceptable because of the visibility of the material. Vibration glass alarms or one of the glass breakage alarms might be employed.

Perhaps the best of several possible solutions is the use of ultrasonics. The energy generated is reflected within the showcase. An alarm is triggered by any change in the reflecting surfaces as when the glass is broken or simply by the removal of one of the objects if the case is entered any other way. This system is effective in that the units are silent and, most importantly, they are concealed in the showcase and are not visible to the public.

CCTV

Closed circuit television (CCTV) is, of course, useful throughout the facility and installation of cameras and the concomitant cable run can be made without interfering with the aesthetic integrity of the exhibit rooms. An installation of this kind can be flexible and adaptable to any kind of exhibit if the initial installation is made with the thought that any CCTV system will have to be able to supervise every part of all covered rooms at some point in its operation. Such a system in conjunction with a well-trained guard staff will be highly effective in protecting the museum's treasures.

Storage

While many objects are on display, there may be many more which are in storage. These items may have to be available to museum staff for study or repair, or they may simply have been temporarily removed from display to provide room for other art objects. In any event, all of these pieces must be protected around the clock. Alarms, inspection, and accountability procedures would be similar to those used in any retail or industrial operation in which storage is a factor. This problem has been discussed in previous chapters.

Closed to the Public

During the hours that the museum is closed to the public, the security system must make provision for full protection, both against surreptitious entry and against theft or damage by anyone who may manage to penetrate the outer defenses.

Here, as in all security applications, the first line of defense is at the perimeter. Since the perimeter usually consists of the walls of the building itself, the defense should concentrate on those points where entry is most likely to be made. Doors, windows, skylights, ventilation or elevator shafts, the roof, adjoining walls, cellar entrances, trash elevators, service tunnels—nothing can be overlooked. Lower windows may be made of burglar-resistant polycarbonate (Lexan) glazing. Upper windows may be protected by architectural grills or by vibration sensors if grillwork is not feasible for any reason. Windows (especially long window fronts) may be further protected by the installation of infrared photoelectric beams. Exterior walls enclosing display rooms may be alarmed with vibration detectors. These alarms are particularly suitable for the protection of sensitive areas since they will respond to the first attempts to breach the wall, rather than reacting later to an actual penetration.

Internal Protection

The interior spaces in a museum should be protected by some sort of motion detector alarm system. A microwave, ultrasonic, or passive infrared system each has a place in museum protection. Tests have been conducted that show that paintings and other art objects are unaffected by either the ultrasonic or the microwave transmitter, so there can be no concern that the objects protected will be injured by the equipment designed to protect them. These space protection devices are especially valuable in larger museums where the space involved makes guard patrol impractical. Such patrol is, however, extremely important at closing time when the premises must be thoroughly examined for hide-ins and any suspicious items such as packages and briefcases which may carry explosives.

Pressure mats and photoelectric beams protecting corridors, interior staircases, and office areas can also be useful in the overall system.

In today's highly developed technology, there are few security needs that cannot be satisfied, and a well-developed system will employ those elements that are appropriate to each application. Furthermore, the system will not depend totally on any one type of protection but will combine several for the greatest effectiveness. The combination of peripheral protection, area protection and alert, well-trained guard supervision will provide the highest level of protection.

Only through the close cooperation of enlightened and concerned museum management and a knowledgeable, experienced security staff can the safety of the irreplaceable treasures of a museum be assured.

HOSPITAL SECURITY

Hospital security is a particularly exacting field for the security professional because it encompasses virtually every aspect of security. Assuming a 500 bed facility, there must be a feeding operation which will serve between 1,000 and 1,500 meals a day; there will be more than 20,000 patients with an average stay of slightly over a week each; there will be from 500 to 1,000 prescriptions filled daily, and these prescriptions will be administered from numerous nursing stations where substantial stores of narcotics are kept. Linens, surgical gowns and many of the staff uniforms will be laundered in the hospital laundry and then distributed to as many as 30 or 35 linen lockers, where they are further distributed as needed. Nursing supplies, from bedpans to water carafes, must be distributed throughout the hospital to be used as necessary. With an average of four visitors per patient per day, there is a regular daily traffic volume of from 1,500 to 2,000 unaffiliated outsiders moving through the facility, and they are present largely in concentrated groups during visiting hours. Such a hospital would employ a staff of nearly 2,000 men and women. Add to this the 500 or so doctors who are affiliated with the hospital, the 400 to 800 volunteers, and perhaps 200 or 300 student nurses. All of this activity is contained within a complex which must be protected from unauthorized intrusion, and yet which must be in operation on a 24-hour basis—an operation which must have on hand substantial supplies of substances such as anesthetics or oxygen, which are highly flammable.

In many respects a hospital's security program will follow along the same broad outlines that guide the security efforts of every facility. Due consideration must be given to parking areas, proper perimeter controls, adequate lighting, entrances and their uses, guard usage, security control points, communication, regular inspection procedures and employee and visitor identification and control. These considerations, along with fire and safety procedures and the institution of systems to protect against internal theft, are analogous to basic security needs generally applied throughout the business and industrial community, but many of them must be shaped to the specific needs of hospital operation.

Administratively, the security function must be based upon broad—sometimes difficult—policy decisions by top management. As implied above, implementing necessary security policies is difficult for many hospital administrators, since they are frequently convinced that a total security program is not compatible with a patient care facility—an attitude that originates in an underlying conviction that such a program interferes with the primary mission of the institution and ultimately is unnecessary in the facility.

Unfortunately, such an attitude can lead to enormous waste as a result of employee theft or carelessness, theft from and even assault on patients, as well as countless other attacks upon institutions and their occupants. It is essential that the hospital administration be shown the need for a security system that will assist in smooth, safe, and efficient operation rather than hindering it by restrictive authoritarian policy. It must also know that the elimination of theft, either internal or external, will allow the hospital to survive or expand rather than be forced into unwanted economies that could serve to reduce the services the institution would otherwise provide.

The objective of hospital security is a broad-based, integrated program that deals with security problems as a whole rather than attempting to fragment the effort by compartmentalizing it into areas where fire and safety, theft, pilferage, crimes of violence, disaster reaction, traffic control, and security administration are treated separately. A well-integrated, properly coordinated program will be infinitely more effective and more efficient—both in terms of performance and economy—than an effort which attempts to deal with these problems under several supervisory areas.

Security Personnel and Patrols

It would be virtually impossible to provide even rudimentary security for a hospital complex without a security force. The diverse nature of the necessary security program and the conditions under which the program must function make it mandatory that there be trained personnel to inspect, patrol, assist and administer a long and varied list of duties, areas and responsibilities. They cannot be replaced by locks, alarms or barriers.

The flow of personnel and supplies and the admission of patients continue around the clock. There is never time when the facility is locked up and secured. This traffic must be supervised and controlled—usually at more than one point. Such control can be effectively managed only by personnel—and by personnel specifically assigned to this function.

Patrol starts with the exterior areas, including parking lots, which must be patrolled periodically during the day and even more frequently at night. Since most hospitals are located in urban areas where crime rates are more than 800 percent higher than in non-urban areas, such patrols will play an important role in reducing vandalism, surreptitious entry, and damage or theft or cars in the hospital parking area.

Since many hospital facilities consist of numerous buildings, including living quarters for staff employees within a single complex, it is important that the entire area be well-lighted and that landscaping be so designed that shrubs, hedges and other ornamental plantings be located well away from walk-ways (especially those that might be used at night), and that approaches to such possible concealment be particularly well-illuminated. In the case of those hospitals where the complex is broken up into two or more units and where hospital personnel must use public thoroughfares to travel from one unit to another, a more extensive patrol will be required.

Many hospitals today routinely provide security escort service for nurses going off duty late at night. The exact nature of this service depends, of course on the location of the hospital; whether the nurses generally leave singly or in groups; whether they are housed in a nearby nurses' residence or are taking public transportation to their own homes. In some cases a regular patrol along a short route to the nursing residence will suffice, in others it will be necessary for a security officer to walk nurses singly or in groups to the nearest bus stop. In any event, this is an important aspect of personnel protection in the security system of every large urban hospital.

Entrances and Other Perimeter Openings

A few modern hospitals have been built in such a way that traffic control is considerably simplified by the design of the buildings themselves. Unfortunately this is very rare. Most of today's hospitals make no architectural concessions to security considerations. A high percentage of them have been expanded over the years by the addition of new wings or annexes. This has frequently complicated the exposure of the facility by creating new entrances, passages and receiving areas which can be difficult and costly to control.

It is of prime importance that a careful survey be made to determine how entrances can be reduced to an absolute minimum. The ideal arrangement, which would consist of a main entrance essentially for ambulatory patients, attending medical staff, and visitors, an emergency entrance for non-ambulatory patients, and an employees' entrance, can rarely be achieved, but efforts should be made in that direction. Greater efficiency and control might be achieved by allowing use of certain employee entrances only during specified periods of the day and preventing their use at other than authorized times. It might be possible to close off various entrances at night and thus reduce the security exposure during this potentially dangerous period.

Each hospital is unique in design, each requires a thorough examination to determine the optimum number of entrances to satisfy the needs of efficient and effective operation, as well as to provide the maximum in protection from intruders and from employee theft. All such entrances should be manned be security personnel when they are open and in use.

This same evaluation should be made of the perimeter if the hospital grounds are protected by a fence or other barrier. These entrances may not require a guard station unless there is a need to direct visitors entering the main gate to various buildings in the complex, but they should be reduced to the absolute minimum necessary for operation. In most cases where the facility occupies sizable grounds within a single unit, the main gate alone will suffice to handle all entry, including ambulance and pedestrian traffic. Additional gates which are normally locked should be provided for fire equipment in the event of an emergency.

Command and Patrol Posts

Most hospitals have two posts of particular importance in the security system, the main entrance and the communications center. The main entrance is manned by guards, at least during daylight hours, to issue passes, help visitors, and be on call for assistance in various areas. Large hospitals may also have a 24-hour command post as the nerve center for communications, administration and CCTV console if there is one.

Security personnel patrolling the hospital interior should be reachable by a pagemaster system. These units are simply palm-sized portable receivers which emit a tone when one is activated from the central command post. The security officer responds by phone. These are preferable to transceivers, since there is more privacy in a phone call. (Security patrols outside the main building or in areas away from patients and visitors use transceivers, since they are not concerned with the problem of privacy and since phones for fast response are not available in many of the areas being patrolled.)

During these interior patrols, the security officer must be on the alert for any number of situations or conditions that could prove dangerous. He will watch for intruders, unauthorized visitors, unidentified employees, sneak thieves, disorderly conduct, vandalism, and any signs indicating an attack on locked spaces. He will assure himself that all doors that should be locked are properly secured. He will check fire hazards such as accumulations of trash, improperly stored flammables, smoke or the odor of smoke, smoking in no smoking areas, blocked exists, improper stacking in sprinklered storage areas. He will check fire extinguisher tags for inspection dates. He will note any safety hazards such as slippery floors or stairways, or holes in floors or stairs. He will watch for running water, machinery, lights not turned off, or anything of a like nature that is wasteful of hospital resources and ultimately potentially dangerous.

Uniforms and Weapons

In this same regard, many hospitals have found that the overall security effort has benefited from a positive effort to deemphasize or to totally eliminate the authoritarian police concept of the security service. Some have chosen to discard the police-type uniforms and outfit security personnel in blazers with an identifying pocket patch. While this is a uniform and identifies the function of its wearer, it does not suggest "police."

Many hospitals disagree with this policy and prefer instead the traditional uniform for its psychological deterrent capabilities. They argue that the presence of a uniform instantly suggests to the patient and the intruder alike that the facility is secure—that the premises are protected.

There is, at the same time, and largely from similar considerations, disagreement on the question of arming security personnel. Some hospitals feel that it is desirable to issue sidearms to all security officers during their duty hours; others arm only gate personnel and exterior patrols; and some issue no arms at all. Few, if any, hospitals, however, fail to have some weapons available for issue in the event of an emergency that would require them.

Employee Education

Whatever the policy governing uniforms and sidearms, it could be said that there is general agreement among hospital security directors that their job is to create a total, integrated prevention program—a program in which the security department spearheads the effort and bears the major responsibility for its effectiveness, but also one which calls upon the entire hospital organization for active participation. It would be wrong to consider the security department the sole element in the security of the organization. Rather, through education and training, every employee must become, to some degree, a part of the system.

By the same token, every security officer must be made aware that he is part of the total health-care function of the hospital. This can only be accomplished through continuous training in which security personnel are exposed to every facet of the hospital operation. Such training should go well beyond security procedures into all areas of concern and interest in the hospital. It could be of great value if such training sessions were addressed by doctors, nurses, dieticians, pharmacists, and others on the hospital staff who can acquaint the security department with the functions and problems in all areas of the hospital's operation.

Employee Identification Badges

As a control factor it is recommended that all personnel be instantly identifiable by some means. Specific techniques vary from place to place, but the most widely accepted practice is the requirement that each employee wear an identification badge carrying his name, department, employee number, his signature and, in most cases, his picture.

The nature of the information on the badge will vary with the policy of the hospital issuing it. In some cases the badge may designate only the employee's name and department, and he will be required to carry on his person an I.D. card with more complete information as to his status. Badges may be color-coded to indicate department or authorization of the wearer. For example, volunteer workers will generally be asked to wear a badge, but the badge should be color-coded to indicate their volunteer status. Contractor personnel who will be on the premises for any appreciable length of time should be issued badges to be prominently displayed, but they too should be color-coded to immediately indicate the wearer's position.

Attending medical staff is usually required to wear some identification, even if it carries no more information than the name of the wearer. They should, however, be required to carry full identification on their person. Doctors are not always amenable to such procedures as badging, but a well-planned public relations program can go a long way toward convincing them of the necessity of such controls, which must include everyone to be effective.

For such a program to work, it must be enforced. Supervisors in every department must be convinced of its need, and their full cooperation must be sought in order to maintain the program's integrity. In addition, the security department should make spot checks from time to time to see that badges are being worn.

Visitor Control

The continual flow of traffic which is organic to the function and the operation of a hospital is one of the principal factors which makes hospital protection unique in the security profession. Visitors, volunteers, student nurses, consulting physicians, and the regular discharge and admittance of patients make for a continually changing sea of faces. The great majority of this traffic is unfamiliar with the hospital. Coupled with this is the atmosphere of urgency and emotion so frequently present.

Visitors are concerned about the condition of a close friend or a member of the family, and the patients themselves may be nervous and apprehensive—in any event, not completely themselves. Allowances must be made for behavior under these conditions. Every hospital security program must stress the importance of understanding and tolerance of certain behavior. Training programs should include discussions of this problem for regular reindoctrination of the security staff.

Many hospitals have found it advisable to issue passes to visitors in order to maintain control throughout the facility. There is no general agreement on the need for such passes during regular visiting hours, but the hospitals which have adopted the policy have generally been pleased with the increased efficiency in traffic control they achieve without materially interfering with those who are visiting patients.

It is important to create an effective balance between security needs and a concern for the patient, whose recovery can be speeded, or at least whose stay can be made more pleasant, by visits from friends and family.

Since there is the likelihood of a brief jam-up in the lobby at the beginning of visiting hours in any hospital where the use of visitor's passes is a policy, it is desirable to work out a pass design and issuance procedure that can be handled with the greatest speed and efficiency and that will cause the least inconvenience to the visitor.

A number of different visitor pass systems are in use in hospitals around the country, but a general profile would include many or all of the following features:

1. Color coding by floor or department to be visited.
2. Two cards made up for each bed.
3. Both cards clearly marked with name and address of hospital, room number, and bed assignment, and a recap of visiting hours. Special color-coded cards may be issued for other than regular visiting hours authorization.
4. Patient's name can be written in pencil on ½-inch masking tape attached to the two passes assigned to that bed. When that patient leaves, his name can be erased or the masking tape replaced with another name.
5. The reverse side of the card pre-addressed for mailing with return postage guaranteed.
6. The pass of sufficient size to make it too bulky to put easily in handbags or pockets, thus discouraging visitors from taking it home with them after the visit.

Such a system has certain disadvantages, in that some visitors and patients may feel that the procedure is too restrictive, since only two visitors per bed at any one time are permitted. There is also the problem already mentioned of lines forming, with some inevitable delays at the start of visiting hours, particularly in the early evening.

The positive aspect of these controls will probably outweigh the disadvantages, however. Floor traffic in patient areas is diminished. Doctors and nurses alike will find that the conditions for treatment of patients are considerably improved with the reduction of milling visitors. These benefits accrue from a traffic control system that has as its overall effect the reduction of thefts from patients and from the hospital by otherwise unidentified opportunists.

Fire Prevention

No occupancy is entirely free from the threat of fire. Hospitals are by no means an exception. In fact, Stephen Barlay, in his disturbing book *Fire*, points out that, statistically, the chances of fire occurring in a hospital are 100 times greater than in any other building in New York.[69]

This threat to a hospital is particularly severe because many of its occupants are in such a condition that they are unable to fend for themselves in an emergency. In addition, few of the patients, even those who are well on the way to convalescence, are familiar with the hospital beyond their own limited area. If forced to make a rapid evacuation, they would have a difficult and dangerous trip through totally unfamiliar territory. The potential for disaster is obvious.

Although the major ignition source of hospital fires (as well as those in most other buildings) is electrical equipment, with cigarettes and matches close behind, it is in the various fuel sources that hospitals are unique. Oxygen, anesthetics, volatile liquids of various kinds—all can be dangerous under the proper circumstances.

Flammable Waste

In addition, Mr. Barlay points out that, whereas the average American office building generates about two pounds of waste paper per hundred square feet of floor space a day, or several tons a week for the entire building, waste material in a hospital averages about eight pounds per bed each day.[70] With offices and laboratories included, the total waste generated by a thousand-bed hospital would amount to about four tons each day.

This waste is highly flammable. With this amount of fuel generated daily, there is a very real hazard unless the tightest controls are maintained over the movement, holding and collection of all rubbish. Maintenance personnel and other staff members must be taught to deposit all waste material in metal containers, which must be emptied at least once daily. Trash outside the hospital awaiting pick-up must be held in metal containers sufficiently far from buildings so that they will not represent a hazard.

Flammable Liquids

Flammable liquids, which pose a special problem to hospital safety, should be stored, dispensed and disposed of in appropriate safety containers and storage cabinets. Flammable liquids with a flash point above 20 °F must be handled under the strictest supervision and should be limited in the quantity permitted on the premises at any one time.

Electrical Hazards

Electrical wiring must be inspected regularly for any signs of wear or malfunction. Normally extension cords should be run overhead—avoiding any contact with nails, pipes, hooks, grilles or other conductors. In cases where extensions must run along the floor, they must be heavily insulated and protected against the water with which they would come in contact when the floors are scrubbed.

Vulnerable Areas

Elevator and dumbwaiter shafts must be cleaned periodically to prevent the inevitable build-up of oil, grease and rubbish which can create stubborn and dangerous fires that are difficult to combat. Linen and rubbish chutes should be washed weekly or more to prevent dangerous dust hazards.

Operating and delivery-room floors should be tested regularly for static electricity. The importance of such testing cannot be overstated. Even though today less than 20 percent of the anesthetics in use are flammable, accidents can occur. A spark from static electricity in an atmosphere containing cyclopropane or ether—both of which are still in limited use as anesthetics—can be deadly.

In the November, 1969, issue of the *Fire Journal* of the National Fire Protection Association, Dr. James M. McCormick points out that, while pre-1941 reports suggest that the incidence of fire and explosion was in the area of 1 in 300,000 anesthetics, it would now appear that such incidence is in the range of 1 in 600,000 anesthetics. He goes on to indicate that it is impossible to determine the actual frequency because medical reporting of such cases is inhibited by adverse publicity and the fear of malpractice liability. In any event, it is a very real danger and one that must be dealt with by the hospital.

Lavatories must be cleaned regularly, not only for purposes of sanitation but also to avoid the accumulation of refuse in those areas. Overstuffed waste baskets are always a hazard and must not be permitted. Since approximately 20 percent of all hospital fires are caused by smoking and careless use of matches, any area where paper might accumulate is a potential danger. Fireproof receptacles for paper can provide an extra margin of safety in lavatories or other areas where combustible refuse might accumulate.

Public areas such as lounges, cafeterias, waiting rooms or lobbies must be amply supplied with appropriate receptacles for use by smokers. As long as smoking is permitted in areas where people congregate, there is the danger of fire. Fires from a cigarette or cigar butt are particularly treacherous, since it may smoulder in a cushion or on a rug under a couch and not become evident until long after the area is deserted. It is advisable to supply containers of sand (which are regularly cleaned of gum wrappers and other scraps of paper that find their way into them) and ashtrays of adequate size which are equipped with grooves or snuffers long enough to hold the entire cigarette safely.

Inspection

It should be the responsibility of the Security Director to make a daily inspection of the facility to determine the effectiveness of fire prevention procedures. He should assure himself that all specified inspections have been made and that all directives relative to fire safety have been followed out.

- Extinguishers, fire hose, and the water supply for fire lines must be checked regularly.
- All oxygen shutoff valves must be checked and tagged.
- Exit signs must be checked regularly for visibility and for burnt-out bulbs.
- Alarm systems must be checked regularly.
- Signs designating no smoking or special hazard areas must be checked for accuracy and applicability as the needs for them arise or dissipate.
- Fire stairs or smoke towers must be checked for any obstructions or any other factors which could possibly impede a smooth and orderly evacuation if the need should arise.

All of the above procedures must be observed in order to establish the beginnings of a fire prevention program. Obviously, each hospital has its own problems. They must be dealt with after a careful inspection and evaluation of its own circumstances.

Unfortunately, many hospitals are still being built without adequate thought being given to lessons that might have been learned from tragic fires in recent history. Too many hospitals exist with inadequate or badly planned exit arrangements, or with highly combustible interiors and inadequate fire-fighting equipment. In such a facility every effort must be made to correct or overcome these inadequacies before a tragedy occurs. Every administrator must be made aware of these dangers, and a plan must be devised to overcome them. In most cases, it will be the responsibility of the Security Director to so advise the hospital administrator.

Emergency Exits

As a measure of adequacy of the exits in a hospital, we might refer to the National Fire Protection Association's *Life Safety Code*, which indicates the types of permissible exits and their capacity:

Types of Exits	Maximum Capacity Per Unit
Doors directly outside	30
Stairs and smokeproof towers	22
Ramps	30
Horizontal exits	30
Outside stairs	22

The *Code* further specifies that each floor or fire section of a hospital have a minimum of two exits, remote from each other, one of which must be a door directly to the outside or a stair or a smokeproof tower.

Access to these exits is summarized in the following table:

Travel distance to exits	100 feet from the door of any room or 150 feet from the farthest point in any room
Minimum width of access to exits	8 feet
Maximum dead end in corridor	30 feet
Minimum width of doorways from sleeping rooms and diagnostic and treatment areas to exits	44 inches
Minimum width of doorways to nurseries	36 inches

The capacity of exits in hospitals is considerably lower than in other types of occupancy. Not only can hospital patients be expected to move more slowly than others, but it can also be expected that some evacuees will be in beds or on stretchers, which will considerably reduce the movement possible in a given period of time.

A careful study of the NFPA's *Life Safety Code* with regard to institutional occupancies is an essential part of the providing of adequate fire protection to a hospital. Since many hospitals fail to conform to the *Code* in some respects, it is particularly important to be aware of those shortcomings and to make provisions to accommodate those needs.

Supervision of the Fire Program

In this section on fire prevention and protection in a hospital, we have assumed that the supervisor of this function was the Security Director. This is generally the case. Many hospitals, however, have employed full-time fire marshals who report to the administrator and are not involved in or answerable to Security in other respects. In other situations the responsibility for fire safety is assigned to the engineer. While many engineers are or become highly proficient in fire safety, just as many are not; and they are often too busy with their other duties to become even routinely knowledgeable.

It would appear that the most satisfactory arrangement is to assign the total job of protection to the Security Department. In this way the problems of fire, violence, theft, safety, disaster control, and security services are integrated into a broad program, each element of which complements and supports the other. Experienced personnel from city fire departments can be of inestimable value as members of such a total protection program, but it could be argued that their value would be diminished if their efforts were isolated from the total security program.

No matter where he fits into the organizational chart or whatever his background, however, every hospital must have someone who acts as Fire Marshal. He will be directly responsible for developing training programs, fire reporting systems, and evacuation procedures, in addition to establishing and administering the vital area of fire prevention. He will establish and maintain liaison with the local fire department, and he will be responsible for keeping abreast of the latest developments in every aspect of fire safety. He will assume complete charge during a fire emergency until the arrival of the fire department. In a large hospital complex he will undoubtedly be a very busy man.

Fire Alarm Systems

The fire step in the marshal's plan is to survey the area carefully and to pinpoint problems peculiar to specific areas. He must next set up a fire reporting system by alarm or phone or both. This system must report the fire and its location to the Fire Marshal and to the city fire department; and it must be announced in some way over the hospital paging system so that all personnel can respond appropriately.

The usual alarm system in larger hospital complexes is set up to sound various combinations of bells, each combination coded to indicate a different location in the facility. Phones throughout the hospital can also be used to call in the location of a fire. The paging system should be used to announce the fire and its location. Most hospitals use a code name for a fire condition to avoid alarming visitors and patients. In the event the fire becomes more serious and requires some evacuation, the condition might then be announced in plain talk over the system.

Fire Planning and Training

The next step in the fire safety program is the establishment of a plan for various fire situations. Everyone in the fire area must have specific assigned duties which are to be carried out under the direction of floor captains.

In order for such a plan to work at all effectively, all personnel must receive fire training as part of their orientation. This includes chief residents, interns and residents. Each employee must know how and where to turn in an alarm; he must know the location and the proper use of fire extinguishers and fire hoses; he must know the location of *all* exits and be familiar with the alternative actions necessary if any of them are blocked. He must be familiar with the potential fire hazards, his specific duties in given situations, and evacuation procedures. Above all, he must be thoroughly indoctrinated in the need for immediate action. He must be persuaded that prompt and proper fire fighting immediately after reporting the fire is essential. He must be taught that the first two or three minutes of a fire are the most important and that it is in this brief period that training pays off or tragedy can occur.

Only frequent fire drills can drive these lessons home. For these drills to be fully effective, they must be observed, and the performance of each participant must be evaluated. Through the repetition of drills and the follow-up critique, nurses, interns, and hospital employees can become an effective emergency team.

It is important to remember, however, that this training program cannot be a routine, *pro forma* performance of some sort of a charade. A hospital staff consists of many dedicated, hard-working professionals whose focus is on service to the patients. Unless their training in fire safety is presented in an engaging, dramatic way, they may well overlook its significance and its immediacy, and treat it simply as a necessary but unpleasant chore. If that is their reaction, the training program must be re-thought and re-designed at once for the safety of everyone in the facility.

Training in fire prevention never ends. New employees need full indoctrination and old ones need constant review of fire emergency procedure. Only by effective training can a hospital mobilize its staff into an efficient force for safety.

REVIEW QUESTIONS

1. Discuss the implications that the Bank Protection Act of 1968 has had on the banking industry.
2. What factors need to be considered when deciding on appropriations for a bank security program?
3. What things must a security manager consider when developing a security program for a museum?
4. What must be considered when attempting to implement an effective access control system for a hospital?
5. Why is fire control so important in a hospital?

PART V
THE FUTURE OF SECURITY

Chapter 17

Career Opportunities in Loss Prevention

Viewed from the perspective of the late 1970's, security has become a major management function in American business. Where they were almost unheard of 20 or even 10 years ago, there are now vice-presidents of loss prevention reporting directly to the presidents of many companies, and having the same impact upon management decisions as, for example, the vice-presidents of operations or distribution.

Career opportunities in different areas of business, industry, and government security vary; the perceived need for an integral and integrated security function in the management of the widest variety of enterprises can be anticipated as the norm in the near future.

An indication of the level of heightened interest in the broad-based security professional can be seen in the results of the diligent efforts of the American Society for Industrial Security (ASIS). This society has long been interested and involved in creating standards of competence and professionalism to identify those security practitioners who have shown a willingness to devote their attention to achieving higher goals of education and training in their chosen career. As has been noted earlier, the security profession has, in the past, been characterized by the transitory nature of much of its personnel. Training standards have frequently been low and even many executives in the field were generalists without either specific work-related experience or specific training in security. Many factors have been brought to bear on this problem and changes have been and are being made.

The ASIS program is designed to upgrade those career security persons who are willing and able to qualify for certification as a Certified Protection Professional (CPP). The certification board in this program was organized in 1977 and, since that beginning, management in the broadest sense, has sufficient evidence of professional performance capability to stress the importance of a CPP being the desired object in search for management security personnel. The ASIS pamphlet describing certification procedure notes that "Positions Available" announcements in the *Wall Street Journal*, and other publications, have included requirements that state: "Certification as a Protection Professional by the ASIS desirable," or "must have certification as a CPP.'' This trend will become more evident during the next five to ten years, as employers and the public become more aware of the Certified Protection Professional Program.

Certification in this program is far from pro forma. It requires both educational and work experience before a candidate can be considered. If the candidate meets the basic standards, he then must take an examination on both mandatory subjects and on four out of thirteen optional subjects. It is through this program and those given by colleges and universities across the country that the highest standards of professionalism in the practice of security will be achieved.

Factors Increasing Security Opportunities

Among the factors tending to create inviting career paths in security, none is more significant than the explosive growth of the protection function, as briefly described in the first chapter of this book. The number of personnel engaged in private security has doubled in less than a decade to more than a million persons. Various studies place the growth rate generally for security products and services at from 10 to 12 percent annually—and there is no sign of slowing. Rapid advances in electronic technology almost daily create new opportunities.[71]

Other positive considerations for the future not only of jobs in security but also the potential for advancement or growth include the following:

- The increasing professionalism of security is reflected in higher standards of educational criteria and experience, and correspondingly higher salaries, especially at the management levels.
- The rapidity of the growth of the loss-preventive function has created a shortage of qualified personnel with management potential, meaning less competition and greater opportunities for advancement for those who are qualified.
- The shift in emphasis to programs of prevention and service, rather than control or law enforcement, has broadened the security function within the typical organization.
- The acceleration of both two-year and four-year degree programs in criminal justice and/or security at the college level is creating at the corporate management level a new awareness of a rising generation of trained security personnel. Many companies, especially the larger corporations, are actively emphasizing the degree approach in hiring.

It should be noted that, as in many other areas of a society belatedly recognizing the needs and the potential contribution of women, blacks and other minorities, opportunities for these groups in security are particularly good. Significantly, the Private Security Task Force survey of licensed security personnel in St. Louis, Missouri, for example, found that, whereas in 1960 only 10 percent of security personnel were black, with 90 percent Caucasian, by 1975 this ratio was 50/50.[72] Women are also being welcomed in such diverse areas as industrial, retail and hospital security.

Opportunities in Industry

Typically, the greatest opportunities in industrial security exist in larger companies employing proprietary security forces. Here the career-oriented person with a certificate or degree in a recognized security or criminal justice program is actively recruited by many firms.

At one major aircraft company, the salary range for administrative security representatives with a college degree is from $14,000 to $25,000, with the potential for advancement within the company's various divisions. Someone with a B.S. or equivalent degree without experience would start at $250 to $275 per week; with experience, at $300 to $350 a week. Investigators (usually with experience either in military or law enforcement investigation) earn from $325-$400 weekly.

The security director at one major automotive manufacturing facility with a highly progressive security program reports being interested in hiring *only* those applicants with at least two-year credits from a recognized security or criminal justice college program.

J. Kirk Barefoot, Director, Corporate Loss Prevention for Cluett, Peabody & Co., Inc., sees corporate job opportunities continuing to increase in the industrial security field, both in retailing and non-retailing areas, as more and more companies begin to view their corporate security departments as profit-making centers. "As this happens," Barefoot comments, "there will be more importance placed on security, and financial rewards to the individuals so involved will increase accordingly."

Barefoot cites the categories of investigator, guard supervisor, store detective, etc., as typical entry level positions available to those without experience but with varying amounts of formal education. Persons with more education or experience can reasonably expect to start at a higher position. Salary levels are comparable to those found in municipal police departments, with the probability that executive security positoins will command salaries correspondingly higher than those in police departments.

Cluett actively recruits personnel from college campuses. Beginning positions are traditionally those of undercover investigator or store detective. The company's first attention in hiring goes to those college and universities offering four-year programs in police science, industrial security or some allied field. Preferential consideration is given to students who have supplemented the security curriculum with some courses in accouting and business administration.

Opportunities in Retail

The retail field provides a diversity of job opportunities in security, from the entry level position of the uniformed security officer (or blazer-jacketed ''host'') to the shoplifting investigator. Positions are available both with retail stores and chains and with security service companies which provide such services as undercover and shopping investigations. While many of these positions have traditionally been filled by persons with law enforcement experience, there are today many openings for those without experience but with the education, ambition and aptitude that might make them successful in retail security. Many companies today provide their own training for shoppers and other investigators, even though the employees have no investigative experience. Alertness, resourcefulness, courage and self-assurance are often more important than specific experience.

There are many different types of operations in the retail industry. Security has had its impact in virtually every operation, from the discount store to the department store to the supermarket. The recognition of the importance of inventory shrinkage to the company's profit picture, and the necessity for loss prevention, is or soon will be almost universal. Those companies which do not accept this necessity, in the words of one ranking retail executive, simply will not be in business.

One of the nation's largest retail chains of department stores reported a dollar loss of $5,800,000 in ''inventory shrinkage'' for 1977. This figure represented 1.2 percent of sales—a remarkably successful record of shrinkage control. Significantly, this chain has a strong security program.

The entry-level position in this company includes many students recruited from criminal justice and security programs, as well as sales personnel ''crossing over'' to security. Many of these employees work part-time while going to school. Full-time security officers—including many of the formerly part-time students who have completed their college programs—earn approximately $12,000 per year. Investigators earn about $14,000 and up.

These figures apply to a successful retail program—but its very success in indicative of the direction which retail security must take in the future.

Health Care Opportunities

The hospital security officer makes up the vast majority of persons employed in hospital security. According to Russell Colling, a nationally known health care security authority and author of *Hospital Security,* the officer who prepares himself for advancement (through a combination of education beyond high school and field experience) can look to numerous supervisory, investigative, training, fire prevention and safety positions in the field.

Hospital security officers generally earn more than their counterparts in other industries because of the variety of duties requiring a higher than average amount of training. The officer must also be able to interact effectively with the medical community as well as patients and visitors under conditions of frequent stress. Salaries, however, do vary with different locations.

Security directors in the health care field can earn salaries in the $25,000 and $35,000 range. Such positions generally require at least four years of college preparation and considerable field experience.

Like so many other areas of security, as Colling observes, hospital security is just coming into its own.

Airport/Airline Security

The airport and airline security field at the level above the line security officer is one which is still heavily dominated by former law enforcement personnel, particularly former special agents of the FBI. The security director at almost every airline is a former FBI agent, and this is true of many investigators.

This situation is not unique to airlines, of course. Ex-agents of the Bureau can be found in a great many corporate security jobs throughout U.S. business and industry. Both the experience and the qualifications required by the Federal Bureau of Investigation have generally been highly regarded in the private sector. The ambitious, career-minded security aspirant could do far worse than consider a period of service in the FBI as a springboard to a promising position in industry—including the airline industry.

Pay for investigators with major carriers is good, starting at or near the $22,500 level. But qualifications are also high. Almost all have college degrees. Many have law degrees and five years or more of FBI experience.

In airport security, a wide variety of entry-level positions exist for the line security officer, especially at major airports. Here again, the field is relatively new, mushrooming especially since the hijacking scares which began in the late 1960s. It seems clear that, with mandated security requirements including physical security and access controls, baggage screening, 100 percent screening of air passengers and carry-on luggage, cargo security, and other controls, the demand for personnel to fill these needs will continue to rise.

Hotel Security

The hotel-motel industry has been characterized in the past by serious neglect of many security responsibilities, an attitude that has only slowly been changing in spite of a number of very large awards by the courts in recent years against hotels or motels charged with negligent security, particularly in the area of protecting guests. However, this very neglect, coupled with court-mandated responsibility, has created a situation of opportunity for the security professional.

In the words of Walter J. Buzby, co-author of *Hotel & Motel Security Management*, "Opportunities in the hotel industry exist in great numbers both for 'on-the-site' positions and at the corporate home office level." Except at the corporate level or the management level in large hotels, however, the salary range is relatively low in relation to the security industry as a whole. On the other side of the coin, the entry level for the person with any combination of hotel experience and security education or experience can be quite high, with clear opportunities for advancement.

In a related area, the future of security in high-rise apartment buildings and housing complexes offers great potential for the security professional because of the growing emphasis on the concept of total environmental protection.

Campus Security

John W. Powell, nationally known campus security consultant and former security director at Yale University, observes that the rapid progress of campus security during the past 15 years has created excellent opportunities for career positions in the field. Openings in many progressive and professional campus departments provide not only challenge but good salaries and fringe benefits, as well as chance for advancement. Such departments are looking for the young, career-minded individual, with particular interest in those enrolled in or graduate of a criminal justice degree program. Interestingly—unlike many areas of modern security—campus security has generally been evolving from a low visibility operation in the direction of a highly visible, police-oriented image in response to rising crime problems of the 1970s.

A good-sized department will include line officers, field supervisors, shift commanders, coordinator of line operations and director. Many departments also have specialized positions such as investigator and training officer. While salaries vary from department to department, and from one area of the country to another, the average salary for campus officers would be between $10,000 and $16,000 per year. Directors of campus security departments command salaries from $20,000 to $38,000.

Banking Security

Banks must comply with minimum federal regulations on security, as promulgated in the Bank Protection Act of 1968. There is a heavy reliance upon electronic technology and physical security rather than large numbers of personnel.

Uniformed guards in one major bank consulted for the purpose of this chapter earn $9,000 annually. Guard supervisors earn $10,500, and senior operations officers earn from $14,800. Most have some military or law enforcement experience. The latter qualification is particularly true of investigators, who earn from $18,000 to $24,000 and require experience in forgery or related investigations; most are former police officers.

Recreational Park Security

The field of parks and recreational security is another one showing rapid growth. One large Southern California facility employs a cadre of 25 to 30 permanent security personnel whose salary, after initial training is from $5.25 an hour. Extensive company benefits are offered in addition to basic wages. Senior officers earn up to $7.75 hourly, and there is opportunity for advancement through additional ranks, which include sergeant, lieutenant, captain and assistant director, as well as the director.

This facility's needs rise dramatically during seasonal operations, as do those of most parks and recreational centers. In season the security force numbers approximately 100. Most of the temporary officers are recruited from local colleges (especially from administration of justice, police science and security programs), and start at $3.40.

Security Services

In general, security personnel at the lower operational levels earn less in contract security organizations than in proprietary guard forces. This is not necessarily true for investigators and other personnel at higher levels. In the words of Saul Astor, president of Management Safeguards, Inc., "Young people should seek opportunities in security service organizations since the growth of security services has been meteoric and there is no leveling off in sight. The demand for good executives is insatiable. Very high salaries are being paid by security service organizations to the young 'comers.'"

Because the good loss prevention or security executive is much less a policeman than a systems expert, auditor and teacher, Astor recommends broad-based education and experience in such areas as accounting, industrial engineering, management, personnel, law, statistics, labor relations and report writing.

On another level, the doubling of the number of security firms in less than a decade is reflective of the demand for technically qualified individuals capable of providing specialized security services, ranging from alarm sales, installation and service to alarm systems consulting. Continuing changes in the application of security hardware and systems will bring an increasing demand for the services of those who can advise users on their selection and implementation.

Computer Security

Computer security is an important example of the new frontiers opening up in the loss prevention field in response to social and technological change. Table 17-1, developed by John M. Carroll, a leading computer scientist and security consultant and the author of *Computer Security,* provides an analysis of opportunities in computer security, calling for a blend of education and experience in computer science and security. The salary ranges mentioned in the table have increased appreciably in the few years since this data was gathered.

Table 17-1
OPPORTUNITIES IN COMPUTER SECURITY

Title	Employer	Salary ($K)	Duties
EDP Security Coordinator (Administrator)	EDP Centers	11 - 17	Works with conventional security forces, representatives of computer manufacturers and software houses, and local systems programmers to implement and maintain computer security systems.
EDP Security Analyst	Government, Large user companies	14 - 30	Prescribes, reviews and evaluates computer systems. Conducts security inspections, surveys and threat evaluations.
EDP Security Consultant	Government, User companies, Computer mfrs., Software houses, Self employment	26 - 42	Designs and integrates computer security systems and programs. Participates in formulating EDP security policy. Develops innovative solutions.

Sources	Education
Programmers or systems analysts with training in security.	Combination of education and experience in EDP equivalent to a Masters in Computer Science (e.g., 2 years Community College and six years progressive experience in general EDP: or 3 years Community College and four years experience: or BS in Computer Science and 2 years experience): and part-time or full-time training in computer security—30 classroom hours with appropriate preparation and practicum.
EDP security coordinators or junior security analysts with larger firms.	Same as above with 1-5 years experience in computer security.
EDP security analysts; junior consultants with larger firms; teachers or researchers in computer security.	Same as above with record of proven accomplishment.

CONCLUSION

The Changing Picture

While it is dangerous to generalize from the particular examples cited in this chapter, especially since salary scales and security applications vary in different parts of the country as well as within the different areas of business and industry—or even within the same type of business or industry—it is nevertheless possible to perceive the coming of age of security during the 1980s.

The aging, poorly trained and underpaid guard portrayed in the Rand Report is becoming the exception not the rule. The composite portrait drawn by the Private Security Task Force survey in St. Louis and New Orleans, for example, describes the typical security officer in 1975 as a man of 40 years of age, with a high school education, married, 5′9″ tall and weighing about 180 pounds—hardly the picture of an "over-the-hill gang." He is better paid and better trained than he was at the beginning of the 1970s. He is also just as likely to be black as white—and nearly one out of 10 is a woman.[73]

More universally accepted standards of training are needed. Higher wage scales are needed. The opportunity for vertical movement within the security structure must be both present and perceived. But even in these areas there are encouraging signs.

Perhaps most encouraging is the rise in salaries of supervisors and middle management—for these are the leaders and standard bearers of security's future. The 1975 ASIS survey found that these members of middle management in proprietary security organizations have moved into a respectable salary range. Twenty-seven percent earn between $1,000 and $1,250 per month, while 36 percent earn more than $1,250.[74] And it should be noted that at these operational levels the differences in pay between proprietary and contract personnel are not significant.

And at the top of the pyramid there is also evidence of a breakthrough. Two recent cases are revealing. One involved a position as director of security for a Midwest supermarket chain; the salary offered was $50,000. On the West Coast, a position as head of security for a California drug chain was filled at a salary of $49,500. These salaries can obviously be interpreted as management's recognition of the increasing importance of their loss prevention programs. Such recognition is not yet universal by any means, but these examples should offer encouragement to security professionals at every level.

Standards of Professionalism

In the first chapter of this text, the growing professionalism of the private security industry was briefly discussed, as reflected in higher standards of training, better pay, increasing management acceptance, and the growth of professional organizations.

In many areas of security these desirable indications are still only goals to be pursued. There is even some resistance to the establishment, for instance, of mandatory training. Security today, as we have seen, is a rapidly growing and generally lucrative field, and there is little inclination in some quarters to disturb the *status quo*. Other responsible security leaders, however, have expressed their concern over the lack of regulation of the industry, a situation that leads to cutting corners and lowering standards to meet the competitive threat.

The day may already be at hand—witness the widespread promulgation of the Report of the Task Force on Private Security produced through the efforts of the National Advisory Committee on Criminal Justice Standards and Goals—when the need for regulation and generally accepted standards is so clearly perceived that it can be effected, to the benefit of supplier, user and the general public.

Such regulation need not be totally imposed by government agencies. Indeed, to be genuinely effective it should be self-imposed by the security industry. Tighter government controls are clearly indicated, but the thrust toward the achievement of professionalism should properly come from within.

As minimum standards, there should be

- A universally accepted code of ethics, establishing a high standard of professional conduct for security officers.
- The development of a consistent body of theory governing the application of security systems—procedures, equipment, techniques and skills.
- The regular and free dissemination of information to all members of the industry.
- Minimum requirements for entry into the field, such as minimum training, education, licensing by some regulatory body, and certification of attainment by the industry.

Ultimately, the establishment of an organization that would draw up guidelines, set policy, and sponsor research or act as a central clearing-house for the distribution of the results of such research, would create the framework for true professionalization of private security, an industry that in so many ways touches the lives of all of us.

NOTES

1. *An Analysis of Criminal Redistribution Systems and Their Economic Impact on Small Business.* (Oct. 26, 1972), pp. 21-29.
2. *Ibid.*, p. 25.
3. *Private Security: Report of the Task Force on Private Security,* (Washington, D.C.: National Advisory Committee on Criminal Justice Standards and Goals, 1976), p. 30.
4. Kakalik, James S. and Sorrel Wildhorn, *The Rand Report*. (Santa Monica, Calif.: The Rand Corp., 1971).
5. *Ibid.*, Vol. II, pp. 94-95.
6. "The Losing Battle Against Crime in America," *U.S. News and World Report*, Dec. 16, 1974, p. 43.
7. *Report of the Task Force on Private Security*, p. 1.
8. *Ibid.*, p. 18.
9. Reported in the *Los Angeles Times*, Sept. 29, 1975, Part III, p. 12.
10. *A Handbook of White Collar Crime*. (Washington, D.C.: Chamber of Commerce of the United States, 1974). p. 5.
11. *Report of the Task Force on Private Security*, p. 18.
12. *Ibid*.
13. *U.S. News and World Report, op. cit.*, p. 32.
14. *Report of the Task Force on Private Security*, p. 35.
15. *Rand Report*, Vol. II, pp. 34-36.
16. *Report of the Task Force on Private Security*, p. 35.
17. *Rand Report*, Vol. II, p. 38.
18. *Report of the Task Force on Private Security*, p. 1.
19. Colling, Russell L., *Hospital Security*, (Los Angeles: Security World Publishing Co., Inc., 1976). p. 87.
20. Astor, Saul D., *Loss Prevention: Controls and Concepts*. (Los Angeles: Security World Publishing Co., Inc., 1978), p. 246.
21. *Report of the Task Force on Private Security*, pp. 39-40.

22. *Rand Report*, Vol. I, p. 3.
23. *Report of the Task Force on Private Security*, p. 4.
24. *Ibid.*, pp. 32-33.
25. *Security Letter*, Vol. X, No. 14, Part II.
26. *Report of the Task Force on Private Security*, pp. 32-33.
27. *Rand Report,* Vol. II, p. 53.
28. *Ibid.*, Vol. I, p. 30.
29. *Ibid.*, Vol. II, p. 198ff.
30. *Report of the Task Force on Private Security*, p. 87.
31. *Ibid.*, p. 98.
32. *U.S. Government Organization Manual,* Government Printing Office, 1966, p. 60. As cited in *Security Administration*, Post, R.S. and Kingsbury, A.A. (Springfield, ILL: Charles C. Thomas, 1977), p. 173.
33. *Ibid.*
34. *Ibid.*
35. *Ibid.*
36. *Ibid.*, p. 180.
37. For a complete discussion of this question, see *Police Organization and Management*, 3rd edition, Leonard, V.A. and More, H.W. (Mineola, NY; Foundation Press, Inc., 1971), p. 101 et seq.
38. Kakalik, J.S. and Wildhorn, S., *The Private Police: Security and Danger*, Crane, Russak and Co. (the Rand Corp.), New York, NY (1977).
39. *Florida Statutes Annotated*, Section 811.022.
40. Burns, *Indiana Statutes Annotated*, Section 35-3-2-1.
41. *Oklahoma Statutes Annotated*, Chapter 22, Section 1343 (1971).
42. *Minnesota Statutes Annotated*, Section 629.366 (Cumulative Supplement, 1971).
43. *Revised Codes of Montana Annotated*, Section 95-611(3).
44. Kakalik and Wildhorn, *op. cit.,* p. 321.
45. *Ibid.*, p. 364.
46. Specifications of government requirements can be obtained from *Industrial Security Manual for Safeguarding Government Classified Information*, available from the U.S. Government Printing Office.
47. Dumbauld, J. and Porter, H., *Safe Burglars, Part II, A Study of Selected Offenders* (Sacramento: California Dept. of Justice, Division of Law Enforcement, 1971).
48. *Accident Prevention Manual for Industrial Operations*. (Chicago, Ill.: National Safety Council, 1974), p. 3.
49. *Ibid.*, p. 4.
50. Bird, Frank E., Jr., *Management Guide to Loss Control*. (Atlanta, Georgia: Institute Press, 1974).

51. Pratt, L.A., *Embezzlement Controls for Business Enterprises*. (Baltimore, Md.: Fidelity and Deposit Co., Revised Edition, 1966), pp. 30-31.
52. Curtis, Bob, *Security Control: External Theft*. (New York: Chain Store Publishing Corp., 1971), p. 12.
53. Barnash (1971).
54. Curtis, *op. cit.*, pp. 15-16.
55. *Successful Retail Security* (Woburn, Mass.: Butterworth Publishers, Inc., 1974), pp. 188-193.
56. *Ibid*., p. 192.
57. Curtis, *op. cit.*, p. 65.
58. *Ibid*., pp. 75-76.
59. *Ibid*., p. 120.
60. *Ibid*., p. 110, quoting Karl Menninger, M.D., "Verdict Guilty - Now What?" *Harper's Magazine*, Vol. 219 (July, 1958), pp. 131-143.
61. *Successful Retail Security*, pp. 44-49.
62. *Ibid*.
63. U.S. Department of Commerce, Preliminary Staff Report on *The Economic Impact of Crimes Against Business* (February, 1972), p. 22.
64. Davis, Benjamin O., Jr., "How, Where, When and What," *Transportation and Distribution Management* (July, 1972), p. 25. Cited in Ursic and Pagano, *Security Management Systems* (Springfield, Ill.: Charles C. Thomas, 1974), p. 161.
65. Parker, Nycum and Oura, *Computer Abuse* (Menlo Park, Calif.: Stanford Research Institute, 1973), p. 17.
66. *Ibid.*
67. Adelson A., "Embezzlement by Computer," in *Computer Security: Equipment, Personnel and Data* (Woburn, Mass.: Butterworth Publishers, Inc., 1974), p. 43.
68. Hemphill, C.F. Jr., *Security for Business and Industry*. (Homewood, Ill.: Dow Jones-Irwin, 1971), p. 113.
69. Barlay, Stephen, *Fire*. (Brattleboro, Vt.: Stephen Greene Press, 1973), p. 192.
70. *Ibid.*
71. *Report of the Task Force on Private Security*, p. 32.
72. *Ibid*., pp. 358-359.
73. *Ibid*., p. 359.
74. *Ibid*., p. 349.

A SELECTED BIBLIOGRAPHY

Accident Prevention Manual for Industrial Operations. Chicago, Ill.: National Safety Council, 1974.

Applying the OSHA Standards. Long Grove, Ill.: National Loss Control Service, 1973.

Assets Protection. Bimonthly, San Francisco, Calif.: Territorial Imperative, Inc.

Astor, Saul D. *Loss Prevention: Controls and Concepts*. Woburn, Mass.: Butterworth Publishers, Inc., 1978.

Barefoot, J. Kirk. *Employee Theft Investigation*. Woburn, Mass.: Butterworth Publishers, Inc., 1979.

Barnard, Robert. *Intrusion Detection Systems*. Woburn, Mass.: Butterworth Publishers, Inc., 1981.

Berger, David L. *Industrial Security*, Woburn, Mass.: Butterworth Publishers, Inc., 1979.

Byrne, Dennis E. and Jones, Peter H. *Retail Security: A Management Function*. Leatherhead, England: 20th Century Security Education Ltd., 1977.

Cargo Loss Prevention Recommendations. Zurich: International Union of Marine Insurance, 1970.

Carroll, John M. *Computer Security*. Woburn, Mass.: Butterworth Publishers, Inc., 1977.

Carson, Charles R. *Managing Employee Honesty*. Woburn, Mass.: Butterworth Publishers, Inc., 1977.

Cole, Richard B. *The Application of Security Systems and Hardware*. Springfield, Ill.: Charles C. Thomas, 1970.

Cole, Richard B. *Principles and Practice of Protection*, Springfield, Ill.: Charles C. Thomas, 1980.

Colling, Russell L. *Hospital Security*. Woburn, Mass.: Butterworth Publishers, Inc., 1976.

Colling, Russell L., ed. *Hospital Security and Safety Journal Articles.* Flushing, N.Y.: Medical Examination Publishing Co., 1970.

Currier-Briggs, Noel. *Security: Attitudes and Techniques for Management*. London: Hutchinson & Co., 1968.

Curtis, Bob. *Security Control: External Theft*. New York: Chain Store Publishing, 1971.

Curtis, Bob. *Security Control: Internal Theft*. New York: Chain Store Publishing, 1973.

Curtis, Bob. *Modern Retail Security*. Springfield, Ill.: Charles C. Thomas, 1960.

Davis, Keith. *Human Relations at Work: The Dynamics of Organizational Behavior*. New York: McGraw-Hill Book Co., 1967.

Ehrstine, B.I. and Mach, J.A. *Profitability Through Loss Control,* Cincinnati, Ohio: Anderson Co., 1977.

Finneran, Eugene. *Security Supervision*. Woburn, Mass.: Butterworth Publishers, Inc., 1981.

Fisher, James A. *Security for Business and Industry*. Englewood Cliffs, N.J.: Prentice-Hall, Inc., 1979.

Gammage, Allen Z. and Hemphill, Charles F., Jr. *Basic Criminal Law*. New York: McGraw-Hill Book Co., 1974.

Guidelines for the Physical Security of Cargo, Department of Transportation, May, 1972.

A Handbook on White Collar Crime. Washington, D.C.: Chamber of Commerce of the United States, 1974.

Healy, Richard J. *Design for Security*. New York: John Wiley and Sons, Inc., 1968.

Healy, Richard J. and Walsh, Timothy J. *Industrial Security Management*. New York: American Management Association, Inc., 1971.

Hemphill, Charles F., Jr. *Modern Security Methods*. Englewood Cliffs, N.J.: Prentice-Hall, Inc., 1979.

Hopf, Peter S. *Handbook of Building Security Planning and Design.* New York: McGraw-Hill Book Co., 1979.

Hughes, Mary Margaret, ed. *Successful Retail Security*. Woburn, Mass.: Butterworth Publishers, Inc., 1974.

Inbau, Fred E. and Aspen, Marvin E. *Criminal Law for the Layman.* Philadelphia: Chilton Book Co., Inc., 1970.

Industrial Safety and Fire Prevention. London: T. Bell & Son, 1973.

Kakalik, James S. and Wildhorn, Sorrel. *The Rand Report.* Santa Monica, Calif.: The Rand Corp., 1971. Five volumes:

R-869-DOJ *Private Police in the United States: Findings and Recommendations.*

R-870-DOJ *The Private Police Industry: Its Nature and Extent*

R-871-DOJ *Current Regulation of Private Police: Regulatory Agency Experience and Views*

R-872-DOJ *The Law and Private Police*

R-873-DOJ *Special-Purpose Public Police*

Kingsbury, Arthur A. *Introduction to Security and Crime Prevention Surveys*. Springfield, Ill.: Charles C. Thomas, 1973.

Mandelbaum, Albert J. *Fundamentals of Protective Systems.* Springfield, Ill.: Charles C. Thomas, 1973.

Momboisse, Raymond M. *Industrial Security for Strikes, Riots, and Disasters*. Springfield, Ill.: Charles C. Thomas, 1968.

Oliver, Eric and Wilson, John. *Practical Security in Commerce and Industry*. London: Gower Press, 1968.

OSHA Reference Manual. Santa Monica, Calif.: Insuror's Press, 1972.

Parker, Nycum and Oura. *Computer Abuse.* Menlo Park, Calif.: Stanford Research Institute, 1973.

Post, Richard S. *Combating Crime Against Small Business.* Springfield, Ill.: Charles C. Thomas, 1972.

Post, Richard S. *Determining Security Needs.* Madison, Wisc.: Oak Security Publishing, 1973.

Post, Richard S. and Kingsbury, Arthur A. *Security Administration: An Introduction*. Springfield, Ill.: Charles C. Thomas, 1973.

Powell, John W. *Campus Security and Law Enforcement.* Woburn, Mass.: Butterworth Publishers, Inc., 1981.

Pratt, Lester A. *Embezzlement Controls for Business Enterprises*. Baltimore, Md.: Fidelity and Deposit Co., 1952; Revised 1966.

Private Security: Report of the Task Force on Private Security. Washington, D.C.: National Advisory Committee on Criminal Justice Standards and Goals, 1976.

Protection Canada. Bimonthly. Calgary, Alberta, Canada: Protection Canada.

Rosberg, Robert R. *A Practitioner's Guide to Security Risk Management*. Boston: Dorison House Publishers, 1980.

Russell, A. Lewis. *Corporate and Industrial Security*. Houston, Texas: Gulf Publishing Co., 1980.

San Luis, Ed. *Office and Office Building Security*. Woburn, Mass.: Butterworth Publishers, Inc., 1973

Schultz, Donald O. *Principles of Physical Security*. Houston, Texas: Gulf Publishing Co., 1978.

Security Management. Monthly. Washington, D.C.: American Society for Industrial Security.

Security World Magazine. Monthly. Chicago, Ill. Cahners Publishing Co.

Sennewald, Charles A. *Effective Security Management*. Woburn, Mass.: Butterworth Publishers, Inc., 1978.

Sennewald, Charles A. *The Process of Investigation: Concepts and Strategies for the Security Professional*. Woburn, Mass.: Butterworth Publishers, Inc., 1981.

Strauss, Sheryl, ed. *Security Problems in a Modern Society*. Woburn, Mass.: Butterworth Publishers, Inc., 1980.

Strobl, Walter M. *Crime Prevention Through Physical Security* New York: Marcel Dekker, Inc., 1978.

Thorsen, June-Elizabeth, ed. *Computer Security: Equipment, Personnel and Data*. Woburn, Mass.: Butterworth Publishers, Inc., 1974.

Uniform Crime Reports. Washington, D.C.: Federal Bureau of Investigation.

United States Department of Commerce. Preliminary Staff Report on *The Economic Impact of Crimes Against Business*. February, 1972.

Ursic, Henry S. and Pagano, Leroy E. *Security Management Systems.* Springfield, Ill.: Charles C. Thomas, 1974.

Walsh, Timothy J. and Healy, Richard J. *Protection of Assets Manual*. Santa Monica, Calif.: The Merritt Co., 1974.

Weber, Thad L. *Alarm Systems and Theft Prevention*. Woburn, Mass.: Butterworth Publishers, Inc., 1973.

Whitman, Lawrence A. *Fire Prevention*. Chicago, Ill.: Nelson-Hall, 1979.

Wright, K.G. *Cost-Effective Security* (UK) Ltd.: McGraw-Hill, 1972.

INDEX

INDEX